Springer Series in Materials Science

Volume 306

The Springer Series in Materials Science covers the complete spectrum of materials research and technology, including fundamental principles, physical properties, materials theory, and design. Recognizing the increasing importance of materials science in future device technologies, the book titles in this series reflect the state-of-the-art in understanding and controlling the structure and properties of all important classes of materials.

More information about this series at http://www.springer.com/series/856

Jinmin Li · Junxi Wang · Xiaoyan Yi ·
Zhiqiang Liu · Tongbo Wei · Jianchang Yan ·
Bin Xue

III-Nitrides Light Emitting Diodes: Technology and Applications

Science Press
Beijing

Springer

Authors
See next page

ISSN 0933-033X ISSN 2196-2812 (electronic)
Springer Series in Materials Science
ISBN 978-981-15-7951-6 ISBN 978-981-15-7949-3 (eBook)
https://doi.org/10.1007/978-981-15-7949-3

Jointly published with Science Press, Beijing, China
The print edition is not for sale in China Mainland. Customers from China Mainland please order the print book from: Science Press.
ISBN of the Co-Publisher's edition: 978-7-03-065929-3

This Springer imprint is published by the registered company Springer Nature Singapore Pte Ltd.
The registered company address is: 152 Beach Road, #21-01/04 Gateway East, Singapore 189721, Singapore

Jinmin Li
The State Key Laboratory for Solid State Lighting, Beijing, China

Center of Materials Science and Optoelectronics Engineering
University of Chinese Academy of Sciences
Beijing, China

Institute of Semiconductors
Chinese Academy of Sciences
Beijing, China

Junxi Wang
The State Key Laboratory for Solid State Lighting, Beijing, China

Center of Materials Science and Optoelectronics Engineering
University of Chinese Academy of Sciences
Beijing, China

Institute of Semiconductors
Chinese Academy of Sciences
Beijing, China

Xiaoyan Yi
The State Key Laboratory for Solid State Lighting, Beijing, China

Center of Materials Science and Optoelectronics Engineering
University of Chinese Academy of Sciences
Beijing, China

Institute of Semiconductors
Chinese Academy of Sciences
Beijing, China

Zhiqiang Liu
The State Key Laboratory for Solid State Lighting, Beijing, China

Center of Materials Science and Optoelectronics Engineering
University of Chinese Academy of Sciences
Beijing, China

Institute of Semiconductors
Chinese Academy of Sciences
Beijing, China

Tongbo Wei
The State Key Laboratory for Solid State Lighting, Beijing, China

Center of Materials Science and Optoelectronics Engineering
University of Chinese Academy of Sciences
Beijing, China

Institute of Semiconductors
Chinese Academy of Sciences
Beijing, China

Jianchang Yan
The State Key Laboratory for Solid State Lighting, Beijing, China

Center of Materials Science and Optoelectronics Engineering
University of Chinese Academy of Sciences
Beijing, China

Institute of Semiconductors
Chinese Academy of Sciences
Beijing, China

Bin Xue
The State Key Laboratory for Solid State Lighting, Beijing, China

Institute of Semiconductors
Chinese Academy of Sciences
Beijing, China

Preface

GaN-based high brightness blue Light Emitting Diode (LED) marks the milestone of Solid State Lighting (SSL) development which expands a brand-new horizon of illumination. SSL has become one of the most competitive research focuses worldwide. When I was a senior visiting scholar to the USA between 2000 and 2002, I was impressed by the prototype of such device during a photonics exhibition. I realized that LEDs could lead revolution of artificial lighting. At the beginning of 2003, together with my colleagues, we made a draft of SSL research development plan and reported to the Chinese authority in order to emphasize the significance of developing III-nitrides based LED technology. In the same year, Chinese Ministry of Science and Technology (MOST) launched the National SSL Project and three years later, SSL research and development center of Chinese Academy of Sciences (CAS) was established upon Institute of Semiconductors, CAS, where the authors are working.

Thanks to continuous support from the government through R&D funding and hardworking researchers, Chinese SSL technology has made tremendous progress in full technical chain. It has promoted the construction and improvement of Chinese SSL industry. The technology has been successfully employed in various important scenarios, such as Beijing Olympics, Shanghai Expo, etc. The application is a good demonstration and enables the public experience the fascination of SSL.

SSL technology progress is based on the technology breakthrough in fundamental problems of wide bandgap semiconductor materials. Improvement of material quality and quantum efficiency is benefited from low temperature buffer layer and p-type doping technology. Doping mechanism and density are investigated and improved by studying first principle calculation, which enables the performance enhancement of the devices. Although SSL products have been widely used in our daily life, performance of these products can be further improved. It is necessary to strengthen basic scientific research to boost the drive force of SSL technology.

Fundamental concept and material property of III-nitrides LEDs is discussed in Chaps. 2 and 3. Basic concept of material growth and epitaxy technology, concept of InGaN/GaN multiple-quantum-well-structured blue and green light emitting diodes, AlGaN/GaN multiple-quantum-well-structured violet light emitting diodes are presented in Chaps. 4–6. Quantum efficiency enhancement technology, chip fabrication, packaging, reliability analysis, and application of LEDs are discussed in Chaps. 7–12.

Beijing, China Jinmin Li

Acknowledgements

I would like to express my deep gratitude to the respected Chinese academicians including Profs. Wang Zhanguo, Zhou Bingkun, Gan Zizhao, Zheng Youdou, Chen Lianghui, and Xia Jianbai for their continuous guidance and contribution. I would like to take this opportunity to express my appreciation for continuous support from the Ministry of Science and Technology of China. The concept of this book is fulfilled by the research progress of my colleagues at CAS and their input is very much appreciated. Many thanks to Dr. Zhe Liu, Dr. Hua Yang, Dr. Yiyun Zhang, Dr. Yi Lu, and Dr. Yang Li. Furthermore, I want to acknowledge here the contribution of Prof. Quanxi Jia (University at Buffalo, the State University of New York), and thank him for his support and friendship.

I do hope this book could provide assistance to those who are working or interested in solid state lighting.

Beijing, China

Jinmin Li

Brief Introduction of This Book

The concept of this book is based on the research progress of the authors and covers academic and practical content of III-nitrides based light emitting diodes, including epitaxy growth, chip fabrication, packaging, and application. This book consists of 12 chapters, including basic principles of III-nitrides based light emitting diodes, material property and epitaxy technology, InGaN/GaN multiple-quantum-well-structured blue and green light emitting diodes, AlGaN/GaN multiple-quantum-well-structured violet light emitting diodes, quantum efficiency enhancement technology, chip fabrication and packaging technology, reliability and application of light emitting diodes. Last but not least, cutting-edge and hot spots of III-nitrides based light emitting diodes research is presented.

This book will be of interests to scientists, engineers working on LED technology and applications, and also postgraduate students in related subjects.

Contents

About the Author

Prof. Jinmin Li obtained Bachelor degree in semiconductor physics and devices from the Department of Electronic Engineering, Xi'an Jiaotong University in 1982. He was conferred Master's degree in physics of semiconductor materials and devices from the Electronic Research Institute, Ministry of Information of China in 1984, and Ph.D. degree in Xi'an Institute of Optics and Precision Mechanics of Chinese Academy of Sciences (CAS) in 1991, respectively. In 1993, he accomplished post-doctoral research work at the Institute of Semiconductors, CAS (ISCAS). Since then, he started his professional research and was mainly focused on the study of novel semiconductor materials at ISCAS. Between 2000 and 2002, he worked as a senior visiting scholar at UCLA. From 2003 to 2012, he served as director general of the ISCAS. Since 2010, he has been an expert in the first and second technical committees of the International Solid State Lighting Alliance (ISA). At present, he is the director of the State Key Laboratory of Solid State Lighting of China.

Chapter 1
Introduction

Since ancient times, history of artificial lighting sources has been closely related to the development of human civilization. Lighting source is not only the witness of human civilization, but also the driving force for the continuous development and progress of human civilization. In 1879, Thomas Edison invented the incandescent lamp, which brought mankind into the era of electrical lighting and changed the production and life style of human society. Since then, with the rapid development of lighting sources, fluorescent lamps and high-pressure sodium lamps have emerged one after another, acting as an important lighting tool. However, incandescent lamps have low luminous efficiency, mercury and other harmful substances used in fluorescent lamps, and poor color rendering for high-pressure sodium lamps. The fatal weakness of these light sources indicates that they cannot meet the high demand of energy saving and environmental protection for the growing human civilization. While people are devoting themselves to the research and development of new light sources, the development of semiconductor technology has led people into the information age, and people are paying more and more attention to the research and application of semiconductor materials. Under this background, light-emitting diode (LED) emerges, but over time. In 1962, Nick Holonyak Jr. invented the first p-n red LED [1] using GaAsP as luminescent material. In the next forty years or so, GaP (550 nm, green), GaAsP (650 nm, orange and yellow), GaAlAs (680 nm, red), AlInGaP (590–620 nm, yellow-orange) and other materials have been introduced successively. At the same time, the luminous wavelength has been limited to red and yellow-green for a long time, and it is difficult to achieve a shorter wavelength of blue LED [2]. Until around the 1990s, Isamu Akasaki and Hiroshi Amano of Nagoya University, Japan, and Shuji Nakamura of Japan Nichia Company successfully solved the issues related to the growth of high quality GaN film and p-type doping, and realized the commercialization of high-efficiency nitride blue-light LED. The invention of GaN-based blue-light LED is known as "the second lighting revolution". They also won the 2014 Nobel Prize in Physics for "the invention of an efficient blue light-emitting diode (LED) with a bright and energy-saving white light source" [3].

J. Li et al., *III-Nitrides Light Emitting Diodes: Technology and Applications*, Springer Series in Materials Science 306,
https://doi.org/10.1007/978-981-15-7949-3_1

Group III nitride semiconductors, including AlN, InN, GaN and their alloys, belong to all-component direct bandgap semiconductors. They have the advantages of momentum conservation of radiation recombination and high radiation transition efficiency. Their bandgap energy covers a wide spectrum, ranging from near infrared to deep ultraviolet. They are ideal materials for realizing high efficiency light emitting diodes and are also the most attractive in wide bandgap semiconductors [4]. However, the luminous efficiency of light-emitting diodes will be affected by materials, devices, packaging and other processing steps and technologies in actual production. This series of problems has attracted extensive attention of many scientists and researchers all over the world and become a hot field of scientific research.

Substrate material is the basis of the growth of group III nitride epitaxy film, and also the main component of LED devices, which has an important impact on the preparation and performance of LED devices. Taking GaN substrate as an example, GaN homogeneous substrate is the most ideal substrate for growth of GaN epitaxial layer. Its greatest advantage lies in the lattice match between the substrate and the epitaxial layer, which can eliminate crystal defects caused by lattice mismatch to the greatest extent. Furthermore, GaN substrate can improve the vertical conductivity and thermal conductivity, thus greatly enhance the performance of GaN-based LED. However, the melting point of GaN is very high (2800 C) and the equilibrium vapor pressure is very high (4.5 GPa). It is very difficult to prepare GaN single crystal [5]. Although some research institutes have prepared GaN single crystals with good crystal quality through hydride vapor phase epitaxy (HVPE), Ammon thermal method and low-pressure growth at lower pressure method, the high growth cost and small size make it difficult to meet the needs of the market [6]. Therefore, heteroepitaxy is widely used at present. Commonly used substrates are sapphire (a-Al_2O_3), silicon carbide (6H-SiC) and silicon (Si). However, heteroepitaxy has its inherent shortcomings, such as lattice mismatch, thermal expansion mismatch and poor chemical solubility. These problems lead to high dislocation density, mosaic crystal structure, biaxial stress and warping of epitaxy sheets in epitaxy materials, which affect the crystal quality of epitaxy materials and the performance of LED devices [7]. In order to improve the crystal quality of heteroepitaxy materials, some special methods and means are usually adopted. For example, epitaxial lateral overgrowth (ELOG), buffer layer or insertion layer, graphical substrate and non-polar and semi-polar surface epitaxy growth have been explored [8, 9].

The epitaxy methods of group III nitrides mainly include metal organic compounds vapor phase epitaxy (MOCVD), hydride vapor phase epitaxy (HVPE) and molecular beam epitaxy (MBE). Due to the high growth rate of HVPE, it cannot be used to grow low-dimensional structures such as quantum wells and superlattices, while MBE is quite expensive and has a slow growth rate. Considering the cost and efficiency, MOCVD technology is used commercially for heteroepitaxial LED structures along c-axis on sapphire substrates. Since wurtzite nitride semiconductors grown along the c-axis do not have central symmetry and the lattice mismatch between nitride materials and substrates and nitride heterostructures, spontaneous polarization and piezoelectric polarization occur in their crystals. The polarization effect generates built-in electric field at the interface and changes the energy band structure of crystals. This phenomenon occurs in InGaN/GaN and AlGaN/GaN quantum wells. Such a phenomenon is often called quantum confined Stark effect [10]. The built-in electric field tilts the conduction band and valence band edges, which separates the wave functions of electrons and holes in quantum wells. The band bending reduces the spatial overlap of carrier wave functions, the radiation recombination probability, and consequently the luminescence efficiency of LED. In order to effectively reduce the influence of polarized electric field and quantum confinement Stark effect on the luminescence efficiency of LED, it is necessary to control and reduce the total strain in the crystal. Commonly used methods include epitaxy on heterogeneous substrates with specific crystal surfaces (such as non-polar or semi-polar surfaces), optimization of quantum well structure, fabrication of micro-nano graphical substrates and nano-pillar LED, etc.[9, 11]. In addition, in order to inject holes into the quantum well region more effectively in p-type GaN, electronic barrier layer, strain compensation technology and gradient growth technology are usually adopted. These methods can effectively improve the internal quantum efficiency of LED [12].

There are three major technological routes for LED chips: front-loading, flip-flop and vertical structure. Chip technology has an important impact on improving the efficiency of electric injection and light extraction of LED devices. In order to improve the injection efficiency of LED and obtain low working voltage devices, the ohmic contact problem between electrode material and GaN material must be solved. Ohmic contact fabrication of p-type GaN has always been a difficult problem because the current growth technology cannot obtain p-type GaN with high hole concentration [13]. The work function of p-type GaN is large and there is no equivalent metal material. In order to achieve low specific contact resistivity and high optical transmittance, transparent conductive oxide films, metal nanowires and graphene were used as ohmic contact layers [14]. In addition, in order to further improve the injection efficiency of LED and solve the current aggregation effect, current diffusion layer, current barrier layer and other technologies are usually used. Chip technology can not only improve the injection efficiency, but also improve its light extraction efficiency. Because the refractive index difference between GaN material (refractive index 2.5) and air (refractive index 1) is large, the light emitted from the active region cannot be emitted from the interface because the critical angle of total reflection is too small (only 23.5°), resulting in low light extraction efficiency (only 4%). The light that cannot be emitted will propagate repeatedly in the dielectric material until all the light energy is dissipated into thermal energy. This can consequently affect the device performance [15]. In view of this, a series of technologies has been adopted, such as graphical substrates, surface roughening, mirror technology, photonic crystals, nano-column structure and surface plasmon polaritons, etc.[16, 17].

LED packaging technology directly affects the reliability of devices. Due to the uniqueness of the nitride LED, the reliability of LED devices is mainly affected by: heteroepitaxy growth of substrates, materials and compound semiconductors; passivation of p-type doping and doping elements of GaN; metallization of electrodes and ohmic contact; current, temperature and static electricity in device storage and working environment. A large number of studies have shown that for LED light source devices the light source failure caused by package is much larger than chip itself [18]. Packaging process is an important part of LED production. Appropriate packaging can protect the LED chip from damage, improve its mechanical performance and life,strengthen heat dissipation, reduce the junction temperature of the LED chip, improve the photoelectric and thermal performance of the LED,optimize the optical design, and improve the luminous efficiency of the LED. Imperfect packaging can lead to serious light loss, low luminous efficiency, uneven light color, short service life and many other problems of LED devices. From the perspective of packaging technology, LED packaging has gone through the development stages of pin-type (Lamp LED), surface-mounted (SMT LED), on-board chip direct-mounted (COB), system-mounted (SiP) and so on. It is now developing chip-scale wafer-level packaging, IC packaging and other forms. Regardless of the packaging technology, we must take into account the effects of thermal management, packaging materials, optical design, electrostatic protection and other factors on the photoelectric performance, luminous efficiency and lifetime of LED [19].

In 2014, the Nobel Prize Committee stated: "Incandescent lamps will light up the twentieth century, and LED lights will light up the twenty-first century." In the past 20 years, the technology of group III nitride LED has been widely used not only in lighting, backlight, display and other fields, but also in the progress of modern lighting technology. It has gradually changed people's way of life. It has been developing in the fields of biology, agriculture, medical treatment, health care, aviation, aerospace and communication. Nitride LED has played an important role in social progress and economic prosperity [20]. However, with the continuous expansion of LED applications and the emergence of new applications, many new basic scientific problems intersecting with communication, control, sensing, information processing, electronics, software, optics and other fields will emerge. Such new scientific issues need to be solved urgently.

Research on the technology of group III nitride light emitting diodes in China began in the late 1990s. Especially since the launch of "Semiconductor Lighting Project" in 2003, we have carried out a series of basic, prospective and strategic scientific and technological research. The most noticeable achievements are the manufacturing of MOCV equipment, the growth of epitaxy materials, chip manufacturing process and packaging integration technology. We have done original works which have been recognized internationally. The luminous efficiency of white LED developed by us has increased from 20 lumens/W in 2004 to 160 lumens/W in 2014 at 350 mA working current. The development of LED has also made tremendous contributions to energy saving, emission reduction and economic construction in China. With the deepening of basic research and engineering application, the technology of group III nitride light emitting diodes will further promote the development of the third-generation semiconductor technology and industry in China.

References

1. N. Holonyak, S.F. Bevacqua, Coherent (visible) light emission from Ga ($As_{1-X}P_x$) junctions. Appl. Phys. Lett. **1**, 82–83 (1962)
2. Craford MG, Dupuis RD, Feng M, et al. 50th anniversary of the Light-Emitting Diode (LED): an ultimate lamp, in Proceedings of the IEEE (2013)
3. D. Normile, Physicists change the light bulb. Science **346**, 149–150 (2014)
4. F.A. Ponce, D.P. Bour, Nitride-based semiconductors for blue and green light-emitting devices. Nature **386**, 351–359 (1997)
5. T. Paskova, D.A. Hanser, K.R. Evans, GaN substrates for III-nitride sevices. Proc. IEEE **98**, 1324–1338 (2010)
6. K. Sardar, C.N.R. Rao, New solvothermal routes for GaN nanocrystals. Adv. Mater. **16**, 425 (2004)
7. S. Nakamura, The roles of structural imperfections in InGaN-Based blue light-emitting diodes and laser diodes. Science **281**, 956–961 (1998)
8. H. Masui, S. Nakamura, S.P. Den Baars et al., Nonpolar and semipolar III-nitride light-emitting diodes: achievements and challenges. IEEE Trans. Electron Devices **57**(1), 88–100 (2010)
9. J. Jang, S. Woo, D. Min, O. Nam, Recent advances in nonpolar and semipolar InGaN light-emitting diodes (LEDs). J. Nanosci. Nanotechnol. **15**, 1895–1906 (2015)
10. M.H. Crawford, LED for solid-state lighting: performance challenges and recent advances. IEEE J. Selected Topics Quantum Electron. **15**, 1028–1040 (2009)
11. T.W. Yeh, Y.T. Lin, L. Stewart, et al, InGaN/GaN multiple quantum wells grown on nonpolar facets of vertical GaN nanorod arrays. Nano Lett. **12**(6), 3257–3262 (2012)
12. J.H. Park, D.Y. Kim, S.Y. Hwang, et al, Enhanced overall efficiency of GaInN-based light-emitting diodes with reduced efficiency droop by Al-composition-graded AlGaN/GaN superlattice electronic blocking layer. Appl. Phys. Lett. **103**(6), 061104 (2013)
13. J.O. Song, J.S. Ha, T.Y. Seong, Ohmic-contact technology for GaN-based light-emitting diodes: role of p-type contact. IEEE Trans. Electron Devices **57**(1), 42–59 (2010)
14. F. Bonaccorso, Z. Sun, T. Hasan et al., Graphene photonics and optoelectronics. Nat. Photonics **4**(9), 611–622 (2010)
15. J.J. Wierer, A. David, M.M. Megens, III-nitride photonic-crystal light-emitting diodes with high extraction efficiency. Nat. Photonics **3**, 163–169 (2009)
16. M.K. Kwon, J.Y. Kim, B.Y. Kim, et al., Surface-plasmon-enhanced light-emitting diodes. Adv. Mater. **20**(7), 1253 (2008)
17. A.I. Zhmakin AI. Enhancement of light extraction from light emitting diodes. Phys. Rep. Rev. Sect. Phys. Lett. **498**, 189–241 (2011)
18. O. Ueda On degradation studies of III-V compound semiconductor optical devices over three decades: focusing on gradual degradation. Jpn. J. Appl. Phys. **49**(9), 090001 (2010)
19. M. Chang, D. Das, P. Varde, et al., Light emitting diodes reliability review. Microelectr. Reliabil. **52**, 762–782 (2012)
20. E.F. Schubert, J.K. Kim, Solid-state light sources get smart. Science **308**, 1274–1278 (2005)

Chapter 2
Basic Principles of LED

LED is a semiconductor optoelectronic device. In contrast with traditional light sources, LED has a list of advantages, such as high efficiency, long lifetime, not easy to break, fast reaction speed and high reliability [1]. These features make the LED device attractive for luminaire. Its basic structure is a p–n junction. The basic principle is that electrons and holes in semiconductors recombine and emit photons under forward bias. In order to fabricate efficient LED, the radiation recombination probability of electrons and holes in semiconductor must be increased as much as possible, and the non-radiation recombination probability must be reduced. Due to the fact that the light-emitting mechanism of LED is completely different from that of traditional incandescent lamp and other light sources, it also has different optical and electrical characteristics. In view of the unique optical and electrical characteristics of LED, we can use monochrome LED to mix color or combine phosphor to achieve white LED.

This chapter takes p–n junction as the theoretical model, mainly introduces the basic principle of LED, radiative and non-radiative radiation, the optical and electrical characteristics of LED, and the realization method of white LED.

2.1 LED Luminescence Principle

2.1.1 History and Principle of Lighting Source

The history of lighting source is as old as human civilization. The lighting history of human beings has experienced four stages [2]: fire lighting, incandescent lamp lighting, fluorescent lamp lighting and high intensity gas discharge lamp. Among them, people use torches, oil lamps and candles as lighting tools for a long time, but these light sources have low luminous efficiency, lower brightness, incomplete

J. Li et al., *III-Nitrides Light Emitting Diodes: Technology and Applications*, Springer Series in Materials Science 306,
https://doi.org/10.1007/978-981-15-7949-3_2

material combustion, and easy to produce smoke. In 1879, Edison invented the incandescent lamp with carbon filament. From then on, mankind entered the era of electrical lighting. Incandescent lamps emit light by the thermal effect resulted from electric current through the filament. Any object heated above 525°C emits visible light. However, luminous efficiency of incandescent lamps is still very low, because most of the electric energy is wasted by converting it into heat. Unlike incandescent lamp, fluorescent lamp is based on gas discharge. The tube is filled with argon and a small amount of mercury. The inner wall of the lamp is coated with fluorescent powder which converts ultraviolet light into visible light. In the luminous process of fluorescent lamp, the heat generation is small. Furthermore, the emitted light is cold light which makes the luminous efficiency of fluorescent lamp greatly improved. However, fluorescent lamp has drawbacks such as electromagnetic pollution, fragility, and mercury pollution in the waste. In the meantime, people have also developed high-intensity gas discharge lamps which have been used in industrial and street lighting. The lamp is filled with mercury vapor, high-pressure sodium or mixed gas vapors. When current passes through the high-pressure gas vapor, the gas vapor can be ionized, where the collision between electrons, atoms and ions in the discharge tube takes place and emits light. Unfortunately, this kind of light source has shortcomings such as high cost, difficult maintenance, high power consumption, electromagnetic radiation hazards and poor color rendering.

While people are devoting themselves to the research and development of new light sources, the development of semiconductor technology has led people into the information age. At the same time, people are paying more and more attention to the research and application of semiconductor materials in different fields. Under this background, LED emerges, but over time.

The development of LED is closely related to the progress of materials. From the invention of red LED in 1962 to the commercialization of blue and green LED in 1994, the progress of materials has greatly accelerated the development of LED [3]. Although the materials used to fabricate LED are different, the basic structure of LED has not changed, all of them are p-n junctions [4]. For instance, semiconductor materials can be indirect bandgap and direct bandgap. The spectrum can cover red, green and blue.

2.1.2 p–n Junction and the Principle of LED Luminescence

The principle of LED is different from that of conventional light sources. Output from a LED is based on the light radiation (ultraviolet, visible or infrared) produced by the electron excitation transition in the device or material. It excludes any radiation (incandescence) caused by pure material temperature.

The basic structure of the LED chip is p–n junction, as illustrated in Fig. 2.1. The so-called p–n junction is to incorporate p-type (or n-type) impurities into n-type (or p-type) semiconductor single crystals by appropriate technological methods (such as alloy, diffusion, growth, ion implantation, etc.). The different regions of the single

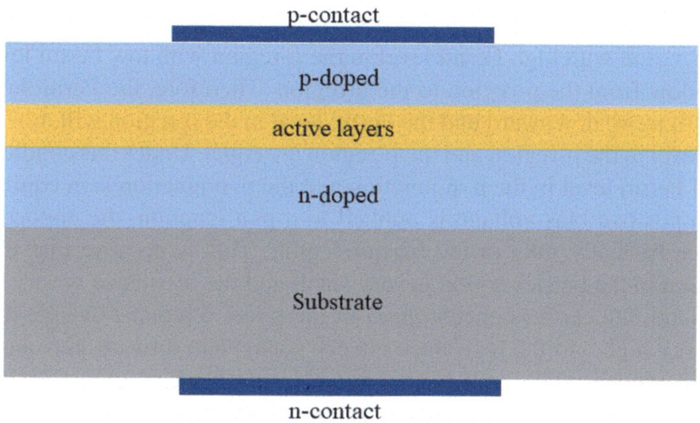

Fig. 2.1 Basic structure of LED chip

crystal have n-type or p-type conductivity respectively. A special region is formed at the interface between the n- and p-type semiconductors.

When two semiconductors combine to form a p–n junction, the carrier concentration gradient between them leads to the diffusion of holes from p to n and electrons from n to p. For the p region, the immobile ionized-acceptors are negatively charged after the holes move away. On the other hand, a positive charge region is formed by ionizing donors on the n side near the p–n junction. The charges of these ionizing donors and ionizing acceptors near the p–n junction are usually called space charges. The region in which they exist is called the space charge region [5].

These charges in the space charge region produce an electric field from n to p, that is, from positive charge to negative charge, which is called built-in electric field. Under the built-in electric field, the carrier drifts. Obviously, the drift motion direction of electrons and holes is opposite to their diffusion motion direction. Therefore, the built-in electric field acts as a barrier to the continued diffusion of electrons and holes.

With the diffusion movement, the space charge increases, and the space charge region expands gradually. At the same time, the built-in electric field increases, and the carrier drift motion increases steadily. In the absence of applied voltage, the carrier diffusion and drift will eventually reach a dynamic equilibrium. In other words, the numbers of electrons have diffused from n to p region will be the same as the electrons will return to n region under the action of built-in electric field. Therefore, the diffusion current and drift current of electrons are equal in magnitude and opposite in direction but cancel each other out. For holes, the situation is quite similar. Therefore, no current flows through a p–n junction under no external electric field, or the net current flowing through the p–n junction is zero. In this case, the space charge region will not continue to expand but maintain a certain width. Under such a condition, a certain built-in electric field is established. It is generally called the p–n junction under such a condition in the state of thermal equilibrium.

When two semiconductors combine to form a p–n junction, electrons will flow from the n-region with high Fermi level to the p-region with low Fermi level, while holes will flow from the p-region to the n-region. Therefore, the Fermi level in the n-region will move downward and the Fermi level in the p-region will move up until the Fermi level in the n-region and the p-region are equal. Under this condition, there is a unified Fermi level in the p–n junction and the p–n junction is in equilibrium.

When a positive bias voltage is applied to a p–n junction, the applied positive bias voltage basically falls in the barrier region. This is because that the carrier concentration in the barrier region is very small and the resistance is very large. On the other hand, the carrier concentration in the p and n regions outside the barrier region is very high and the resistance is very small. The forward bias produces an electric field opposite to the built-in electric field in the barrier region, which weakens the electric field intensity in the barrier region, resulting in the injection of minority carriers from both sides (n and p region) of the junction. The Fermi level in the n region moves up but the Fermi level in the p region moves down, until the difference of Fermi levels in the n and p regions equals to the difference between the built-in electric field and the applied electric field. The non-equilibrium carriers higher than the concentration at the equilibrium state will recombine near the junction and produce photons.

When a reverse bias voltage is applied to a p–n junction, the electric field generated by the reverse bias in the barrier region is in the same direction as the built-in electric field. The electric field in the barrier region is enhanced. A very small number of carriers in the junction region could be driven away by the electric field. The Fermi level in the n region moves down and the Fermi level in the p region moves up until the difference between the built-in electric field and the applied electric field in the n region is equal to the sum of the built-in electric field and the applied electric field in the p region. The current of the p–n junction is also very small and tends to remain unchanged. The LED does not emit light under reverse bias. It can be seen that, like the ordinary diodes, LED has the characteristics of 'ON' under forward bias and 'OFF' under reverse bias.

2.2 Radiation and Non-radiation Recombination

Because of the interaction within the semiconductor, there are always a certain number of electrons and holes in the equilibrium state of any semiconductor. From the microscopic point of view, equilibrium state refers to the balance between the microscopic processes caused by certain interactions within the system. It is precisely these micro-processes that make the system transition from non-equilibrium state to equilibrium state, resulting in the recombination of non-equilibrium carriers. There are two mechanisms for the recombination of unbalanced carriers in LED. In the process of recombination, the excess energy of electrons can be released in the form of radiation (photon emission). This recombination is called radiation recombination, which is an inverse process of light absorption. The excess energy of electrons can also be released in other forms (such as transferring excess energy to the lattice to enhance the lattice vibration, or giving energy to other carriers to increase their kinetic energy, also known as Auger recombination [6] without emitting photons. This recombination is called nonradiative recombination. The LED luminescence is a radiation recombination process, and the non-radiation recombination process is not beneficial to the luminescence of LED [7].

In LED, the non-radiation compound effect can only be reduced and cannot be eliminated. There is always a competitive relationship between radiation recombination and non-radiation recombination. The transition from band to bank and the transition through the trap can be either radiative recombination or non-radiative recombination. For instance, the inter-band recombination in indirect bandgap semiconductors is non-radiative recombination, while the recombination is radiative recombination through traps of isoelectronic levels.

In the process of structure design of LED, there are many ways to increase the radiation recombination probability of unbalanced carriers. For example, the structure of p–n junction is transformed from homojunction to heterojunction and/or double heterojunction to enhance the radiation recombination efficiency of electron–hole in the junction region [8]. In heterojunction structures, the region where electrons and holes recombine is called active region. Thinning the active region can improve the internal quantum efficiency of the LED and reduce the photon reabsorption in the active region. As the thickness of the active region getting close to the de Broglie wavelength of the electron, the energy of the carriers moving in the direction perpendicular to the junction is no longer continuous. This nanoscale active region is called a quantum well. The thin films with multi-layer quantum well structure can be epitaxially grown by MOCVD. At present, the active region of high brightness LED chips is composed of multiple-quantum-well (MQW) structure. The quantum well structure plays an important role in enhancing the radiation recombination probability of unbalanced carriers in LED [9].

2.3 LED Optical and Electrical Characteristics

2.3.1 LED Quantum Efficiency

Quantum efficiency is an important parameter related to radiation in the characteristics of light emitting diodes. It reflects the efficiency of photon generation by recombination of injected carriers. Quantum efficiency can refer to both internal quantum efficiency and external quantum efficiency.

Internal quantum efficiency refers to the ratio of the number of photons produced by the radiation recombination of semiconductors per unit time to the number of injected carriers, which reflects the efficiency of converting carrier current into photons. Ideally, all the photons emitted by the recombination of carrier radiation in the active region can be propagated to the outside. However, due to the existence of various loss mechanisms, the number of photons propagating outside the active region will be less than the number of photons generated in the active region. There are three main loss mechanisms:

(1) Absorption in LED materials: due to the characteristics of semiconductor materials;
(2) Fresnel loss: caused by internal reflection of semiconductor;
(3) Critical angular loss: When photons incident to the surface at an angle greater than the critical angle, they are caused by total reflection.

The ratio of the number of photons radiated to the outside in the active area of LED per unit time to the number of photons produced by the radiation recombination in the active area is called light extraction efficiency. The external quantum efficiency of LED refers to the ratio of the number of photons radiated to the outside of the diode per unit time to the number of carriers injected. It can be seen that the external quantum efficiency of LED is the product of the internal quantum efficiency and the light extraction efficiency [10].

2.3.2 Radiation Spectrum

Because the physical mechanism of LED luminescence is the recombination of electrons and holes in the active region with spontaneous radiation, the optical properties of LED are determined by spontaneous radiation.

The photon energy emitted in the active region of LED is not simply equal to the bandgap E_g of related materials because the energy distribution of conduction band electrons and valence band holes in the material follows certain rules. The following is deduced from the basic theory of semiconductor physics [11].

Suppose that the conduction band and valence band correspond to the energy of E_{Ck} and E_{Vk} at the wave vector k in the K space, respectively.

$$E_{Ck} = E_C + \frac{\hbar^2}{2m_C^*}k^2$$

$$E_{Vk} = E_V - \frac{\hbar^2}{2m_V^*}k^2$$

where m_C^* and m_V^* are the effective mass of electron and hole, E_C and E_V the conduction band bottom energy and valence band top energy, respectively.

There are:

$$\varepsilon(k) = E_{Ck} - E_{Vk} - \hbar\omega = \frac{\hbar^2}{2m_C^*}k^2 + E_g + \frac{\hbar^2}{2m_V^*}k^2 - \hbar\omega$$

$$= E_g + \frac{\hbar^2 k^2}{2}\left(\frac{1}{m_C^*} + \frac{1}{m_V^*}\right) - \hbar\omega = E_g + \frac{\hbar^2 k^2}{2m_r^*} - \hbar\omega$$

Among them, the effective mass of electron and hole is converted into effective mass.

Consider the conservation of energy:

$$\varepsilon(k) = E_g + \frac{\hbar^2 k^2}{2m_r^*} - \hbar\omega = 0$$

The density of states can also be deduced from the above formula, because, $d\varepsilon = \frac{\hbar^2}{m_r^*} kdkdk = \frac{m_r^* d\varepsilon}{\hbar^2 k}$.

so there are:

$$d\vec{k} = 4\pi k^2 dk = \frac{4\pi k^2}{\hbar^2 k} m_r^* d\varepsilon = \frac{4\pi k}{\hbar^2} m_r^* d\varepsilon = \frac{4\pi m_r^*}{\hbar^2}\frac{\sqrt{2m_r^*}}{\hbar}\sqrt{\hbar\omega - E_g d\varepsilon}$$

Among them $d\vec{k}$ refers to the volume element of K space, $d\vec{k} = dk_x dk_y dk_z$.

Furthermore, the density of states: $\rho = \frac{1}{2\pi^2}\left(\frac{2m_r^*}{\hbar^2}\right)^{3/2}\sqrt{\hbar\omega - E_g}$

If the distribution of carriers in the energy band follows Boltzmann distribution, the probability of distribution is a function of carrier energy.

$$f_B = e^{-\hbar\omega/k_B T}$$

From the above formula, it can be deduced that the intensity of photon radiation at different energies is proportional to the density of states of carriers and the Boltzmann function.

$$I(E = \hbar\omega) \propto \sqrt{E - E_g} e^{-\hbar\omega/k_B T}$$

Figure 2.11 shows the spontaneous emission spectrum of the LED. The maximum radiation intensity is

$$E = E_g + \frac{1}{2}k_B T$$

The spectral linewidth of LED emission light is an important parameter to measure LED. The emission spectrum width of LED is defined as the full width at half maximum (FWHM) intensity, which corresponds to the spontaneous emission spectrum of semiconductor. In the process of direct transition, the emission wavelength is basically determined by the bandgap energy. Other transition processes (non-radiative recombination process, Auger recombination process, etc.) will widen the emission spectrum [12].

2.3.3 Basic Photometric Concepts in LED

Photometry is the study of the perception of light intensity by the human eye. LED is a kind of light-emitting element, so photometric parameters are also critical for evaluating the performance of LED.

Luminous Intensity: Also known as photometry or intensity, luminescence intensity refers to the intensity of a point light source in a certain direction, symbolized as I, in candela (cd). It is defined as the quotient between the luminous flux emitted by the light source in this direction and the solid angle. Light intensity is often used to describe the distribution density of light flux emitted by light source in all directions of space or in selected directions. 1 cd = 1 lm/sr (lumen per sphericity).

Luminous flux: The sum of the energy emitted by the light source and received by the human eye is the luminous flux, denoted as Fv, in lumen (lm). Light flux, also known as light beam, is the radiation flux commonly used in the international evaluation of human visual characteristics. When LED is used for illumination, luminous flux is the basic quantity to evaluate the luminous ability of light source. Lumen is a unit of luminous flux. The luminous flux emitted by a point light source with uniform luminous intensity of 1 cd (candela) is 1 lm per unit solid angle, or all the luminous flux received from the source is 1 lm per unit area per unit distance.

Luminance: It refers to the light source or reflected by the object entering the eyes and imaging on the retina so that we can recognize its shape and shade. Brightness is the density of light intensity per unit surface in a certain direction. It is equal to the quotient of the intensity of light in this direction and the projection area of this plane in this direction, which is expressed by the symbol L. The unit of brightness is Candela per square meter (cd/m^2).

Illuminance: It refers to the surface density of the light flux received on the irradiated plane. The unit of illumination is Lux, and the symbol is lx. 1 lx is equal to the illumination produced by the uniform distribution of the luminous flux of 1 lm on the surface of 1m^2, i.e. 1 lx = 1 lm/m^2. Illuminance is an important indicator of illumination design, which indicates how bright the illuminated object is.

Luminous efficiency: It refers to ratio of the luminance of the light emitted by the light source to its power consumption, that is, the luminous efficiency (lm/W) = the luminance (lm)/the power consumption (W). Luminescent efficiency refers to the amount of light converted by each watt of electricity. The larger the value, the higher the efficiency of the light source is.

2.3.4 Electrical Characteristics of LED

The current–voltage characteristic of LED is similar to that of ordinary diode. Therefore, when the device is operated under forward bias, injection current and the luminance of the device increases with increasing voltage. When the forward voltage increases to a certain value, the luminance decreases while the forward current still increases. If there is no protection circuit, the LED device will be burned down due to the accumulation of heat resulted from the increase of the current. LED does not emit light when it is reversely biased. However, the LED does have reverse current which is very small. Usually, the reverse voltage of LED should not exceed 10 V, and the maximum is no more than 15 V. If the reverse bias voltage exceeds this voltage, the reverse breakdown will occur, which will lead to the damage of LED [13].

Current vs. voltage (I–V) characteristic is one of the most important parameters in evaluating the electrical performance of LED. The following important parameters can be obtained from the I-V characteristic curve:

Forward voltage V_F: The voltage of the LED when it is loaded with 20 mA or 350 mA forward current.

Forward current I_F: For low-power LED, its forward working current is generally 20 mA. For high-power chip, the forward working current should be determined according to the size of the chip, which is generally 350 mA.

Reverse leakage current I_R: It refers to the current of the LED when a 5 V reverse bias voltage is applied to the device. Usually the smaller the reverse leakage current is, the better the electrical performance of the LED is.

Reverse breakdown voltage V_R: In general, the reverse voltage V_R can be tested when the reverse current is maintained at a specified value. The reverse current is generally between 5 and 100 μA. Reverse breakdown voltage usually cannot exceed 20 V.

2.4 Principle of White LED

2.4.1 The Principle of Three Primary Colors and Addition of Light

Human eye recognizes color according to the wavelength of light. Most of the colors in the visible spectrum can be mixed by three basic colors in different proportions. The three basic colors are red, green and blue. When these three kinds of light mix in the same proportion and reach a certain intensity, they appear white (white light); if the intensity of the three kinds of light is zero, it is black (dark), which is the principle of additive method [14, 15]. Using the principle of additive method, white LED can be obtained through different ways.

2.4.2 The Realization Method of White LED

White light-emitting diodes have the advantages of small size, high efficiency, long life, energy saving and environmental protection, and will replace traditional light sources in the future [16]. From the packaging level, the main fabrication methods

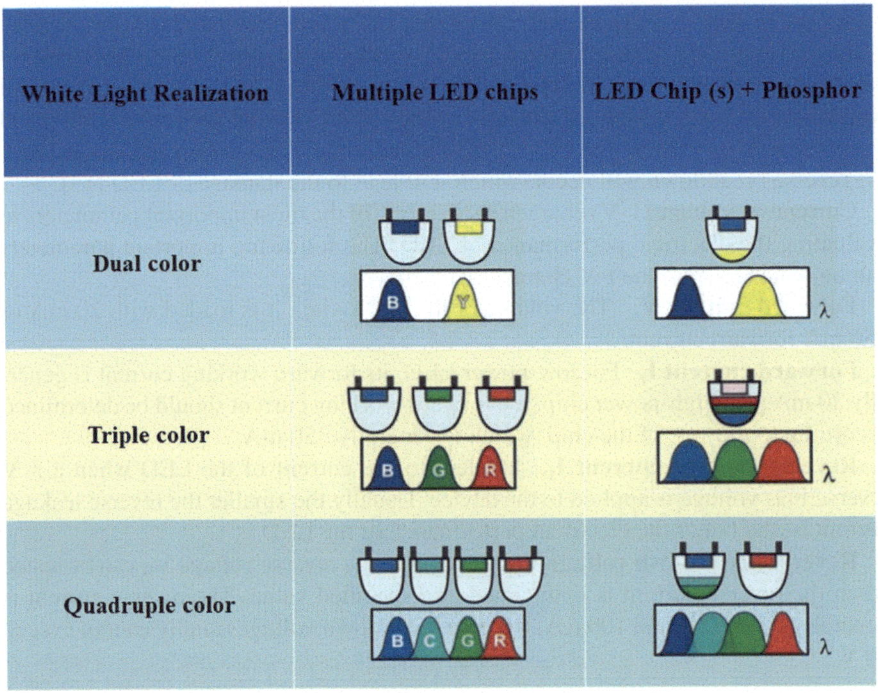

Fig. 2.2 Main implementation method of White LED

can be divided into two categories, as shown in Fig. 2.2: one is the multi-chip type, and the other is the chip type where phosphor is used to generate white light.

White light can be generated from multi-chip white LED thanks to the combination of various monochrome LEDs. It can be adjusted to white light color in many ways, with high color rendering. However, due to the differences in material system, driving voltage, temperature and light attenuation rate of each monochrome LED, and the need for multiple circuit designs to control the current separately, resulting in design difficulties and increased costs. At present, the industrialized white LED and the future development trend are still dominated by single chip type [17].

Single chip type is also named as pc-LED (phosphor-converted LED). In such design, white light is realized by exciting different phosphors by using a monochrome LED. There are many approaches to do so while the combination of a blue LED with yellow phosphor is a straightforward solution. At present, cerium doped yttrium aluminum garnet (YAG) [18] is mostly used as phosphor. The blue light emitting from the LED is partly used to excite the phosphor, and partly penetrates through the gap between the phosphors. The yellow light produced by the phosphor mixes with the blue light to generate blue-white quasi-white light. Since such design is simple and low cost, it is still the mainstream of the market at present. However, due to the lack of red light in its spectrum, it usually comes with relatively low color rendering. In contrast, ultraviolet LED together with red, green and blue phosphors can produce much better color rendering of white light. up to more than 90. Furthermore, various light colors can be combined by different proportions of phosphors, and have the advantages of simple single chip control circuit. Unfortunately, the drawback of such a design is that it is difficult and expensive to mix three phosphors, which also restricts the popularity of this technology. The combination of blue and red LED with cyan and green phosphors can achieve the best color rendering, but the technology is limited due to its complexity, including control circuit and phosphor ratio, and other issues needed to be further studied [19].

Single chip LED with phosphor to produce white light will inevitably cause energy loss, that is, Stokes energy loss which is in the range of 10–30%. In order to further improve the luminous efficiency of single-chip white LED and avoid the energy loss caused by phosphor conversion, researchers have developed novel structure white LED [20–22] through nano-pillar, nano-pyramid and quantum dot technology. These approaches will be introduced in Chap. 12.

References

1. S. Pimputkar, J.S. Speck, S.P. DenBaars, S. Nakamura, Prospects for LED lighting. Nat. Photonics **3**, 179–181 (2009)
2. E.F. Schubert, J.K. Kim, Solid-state light sources get smart. Science **308**, 1274–1278 (2005)
3. M.G. Craford, From Holonyak to today. Proc. IEEE **101**, 2170–2175 (2013)
4. I. Akasaki, GaN-based p-n junction blue-light-emitting devices. Proc. IEEE **101**, 2200–2210 (2013)
5. MorkoH. The p-n junction. Nitride Semiconductor Devices. Wiley. KGaA (2013)
6. Y.C. Shen, G.O. Mueller, S. Watanabe, et al. Auger recombination in InGaN measured by photoluminescence. Appl. Phys. Lett. **91**(14), 141101 (2007)
7. E.F. Schubert, T. Gessmann, J.K. Kim, Light emitting diodes. Wiley (2005)
8. H. Kroemer, The double-heterostructure concept: how it got starte. Proc. IEEE **101**, 2183–2187 (2013)
9. S. Nakamura, N. Senoh, N. Iwasa, S.I. Nagahama, High-brightness INGAN blue, green and yellow light-emitting-diodes with quantum-well structures. Jpn. J. Appl. Phys. Part 2 Lett. 1995, 34:L797-L799.
10. M.-H. Chang, D. Das, P.V. Varde, et al., Light emitting diodes reliability review. Microelectr Reliabil. **52**, 762–782 (2012)
11. M. Razeghi, *Fundamentals of solid state engineering* (Springer, US, 2006)
12. MorkoH. *Optical Processes. Nitride Semiconductor Devices* (Wiley, KGaA, 2013)
13. MorkoH. *Light-emitting Diodes and Lighting. Nitride Semiconductor Devices* (Wiley, KGaA, 2013)
14. H. Ries, I. Leike, J. Muschaweck, Optimized additive mixing of colored light-emitting diode sources. Opt. Eng. **43**, 1531–1536 (2004)
15. J. Orava, T. Jaaskelainen, J. Parkkinen, V.P. Leppanen, Diffractive CIE 1931 chromaticity diagram. Color Res. Appl. **32**, 409–413 (2007)
16. R. Mueller-Mach, G. Mueller, et al., Highly efficient all-nitride phosphor-converted white light emitting diode. Phys. Status Solidi Appl. Mater. Sci. **202**, 1727–1732 (2005)
17. D. Eisert, U. Strauss, S. Bader, et al., White light sources based on InGaN, in Proceedings of the IWN 2000, Sep 24–27, 2000, Nagoya, Japan (2000)
18. L. Chen, C.C. Lin, C.W. Yeh et al., Light converting inorganic phosphors for white light-emitting diodes. Materials **3**, 2172–2195 (2010)

19. R. Mirhosseini, M.F. Schubert, S. Chhajed et al., Improved color rendering and luminous efficacy in phosphor-converted white light-emitting diodes by use of dual-blue emitting active regions. Opt. Express **17**(3), 10806–10813 (2009)
20. K. Wu, T.B. Wei, D. Lan, et al., Phosphor-free nanopyramid white light-emitting diodes grown on {10-11} planes using nanospherical-lens photolithography. Appl. Phys. Lett. **103**(24), 241107 (2013)
21. S. Albert, A. Bengoechea-Encabo, M.A. Sanchez-Garcia, et al., Selective area growth and characterization of InGaN nanocolumns for phosphor-free white light emission. J. Appl. Phys. **13**(11), 114306 (2013)
22. H.P.T. Nguyen, Q. Wang, Z.T. Mi, Phosphor-free InGaN/GaN dot-in-a-wire white light-emitting diodes on copper substrates. J. Electron. Mater. **43**(4), 868–872 (2014)

Chapter 3
Properties and Testing of Group III-Nitride LED Materials

The group III nitride semiconductor material mainly includes the compounds GaN, AlN, InN, and their alloys such as AlGaN, GaInN, AlInN, and AlGaInN. The band gap (0.7~6.03 eV) of these materials with the different components covers the whole visible spectrum, which makes the nitride semiconductor materials attractive for a range of applications. However, due to the nature of the nitride material, there are many difficulties in the development process that need further research. This chapter mainly introduces the basic properties of nitride semiconductor materials, including crystal structure, band structure, polarization effect, doping, crystal defects and certain characterization methods to measure the physical properties of the materials.

3.1 Crystal Structure and Band Structure of Group III Nitride

3.1.1 Crystal Structure

The III-nitride semiconductor material has three crystal structures: hexagonal wurtzite, cubic sphalerite, and sodium chloride-type rock salt. The rock salt structure is generally formed only under extreme high pressure and will not be specifically introduced here.

The sphalerite structure is a cubic close-packed structure formed by nesting 1/4 diagonal lengths of two face-centered cubic lattice structures along the diagonal line [1]. A unit cell contains 8 atoms, of which 4 are Ga atoms and 4 are N atoms. This structure belongs to the thermodynamic metastable structure and is mostly prepared at low temperatures.

The wurtzite structure is a hexagonal symmetry structure formed by two hexagonal dense stack lattices translating 3/8 cell heights along the c-axis. A unit cell contains

J. Li et al., *III-Nitrides Light Emitting Diodes: Technology and Applications*,
Springer Series in Materials Science 306,
https://doi.org/10.1007/978-981-15-7949-3_3

12 atoms, of which 6 are Ga atoms and 6 are N atoms. The wurtzite structure is a thermodynamically stable structure of the nitride semiconductor material. It has two lattice constants, an in-plane lattice constant and an out-plane lattice constant c. The hexagonal system usually uses a four-axis coordinate system to describe the crystal orientation and crystal plane of the crystal lattice. On the bottom surface, the three axes a_1, a_2, and a_3 form an angle of 120° with each other, and the unit of measurement is a lattice constant of a. The c-axis is perpendicular to the bottom surface, and the unit of measurement is the lattice constant c. The crystal orientation and the crystal plane are expressed as [uvtw] and (hkil), respectively, and there are only three independent parameters in the four coordinates, satisfying $u + v = -t$ and $h + k = -i$, and sometimes the third coordinate can be omitted and expressed as [uvw] and (hkl).

Both structures of the nitride semiconductor are bonded to four N atoms or Group III atoms nearest to each other to form a tetrahedral structure. The difference is in the stacking order. The stacking order of wurtzite along the c-axis <0001> direction is ABABAB..., and the stacking order of sphalerite in the <111> direction is ABCABC.... The physical properties of the materials with two different crystal structure are also very different, as shown in Table 3.1. Due to the difficulty in material preparation and the quality of the crystal, the wurtzite structure of nitride semiconductor is the mostly investigated and widely used. The material properties and devices involved in this book are based on this structure.

The wurtzite structure GaN, AlN, InN compound semiconductors can be alloyed with each other to form a solid solution in different proportions. In this case, the lattice structure is unchanged, but the lattice parameters change with the composition. Their alloys can be expressed as $Al_xIn_yGa_{1-x-y}N$ $(0 < x + y < 1)$, where x and y are the contents of Al and In, respectively. The lattice constant of an alloy can be calculated by the following formula:

Table 3.1 Structural parameters of GaN material (300 K)

	Wurtzite structure	Sphalerite structure
Lattice constant (nm)	a = 0.3189 c = 0.5185 [2]	a = 0.452
Band gap (eV)	3.39 [3]	3.2
Effective mass of electron (m_0)	0.20	0.13
Effective DOS of the conductive band Nc/cm^{-3}	2.3×10^{18}	1.2×10^{18}
Effective DOS of the valence band Nv/cm^{-3}	4.6×10^{19}	4.1×10^{19}
Refractive index	2.67	2.5
Dielectric constant	$\varepsilon_r = 8.9$ $\varepsilon_\infty = 5.35$	5.3
Coefficient of the thermal expansion ($10^{-6}K^{-1}$)	$a_a = 5.59$ $a_c = 3.17$	–

Note m_0 is the resting mass of electron

$$a_{Al_xIn_yGa_{1-x-y}N} = \mathrm{x} \times a_{AlN} + y \times a_{InN} + (1 - x - y) \times a_{GaN} \quad (3.1)$$

$$c_{Al_xIn_yGa_{1-x-y}N} = \mathrm{x} \times c_{AlN} + y \times c_{InN} + (1 - x - y) \times c_{InN} \quad (3.2)$$

wherein a_{AlN}, c_{AlN}, a_{InN}, c_{InN}, a_{GaN}, c_{InN} represent lattice constants of the AlN, InN, and GaN binary compounds, respectively.

3.1.2 Band Structure

In the energy band theory, the problem can be described and analyzed more clearly using the inverted lattice. The Brillouin zone of the wurtzite structure of the III-nitride semiconductor material is shown in Fig. 3.1. Where k_x, k_y, and k_z are three coordinate axes orthogonal to each other in the inverted lattice. Γ is the center of the Brillouin zone at $k = 0$.

The energy band structure of AlN and InN is similar. The conduction band energy minimum and the valence band energy maximum are located at the Γ point of the center of the Brillouin zone, where $k = 0$. GaN, AlN, and InN are direct band gap semiconductor materials and are suitable as luminescent materials. Their band gap width at room temperature and related parameters are shown in Table 3.2. In addition to this, the second energy valley M-L valley and the third energy valley A valley are also found in the energy band. Due to the action of the crystal field and the spin-orbit coupling, the valence band splits into a heavy hole band, a light hole band, and a spin-coupled split band. E_{so} is the spin-coupled splitting energy, and E_{cr} is the splitting energy of the crystal field.

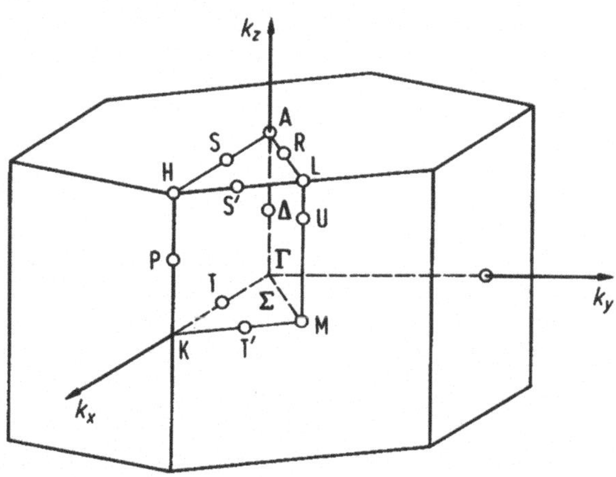

Fig. 3.1 The first Brillouin zone of the wurtzite crystal [4]

Table 3.2 Structural parameters of the wurtzite GaN, AlN and InN (300 K) [4]

	GaN	AlN	InN
Lattice constant (nm)	a = 0.3189	a = 0.3112	a = 0.3545
	c = 0.5186	c = 0.4982	c = 0.5703
Density (g/cm^3)	6.15	3.23	6.81
Band gap (eV)	3.39	6.026	0.641
Effective DOS of the conductive band Nc/cm^{-3}	2.3×10^{18}	6.3×10^{18}	9×10^{17}
Effective DOS of the valence band Nv/cm^{-3}	4.6×10^{19}	4.8×10^{20}	5.3×10^{19}
Coefficient of the thermal expansion (10^{-6}K^{-1})	$a_a = 5.59$	$a_a = 4.2$	$a_a = 3.8$
	$a_c = 3.17$	$a_c = 5.3$	$a_c = 2.9$
Effective mass of electron (m_0)	0.2	0.4	0.11
Effective mass of hole (m_0)			
Heavy hole	1.4	k_x:10.42 k_z:3.53	1.63
Light hole	0.3	k_x:0.24 k_z:3.53	0.27
Band splitting	0.6	k_x:3.81 k_z:0.25	0.65
Thermal conductivity (W·cm^{-1}·K^{-1})	1.3	2.85	0.45

In general, as the temperature increases, the energy band of the semiconductor material changes. The dependence of the GaN bandgap on temperature can be expressed by Eq. (3.3) [3]:

$$E_g(T) = E_g(0) - \frac{\alpha T^2}{T + \beta}(\text{eV}) \tag{3.3}$$

where Eg(T) represents the band gap width when the temperature is T, $E_g(0)$ is the band gap width at absolute zero, and α and β are the corresponding temperature parameters. The results of these parameters obtained by different studies can be different [5].

For the rough estimation of the forbidden band width of the alloy material, a linear relationship similar to the calculation of the lattice constant can be adopted. For the accurate calculation, the nonlinear factor needs to be considered, and the forbidden band width is expressed as:

$$E_{g\text{Al}_x\text{In}_y\text{Ga}_{(1-x-y)}N} = xE_{g\text{AlN}} + yE_{g\text{InN}} + (1-x-y)E_{g\text{GaN}} - b_{\text{Al}}x(1-x) - b_{\text{In}}y(1-y) \tag{3.4}$$

where b_{Al} and b_{In} are the bending indexes. Materials with different compositions have different bending indexes. Empirical values are usually used for some specific compositions.

3.2 Polarization Effect of Group III Nitride Materials

3.2.1 Polarization Effect

The wurtzite structure nitride semiconductor crystal does not have central symmetry. Taking GaN as an example, in the c-axis direction, the plane formed by Ga atoms (Ga plane) is separated from the plane formed by N atoms (N plane). This results in the spontaneous polarization effects due to the misalignment between the positive and negative charge centers. The polarization direction is along the c-axis [0001] direction. Usually, the (0001) plane is called the *c*-plane, the {1–100} plane is the *m*-plane, the {11–20} plane is the *a*-plane, and the {1–102} plane is the *r*-plane [6]. The c-plane is a polar plane. The epitaxial growth of GaN on the sapphire substrate along the c-axis direction is a relatively mature method for obtaining a better-quality crystal. The most studied group III nitride semiconductor devices are based on the *c*-plane Ga-polar face. In order to reduce the influence of the polarization effect on the performance of the device, nitride devices built on non-polar surfaces such as *a*- and *m*-plane or semi-polar *r*-plane at a certain angle with respect to the *c*-plane have also been studied. Although there are still many difficulties in the quality of epitaxial crystals, many studies have reported the high-performance devices on non-polar or semi-polar crystal planes, where the performance of the devices is close to that built on polar materials.

The spontaneous polarization of Group III nitrides is negative, indicating that the spontaneous intensity is in the opposite direction of [0001]. The spontaneous polarization of GaN, InN, and AlN are $-0.034\,\mathrm{C/m^2}$, $-0.042\,\mathrm{C/m^2}$, and $-0.090\,\mathrm{C/m^2}$, respectively. The Vegard interpolation method is used to calculate the spontaneous polarization of the ternary compound $A_xB_{1-x}N$:

$$P_{sp}(A_xB_{1-x}N) = xP_{sp}(AN) + (1-x)P_{sp}(BN) + b_{ABN}x(1-x) \tag{3.5}$$

The calculation formula for the spontaneous polarization of quaternary alloys is:

$$\begin{aligned} P_{sp}\left(Al_xIn_yGa_{1-x-y}N\right) &= xP_{sp}(AlN) + yP_{sp}(InN) + (1-x-y)P_{sp}(GaN) \\ &\quad + b_{AlGaN}x(1-x-y) + b_{\mathrm{In}\,GaN}y(1-x-y) + b_{Al\,\mathrm{In}\,N}xy \\ &\quad + b_{Al\,\mathrm{In}\,GaN}xy(1-x-y) \end{aligned} \tag{3.6}$$

where b is the correction factor, $P_{sp}(AlN) = -0.09$, $P_{sp}(GaN) = -0.034$, $P_{sp}(InN) = -0.042$, $b_{AlGaN} = 0.019$, $b_{InGaN} = 0.038$, and $b_{AlInN} = 0.071$ [7, 8]. Since the value of $xy(1-x-y)$ is small, the last item in the calculation is usually ignored.

In addition to spontaneous polarization, the wurtzite nitride semiconductor material also shows a piezoelectric effect. The lattice constants of AlN, GaN, InN materials and their alloys vary with composition. When heterojunction is formed by different materials, the epitaxial materials are subjected to tensile or compressive stress due to lattice mismatch. The lattice constant will change when strain is generated by stress in the critical thickness range. Polarized charges or piezoelectric effect will appear at the interface resulted from the lattice strain. The total polarization of the material is the sum of spontaneous and piezoelectric polarization.

The piezoelectric polarization is related to the strain ε of the material and is expressed as the ratio between the difference of the lattice constants of the substrate and the epitaxial layer to the lattice constant of the substrate:

$$\varepsilon = (a_{\text{subs}} - a_L)/a_L \tag{3.7}$$

where a_{subs} is the substrate lattice constant, and a_L is the epitaxial layer lattice constant.

The piezoelectric polarization of AlN, GaN, and InN are:

$$P_{pz}\text{AlN} = \begin{cases} -1.808\varepsilon + 5.624\varepsilon^2 \varepsilon < 0 \\ -1.808\varepsilon - 7.888\varepsilon^2 \varepsilon > 0 \end{cases} \tag{3.8}$$

$$P_{pz}(\text{GaN}) = 0.918\varepsilon + 9.541\varepsilon^2 \tag{3.9}$$

$$P_{pz}(\text{InN}) = -1.373\varepsilon + 7.559\varepsilon^2 \tag{3.10}$$

The piezoelectric polarization strength of the alloy $Al_xIn_yGa_{1-x-y}N$ [7] can be expressed as

$$P_{pz}\left(Al_x\,\text{In}_y\,Ga_{1-x-y}N\right) = P_{pz}(\text{AlN})x + P_{pz}(\text{In N})y + P_{pz}(Ga\text{N})(1-x-y) \tag{3.11}$$

The total polarization is the sum of spontaneous and piezoelectric polarization:

$$\text{P}\left(Al_xIn_yGa_{1-x-y}N\right) = P_{pz}\left(Al_xIn_yGa_{1-x-y}N\right) + P_{sp}\left(Al_xIn_yGa_{1-x-y}N\right) \tag{3.12}$$

3.2.2 *Influence of Polarization Effect*

The active region is usually a multiple quantum well structure when nitride semiconductor material is used as the light-emitting device. In the quantum well structure, a polarization electric field is generated at the interface of the heterojunction due to the polarization phenomenon. The energy band in the quantum well is tilted under the action of the polarization electric field. The spatial distribution of the electron and hole wave functions changes, the electrons and holes are spatially separated, and the recombination probability decreases that results in the reduction of luminous efficiency [8]. This is the Quantum Confined Stark Effect (QCSE) [9]. It is precisely because of these effects of the c-plane polarization in light-emitting devices that researchers are continually trying to improve and search for non-polarized surfaces (*a*-plane, *m*-plane) or semi-polarized surface devices [10, 11], hoping to avoid or reduce the polarization effects [6].

3.3 Doping of Group III-Nitride LED Materials

3.3.1 Doping of Nitride LED Materials

The doping of the Group III nitride semiconductor is mainly exemplified by wurtzite GaN. Since undoped GaN has N vacancies, it exhibits a weak n-type [12, 13]. The electron concentration is on the order of $10^{16}/cm^3$. Table 3.3 shows the ionization energies of the donor doping elements in GaN. It can be seen from the table that the Si doping in the place of the Ga site is a shallow donor impurity. As the minimum ionization energy is 0.012–0.02 eV, the activation efficiency is quite high. In the actual process, silane (SiH_4) is usually used for Si doping of the nitride material to form an n-type material, which can be simultaneously completed during the epitaxial growth of the material. Many studies have shown that Si doping can reduce point defects in GaN materials, improve interface free energy, change the atomic mobility during material growth, and improve the crystal quality of GaN materials. In the multi-quantum well structure, the doped Si atoms in the GaN material are activated to generate interface charges, partially shield the polarized electric field and attenuate the quantum-confined Stark effect.

The n-type doping of GaN is easy to implement. On the other hand, p-type doping is difficult to achieve. This has been a challenge to realize GaN-based semiconductor material LEDs. Some of the important acceptor doping materials in GaN materials are listed in Table 3.4. It can be seen from the table that the Mg has the smallest ionization energy. The absolute value is, however, still large that leads to the low activation efficiency. It is difficult to obtain a p-type GaN material with high hole concentration. At the same time, the Mg is passivated by the H impurities during the growth of nitride and cannot be activated effectively, resulting in high resistance characteristics. Until H. Amano et al. successfully realized the p-type doping of Mg in GaN by low energy electron beam irradiation (LEEBI), the Mg-doped GaN with a Mg concentration of $10^{20}/cm^3$ can only achieve a hole concentration of $2 \times 10^{16}/cm^3$. Although the

Table 3.3 Common donor doping in wurtzite Ga [4]

Donor	Substitutional site	Impurity ionization energy (eV)
Si	Gasite	0.012–0.02
V_N	Nvacancy	0.03–0.1
C	Ga site	0.11–0.14
O	N site	0.03

Table 3.4 Common acceptor doping in wurtzite GaN

Acceptor	Substitutional site	Impurity ionization energy (eV)
V_{Ga}	Ga vacancy	0.14
Mg	Ga site	0.14–0.21
Si	N site	0.19
Zn	Ga site	0.21–0.34
Hg	Ga site	0.41
Cd	Ga site	0.55
Be	Ga site	0.7
Li	Ga site	0.75
C	N site	0.89
Ga	N site	0.59–1.09

activation rate of Mg is only 2×10^{-4}, it brings the hope to the possibility of Mg doping. In 1992, S. Nakamura utilized annealing at 700 °C in nitrogen atmosphere to active Mg atoms. This lead to a hole concentration of $3 \times 10^{17}/\text{cm}^3$ [14]. Up to now, the Mg doping, in combination with the annealing process, has become a commonly used p-type standard process for GaN. It is generally believed that Mg doping in GaN materials will affect the material quality due to the large Mg atoms. However, some research results show that Mg doping is beneficial to the reduction of V-type defects in GaN materials and improves the crystal quality of GaN materials.

3.4 Test and Analysis of the Properties of Group III Nitride Materials

The performance of the device is directly affected by the quality of the material. The characterization and analysis of the material can reflect the quality of the material and guide the adjustment of the material growth process. In this section, we briefly describe the common analytical methods for testing the properties of nitride materials.

3.4.1 Structural and Morphological Analysis

The structure and morphology of crystals are fundamental research in the application of materials. Especially in semiconductors, the crystal structure determines the band structure of the material. The band structure of the material is the main factor to be considered for light-emitting device. Since the melting temperature of the GaN material is higher than the decomposition temperature, the preparation of the single crystal is not as easy as that of the first generation and the second generation semiconductors. As such, the quality of the crystal is not very high. In this case, the characterization of crystal structure and morphology is particularly important in improving the quality of crystal growth. The following are some of the commonly used methods for analyzing structures and morphologies in the application of Group III nitride semiconductor materials.

High-resolution X-ray Diffraction (HRXRD): X-ray diffraction can characterize the crystal quality, dislocation density, composition, defects, stresses, strains, and lattice constants in different epitaxial layers. It has the advantages of no damage to the material, no pollution, high efficiency and high precision. The information of the material is obtained by analyzing the diffraction pattern. X-rays with a wavelength of 0.01–10 nm are extremely penetrating. Its working principle is Bragg's law. When the angle between the incident X-ray and the crystal plane meets Bragg's law, the diffraction intensity reaches a maximum value:

$$2\mathrm{d}\sin\theta = n\lambda \tag{3.13}$$

where θ is the angle between the incident X-ray and the crystal plane, that is, the Bragg angle, d is the interplanar spacing, λ is the X-ray wavelength, and n is the diffraction order.

X-ray double crystal diffraction is often used in practical applications. The three commonly used scan modes are:

ω mode or the rocking curve: The detector is fixed, no slit is placed in front of the detector, and the sample is rotated by ω. This mode is used to evaluate the quality of the crystal, the shape and number of dislocations in the crystal, and the dislocation density.

2θ mode: The detector and the X-ray source are fixed. The slit is added in front of the detector, and the sample is rotated by 2θ. This mode can be used to study the dispersibility of the crystal domain of the sample.

2θ-ω mode: On the basis of the ω scan, a slit is added in front of the detector. The sample and the detector are rotated by ω and 2θ, respectively. The rotation angle ratio is 1:2. The lattice constant can be calculated from the position of the diffraction peak. Component information can be obtained for the alloy material. The thickness of the epitaxial layer and the superlattice structure can also be measured.

Scanning Electron Microscopy (SEM): Scanning electron microscopy uses a very fine electron beam to scan a sample to excite secondary electrons on the surface of the sample. The number of secondary electrons is related to the angle of incidence of the electron beam so that the surface structure can be obtained via the number of the received electrons. Since the De Broglie wavelength of electrons is much smaller than the wavelength of visible light, its resolution is much larger than that of an optical microscope. The best resolution can reach 0.1 nm.

Transmission Electron Microscopy (TEM): The principle of transmission electron microscopy is the same as that of optical microscopy. It is a microscopic imaging technique in which an electron beam is emitted by an electron gun, where an electromagnetic beam is used to focus the image. The accelerated and focused electron beam is projected onto an ultra-thin sample with a periodic structure, and the electrons are diffracted to form a diffraction spectrum. After the diffraction spectrum is recombined and imaged in the image plane, different types of TEM images can be obtained. Transmission electron microscopy is currently the highest spatial resolution microscope with a spatial resolution of up to one crystal lattice. It can be used to analyze the fine structure inside the material, to observe the defects such as dislocations in the material and the state of the heterojunction interface.

Atomic Force Microscopy (AFM): Atomic force microscopy is based on an improved scanning tunneling microscope (STM). It overcomes the shortcomings of scanning tunneling microscopes that are not suitable for non-conductive samples. AFM is the most commonly used method for characterizing the surface topography of semiconductor materials. Using a sharp tip to move on the surface of the sample, the surface topography of the sample is imaged by a force test between the tip and the sample. The AFM has a cantilever beam that is sensitive to weak forces. The cantilever beam is fixed at one end and has a sharp tip at the other end. When the needle scanning, the interaction force between the tip and the sample causes the cantilever beam to deform. The back of the tip is machined into an optical mirror that is reflected onto the photodetector when it is incident on the back of the cantilever beam. The deformation of the cantilever beam causes the position of the spot on the detector to shift, reflecting information on the surface topography of the sample. The main parameter of the AFM test is the surface roughness, which characterizes the degree of fluctuation of the surface flatness. With high lateral and vertical resolution, the general lateral resolution can reach 0.1~0.2 nm, and the vertical resolution can reach 0.01 nm.

3.4.2 *Surface and Film Composition Analysis*

GaN materials are mainly obtained by epitaxial growth on a substrate such as sapphire. The analysis methods for the film and the surface are as follows.

Auger Electron Spectroscopy (AES): When the electrons in the atom are excited to form holes, the electrons are unstable at high energy levels and will transition to low energy levels. When the energy is radiated in the form of photons, characteristic X-rays are formed. When the energy of the transition is absorbed by another electron and is emitted, Auger electron is formed. Auger electron spectroscopy uses the interaction between the incident electron beam and the matter to excite the inner electrons of the atom. The energy released by the outer electrons to the inner electron transition causes the other electron outside the nucleus to be excited into free electrons, i.e., Auger electrons. The electron energy analyzer is used to collect and analyze the energy of Auger electrons to form a curve of the number of Auger electrons as a function of electron energy, that is, the Auger electron spectrum. Auger electrons can escape the solid surface with only a few layers of atoms on the surface, so Auger electron spectroscopy is an analytical method for solid surface properties.

The elemental X-ray emission probability of the element with large atomic number is large, but the Auger electron emission probability of the element with small atomic number is large. The Auger electron spectroscopy method is more suitable for the analysis of light elements.

X-ray Photoelectron Spectroscopy (XPS): X-ray photoelectron spectroscopy uses high-energy X-rays as excitation sources to excite the photoelectron spectrum of the inner electrons of the atom. Briefly, a beam of X-rays with a certain energy irradiates a solid sample, where the incident photons interact with the sample. The inner electrons absorb photon energy to overcome the binding energy and work function. The remaining energy is emitted as kinetic energy and becomes photoelectron. This process is the photoelectric effect. Photoelectron energy is:

$$E_k = h\upsilon - E_b - \varphi \tag{3.14}$$

where E_k is the kinetic energy of photoelectrons in the photoelectric process, $h\upsilon$ is the energy of the incident X-ray photons, E_b is the binding energy of the bound electrons of the inner shell, and φ is the work function of the spectrometer.

The inner electron binding energy for a given atom shows unique characteristic because it is very different in different molecules. The photoelectron spectrum is obtained by analyzing the photoelectron with an energy analyzer. X-ray photoelectron spectroscopy has evolved into one of the commonly used methods for surface composition analysis.

Secondary Ion Mass Spectrometry (SIMS): Secondary ion mass spectrometry is a method of mass spectrometry of solid surfaces by ion sputtering combined with mass spectrometry. The working principle is as follows: a primary ion of a certain energy is sputtered on the sample target to generate positive and negative secondary ions. The chemical composition of the atomic, molecular or atomic mass ratio after ionization is analyzed by a mass spectrometer. This method can only obtain information on several atomic layers or even a single atomic layer on the surface. The particles sputtered on the surface of the ion bombardment sample have a large part of neutral atoms, and only those secondary ions that are subjected to the electric field magnetic field can enter the analysis system. Therefore, increasing ion yield is important to secondary ion mass spectrometry. There are many factors that affect ion yield, such as atomic number, chemical environment, primary ion species, and incident energy. Experiments show that the use of electron-negative primary ions can greatly increase the yield of positive secondary ions. The positive primary ions can greatly increase the yield of negative secondary ions. Therefore, when using a positive ion to analyze the surface of a material, O^- is often used as a primary ion. When using negative ions to analyze the surface of materials, Cs^+ particles are commonly used to bombard the surface to improve detection sensitivity.

3.4.3 Other Photoelectric Test Methods

In addition to the above analytical methods, there are also many other methods for analyzing materials using the photoelectric properties of materials.

Photoluminescence Spectroscopy (PL): Photoluminescence, as its name implies, utilize the way that light is used to excite the semiconductor illumination. The PL spectrum [15] is one of the commonly used optical characterization methods in the semiconductor field. Its basic principle is to illuminate the sample with a laser with photon energy greater than the forbidden band width. Such a process will generate electrons and excite them from valence band to the conduction band to form unbalanced carriers. The excited electrons will then transit to a lower energy level and recombine with the holes. Meanwhile, the recombination energy is released in the form of photons. The detector receives photons and converts them into electrical signals. A distribution curve of luminous intensity versus photon energy or wavelength is commonly plotted or so called a photoluminescence spectrum.

The mechanisms of electron-hole recombination luminescence include direct radiation recombined luminescence between bands, indirect radiation recombination between bands, radiation recombination between energy band and impurity level, donor-acceptor recombination, free exciton radiation recombination, and binding sub-radiation recombination and deep level impurity radiation recombination. These luminescence mechanisms all form corresponding spectral lines in the PL spectrum. In addition to radiation recombination luminescence, there are many non-radiative recombination that are detrimental to device illumination. At present, many PL test equipment can be used at low temperatures such that the non-radiative recombination is greatly reduced. Measurement at low temperature is advantageous for observing the fine structure of the luminescence peak. Information on the energy band structure, defects, and impurities of the semiconductor can be obtained by analysis of the photoluminescence spectrum.

Cathodoluminescence Spectroscopy (CL): The principle of cathode fluorescence measurement is similar to that of photoluminescence. It uses a high-energy electron beam as an excitation source to excite a semiconductor material to emit a characteristic fluorescence spectrum with energy less than the band gap. Photoluminescence is non-destructive but cathode fluorescence can damage the surface of the sample. The incident electron energy is generally 1–20 keV. When testing the sample, the sample can be at room temperature or be cooled by liquid nitrogen. The cathode fluorescence spectrum can be used to study the luminescence properties of luminescent semiconductors, especially for the luminescence properties of various semiconductor quantum wells, quantum wires, quantum dots and other nanostructures.

Raman: Raman spectroscopy is a scattering spectrum based on Raman scattering. When light interact with medium molecules, the molecules are forced to vibrate to produce scattering. The frequency of the scattered light is generally the same as the frequency of the incident light. This scattering is called Rayleigh scattering and is an elastic scattering. The scattering of the scattered light frequency different from the incident light frequency is Raman scattering, which is an inelastic scattering of the photon by the molecule. In the Raman scattering, a line having a frequency of $\upsilon_0 \pm \upsilon_i$ (i = 1, 2, 3...) is accompanied around each incident line (υ_0). The line on the long wavelength side is called the red line or the Stokes line, and the line on the short side is called the purple line or the anti-Stokes line. Since υ_i is independent of υ_0, it is determined by the nature of the scattering material. Each scattering material has its own specific frequency, and the Raman scattering spectrum can be used to analyze information such as the composition and structure of the crystal or molecule.

UV-vis Spectrophotometer (UV-vis): The UV-visible absorption spectrum of a molecule is the absorption spectrum produced by the electron-level transition of a molecule or atom that absorbs radiation. Since various substances have different molecules, atoms, and different molecular space structures, the absorption of light energy is also different. Therefore each substance has its own unique absorption spectrum. Many materials have characteristic absorption peaks in the ultraviolet-visible (200–800 nm) region. The amount of the substance can be determined based on the absorbance at certain characteristic wavelengths on the absorption spectrum, which is the basis for qualitative and quantitative analysis of the spectrophotometer. The UV-Vis spectrophotometer is an effective mean to study the composition, structure and interaction of substances according to the absorption line of the substance in the ultraviolet-visible region.

Hall effect test (Hall): When electrons in a conductor or semiconductor material move under a magnetic field perpendicular to the electric field, the orbital shift occurs due to the Lorentz force, creating a lateral potential difference in the sample. This phenomenon is the Hall effect [16]. The Hall electric field is expressed as:

$$E_H = R_H \mathrm{JB} \tag{3.15}$$

where R_H is the Hall coefficient, J is the current density, and B is the magnetic field strength.

By measuring the Hall coefficient and conductivity of the conductor material, the key parameters such as the conductivity type, carrier concentration and carrier mobility of the material can be obtained. If the Hall coefficient is measured as a function of temperature, information such as the ionization energy of the material can be obtained.

Current-Voltage Test (I-V): The I-V measurement can be simply done by applying a voltage across the device, testing the current magnitude, and plotting the current as a function of voltage (I-V curve). This method is capable of measuring the electrical conductivity and other fundamental properties of a semiconductor material.

References

1. J. Pankove, T. Moustakas, *Gallium Nitride (GaN) I*. Academic Press (1998)
2. M. Leszcynski, I. Grzegory, M. Bockowski, X-ray examination of GaN single crystals grown at high hydrostatic pressure. J. Cryst. Growth **126**(4), 601–604 (1993)
3. M. Levinshtein, S. Rumyantsev, M. Shur, *Properties of Advanced Semiconductor Materials GaN, AlN, InN, BN, SiC, SiGe* (Wiley, New York, 2001), pp. 1–30
4. V. Siklitsky, Electronic archive: new semiconductor materials. Characteristics and properties [EB/OL] (2012). http://www.ioffe.rssi.ru/SVA/NSM/Semicond/index.html
5. E. Monroy, N. Gogneau, F. Enjalbert et al., Molecular-beam epitaxial growth and characterization of quaternary III-nitride compounds. J. Appl. Phys. **94**(5), 3121–3127 (2003)
6. S. Dmitry, B. Rajaram, Z. Chung, Gallium indium nitride-based green lasers. J. Lightwave Technol. **30**(5), 679–699 (2012)
7. O. Ambacher, J. Majewski, C. Miskys et al., Pyroelectric properties of Al(In)GaN/ GaN hetero- and quantum well structures. J. Phys. Condens. Matter. **14**(13), 3399–3434 (2002)
8. T. Takeuchi, C. Wetzel, S. Yamaguchi et al., Determination of piezoelectric fields in strained GaInN quantum wells using the quantum-confined stark effect. Appl. Phys. Lett. **73**, 1691 (1998)
9. S.F. Chichibu, A.C. Abare, M.S. Minsky et al., Effective band gap inhomogeneity and piezoelectric field in InGaN/GaN multiquantum well structures. Appl. Phys. Lett. **73**, 2006 (1998)
10. S. Yamamoto, Y.J. Zhao, C.C. Pan et al., High-efficiency single-quantum-well green and yellow-green light-emitting diodes on semipolar (2021) GaN substrates. Appl. Phys. Express **3**, 122102 (2010)
11. I. Koslow, J. Sonoda, R. Chung, et al., High power and high efficiency blue InGaN Light emitting diodes on free-standing semipolar (3031) bulk GaN substrate. Jpn. J. Appl. Phys. **49**, 8R, 080203 (2010)
12. S. Nakamura, GaN growth using GaN buffer layer. Jpn. J. Appl. Phys. **30**(10A), L1705–L1707 (1991)
13. J. Neugebauer, C.G. Van de Walle, Atomic geometry and electronic-structure of native defects in GaN. Phys. Rev. B **50**, 8067 (1994)
14. S. Nakamura, M. Senoh, T. Mukai, Highly p-typed My-doped GaN films growth with GaN buffer layers. Jpn. J. Appl. Phys. **30**, 1708–1711 (1991)
15. J. Lu, C. Chen, *Modern Analytical Technique* (Tsinghua University Press, 1995) ISBN: 9787302018308 (in Chinese)
16. D.R. Yang, *Semiconductor Material Testing and Analysis* (China Science Publishing, 2009) ISBN: 9787030270368 (in Chinese)

Chapter 4
Epitaxial of III-Nitride LED Materials

III-nitrides include GaN, InN, AlN and their ternary and quaternary solid solutions. Since they are all direct band gap semiconductors, they are particularly suitable for fabricating light-emitting devices. However, the bulk single crystal growth conditions of the nitride material are complicated and require high temperature and high pressure. The current nitride material is mainly prepared by heteroepitaxial on other substrates. In 1986, Hiroshi Amano and Akasaki used MOCVD epitaxial low-temperature AlN buffer layer to successfully grow GaN films with good surface structure and good crystal quality. This work laid the foundation for the subsequent growth of nitride LED epitaxial materials. In addition to the MOCVD, LPE, MBE, HVPE, etc. can also be used to prepare nitride LEDs.

GaN-based white LEDs have great advantages compared with traditional illumination sources in terms of energy saving, environmental protection, etc. They are a new generation of illumination sources with broad market prospects. The white LEDs that use blue-chips to emit phosphors are the current mainstream technology. The preparation process can be divided into several parts such as substrate material preparation, material epitaxial, chip fabrication and packaging. The most critical process step is the epitaxial growth of materials. The material epitaxial process determines the basic performance of the LED such as wavelength, voltage, internal quantum efficiency and output power.

4.1 Basic Models of Epitaxial

According to its growth mode, the film can be divided into the following three types

J. Li et al., *III-Nitrides Light Emitting Diodes: Technology and Applications*,
Springer Series in Materials Science 306,
https://doi.org/10.1007/978-981-15-7949-3_4

4.1.1 3-D Growth Mode (Volmer-Weber Mode)

If the lattice mismatch between the substrate and the deposited film is relatively large, or the temperature of the substrate is relatively low, the atoms reaching the surface of the substrate are first aggregated into a core at a location that is relatively easy to nucleate. The atoms that reach the substrate later gather around the core, causing the core to grow up and form a thin film. Films formed in a 3-D growth mode generally exhibit polycrystalline morphology. Low temperature GaN nucleation layers deposited on sapphire substrates generally prefer a 3-D growth mode. The 3-D growth mode can be divided into the following four stages.

(a) Nucleation stage: the source gas that reaches the surface of the substrate by the transport of the carrier gas is decomposed, and the Ga and N atoms are transported and adsorbed on the surface of the substrate. They gather at a place where nucleation is easy. During the physical adsorption process, this part of GaN nucleation is not easily decomposed. Other atoms that undergo chemisorption are discharged with the carrier gas from the reaction chamber.
(b) Nucleus growth into small islands: The nucleation sites adsorbed on the surface of the substrate, where the atoms deposited on the substrate, are relatively easily to adsorb. These cores continue to grow into large grains and become islands one by one.
(c) The islands are connected into a grid: As the deposited atoms increase, the nucleus grows and the adjacent islands come into contact (so called coalescence). The islands which are contacted with each other, can release a part of the energy due to the coalescence of the interface and reduce the surface energy. The islands that are in contact with each other tend to be connected to each other to form a grid.
(d) Grid forming a film: As the deposited atoms increase, the meshes heal each other and form a film.

4.1.2 2-D Growth Mode (Frank-Vander Merwe Mode)

The 2-D growth mode generally occurs when the homoepitaxial or substrate temperature is relatively high, but the atomic deposition rate is relatively low. The atoms deposited on the surface of the substrate are adsorbed to the surface of the substrate during transport on the surface of the substrate to form a two-dimensional core. This requires that the surface of the substrate be relatively flat, free of defects or defective outcrops, and that the surface energy of the substrate is consistent with the surface energy of the film. This is difficult to achieve when GaN film is epitaxially grown on sapphire substrate.

4.1.3 2-D and 3-D Mixed Growth Mode (Stranski-Krastanob Mode)

Epitaxial GaN film grown on sapphire substrate is typically of a 2-D and 3-D hybrid growth mode. This is the intermediate state of the above two growth modes. The low temperature nucleation layer provides a 3-D nucleation island, and the subsequent GaN film epitaxially grows on these islands, resulting in a two-dimensional and three-dimensional mixed growth. The high binding energy position close to the step is a preferred position for the deposited atoms. If there is sufficient migration ability, atom can be diffused to the step position to perform epitaxial growth with a step as a medium.

Epitaxial GaN films on sapphire substrates generally have three steps as nucleation sites for epitaxial growth of layered films.

1. The nucleation of the atomic group forms a step. The nucleation of the atomic group is mainly a nucleation point formed at low temperature. In the high temperature state, the migration rate of the atomic group is high. In other words, the atomic group is easy to decompose, and difficult to form a step for the atomic group;
2. The internal dislocations form a step on the surface of the outcrop. In GaN materials, the density of dislocations can be as high as 10^8 cm^{-2}. The steps in GaN materials are mainly formed by the outcrops of dislocations on the surface.
3. The offcut of the crystal forms a step. For a small offcut angle, the surface of the substrate will form a number of steps with a certain periodic spacing. The sapphire substrate we are using now has a certain offcut angle. When a GaN thin film is epitaxially grown on a sapphire substrate using a (0001) plane, the sapphire substrate is generally in the m [1010] direction where the offcut angle is about 0.2°. The offcut angle is θ, the step height is h, and the step spacing is $L = h/\tan\theta$. In the GaN material, the offcut angle is 0.2°, and the step height is assumed to be the thickness of one atomic layer, and the distance between the steps is:

$$L = c_{(GaN)}/\tan(0.2^\circ) = 5.186 \times 10 - 10\ \mathrm{m}/0.00349 = 1.486 \times 10 - 7\ \mathrm{m} \approx 150\ \mathrm{nm} \quad (4.1)$$

Which growth method is adopted depends mainly on the relative size of the deposition rate and the mobility (mainly the substrate temperature). In principle, if the deposition rate is sufficiently low in a high vacuum system, step flow mode epitaxial growth can be performed at any temperature.

Different substrate temperatures have different growth modes:

Low temperature: amorphous, low atomic diffusion rate, not epitaxial growth, annealing required;
Medium temperature: growth by formation of two-dimensional atomic groups;
High temperature: epitaxial growth with a step as a medium.

4.2 Substrate for Epitaxial Growth of III-Nitride LEDs (Sapphire/Si/SiC/$LiAlO_2$/GaN)

Since GaN materials lack a suitable homogeneous substrate, GaN-based LEDs are generally heteroepitaxially grown on substrates such as sapphire, silicon carbide (SiC), silicon (Si), gallium arsenide (GaAs), and the like. Among them, the most widely used substrate is sapphire which has become the material of choice for commercial application of various LED epitaxial factories. Since 2010, the demand for sapphire substrates has increased tremendously.

There is a large lattice mismatch and thermal expansion mismatch between the sapphire substrate and the GaN-based epitaxial material. The GaN material epitaxially grown on the sapphire substrate has dislocations in the order of 10^8 cm^{-2}. However, sapphire substrates have incomparable advantages over other heterogeneous substrates, and may be replaced by other substrates in the future although the sapphire is still currently the most suitable substrate for epitaxial GaN-based LEDs. Below we briefly introduce the advantages and disadvantages of various substrates.

Si substrate: The wide application of Si substrate in the current microelectronics industry is the basis for the development of integrated circuit technology. At the same time, the development of Si substrates is also the most mature, and commercial 8-inch and 12-inch silicon wafers have emerged. However, the epitaxial GaN material on the Si substrate has a large number of misfit dislocations. The performance of GaN-based LEDs on Si is currently not as good as the LEDs on the sapphire substrate. However, the cost advantage of Si substrate and its wide application in the integrated circuit industry make Si substrate attractive over other substrates. Therefore, the epitaxial growth of GaN-based LED on Si substrate has also been investigated by many national research institutes and the companies. Recently, the GaN-based LEDs on Si substrates released by Bridgelux in the United States have reached 160 lm/w at 350 mA [1], reaching certain level on sapphire substrates. Although it has not been put into large-scale commercial applications, its investment in research and development is very large. It is expected that GaN-based LEDs on Si substrates will be further developed in the near future.

SiC substrate: SiC substrate has relatively small thermal expansion and lattice mismatch with GaN. It is an ideal material for epitaxial GaN-based LED. However, the fabrication of SiC substrates is very difficult and expensive. Furthermore, large-scale applications of SiC substrate for LEDs have not been demonstrated yet. Cree in the United States has made good progress in the processing technology of SiC substrates. Its GaN epitaxial growth on SiC substrates has been widely used. In 2014, they announced that GaN-based LEDs on SiC substrates have been reached 308 lm/w at 350 mA [2]. It is the highest index of GaN-based LEDs at present. Cree's success in development of SiC substrates has also pointed out a way for other companies: if the production of SiC substrates can be greatly improved, the price of the LEDs can be significantly reduced. The use of SiC as the substrate for GaN-based LEDs is undoubtedly more advantageous than sapphire.

GaAs substrate: The processing technology of GaAs substrate is also relatively mature. In addition, GaAs has been widely used in the field of optoelectronics. However, the epitaxial GaN material on the GaAs substrate shows cubic phase that has no polarization effect. Furthermore, the cubic phase GaN material is unstable and can be decomposed at high temperatures. Therefore, there are not so much efforts from companies and research institutes that use GaAs materials as substrates for GaN LEDs.

$LiAl_2O_3$ substrate: The lattice parameters of the lithium aluminate substrate and the GaN film are relatively close. The thermal expansion coefficients are also relatively close to each other. Lithium aluminate substrates are suitable for epitaxial GaN materials, but lithium aluminate is not very stable at high temperatures. The lack of commercially available large-scale lithium aluminate substrates further limits such materials for GaN based LED applications.

ZnO substrate: The (0001) plane zinc oxide and gallium nitride have relatively small lattice mismatch and thermal expansion mismatch. ZnO substrate is transparent and easy to modify the surfaces. Such features make ZnO an ideal alternative to sapphire substrates. However, zinc oxide substrates can be decomposed at high temperatures. The thermal stability is in general not ideal. It is also difficult to achieve p-type doping and to grow large-sized and large-scale zinc oxide crystals (Table 4.1).

Sapphire substrate: Generally, GaN with [0001] crystal orientation can be epitaxially grown on the (0001) plane sapphire. Since the (0001) plane of the sapphire substrate and the (0001) plane of the GaN material have a large lattice mismatch, it has been difficult to grow good crystal quality GaN material on the (0001) sapphire substrate. Furthermore, GaN has high intrinsic electron concentration, it is very difficult to accomplish p-doping in GaN.

In 1986, Amano et al. proposed a two-step epitaxial method to improve the crystal quality of GaN materials. A low-temperature AlN buffer layer is deposited on the sapphire substrate, and a high-temperature GaN is epitaxially grown on the AlN buffer layer. The crystal quality of GaN is improved, and the background electron concentration is also greatly reduced [3, 4]. Since then, Nakamura et al. from Nichia

Table 4.1 Basic parameters of various substrate materials

Parameters	H-GaN	H-AlN	Si	Al_2O_3	$LiAlO_2$	SiC
Lattice constant (Å)	3.189	3.112	5.428	4.748	5.65	3.08
Coefficient of thermal expansion (1/K)	5.59×10^{-6}	4.2×10^{-6}	2.6×10^{-6}	7.3×10^{-6}	5.73	4.2
Thermal conductivity W/m K)	130	285	145	40	46	490
Melting point °C	2217	2232	1420	2050	1240	2830
Band gap	3.45	6.2	1.12		1.43	3.1

Corporation of Japan have used low-temperature GaN materials as buffer layers, and further optimized the growth temperature and thickness of the buffer layer to produce epitaxial GaN films capable of producing high-quality LEDs and LDs [5]. Epitaxial high-quality optoelectronic devices such as LEDs and LDs on sapphire substrates are mostly based on the two-step epitaxy method reported by Amano and Nakamura. To further improve the crystal quality of GaN thin films on the basis of the two-step epitaxy, some other modified processes such as epitaxial lateral overgrowth (ELOG) [6], multi-buffer layer epitaxy [7], and suspension epitaxy [8] have been developed.

4.3 Group III Nitride LED Epitaxial Technology

4.3.1 LPE Method

Liquid phase epitaxy (LPE) is a method in which a solid phase material is precipitated from a solution and deposited on a substrate to form a thin layer of a single crystal. Liquid phase epitaxy was invented by Nielson in 1963 and became the main growth method for thin layers of compound semiconductor single crystals. LPE is widely used in the production of electronic devices. Epitaxial growth of a thin layer on the substrate with the same material is called homoepitaxy. On the contrary, it is called heteroepitaxy. Liquid phase epitaxy can be divided into tilting method, vertical method and sliding boat method. The tilting method is to tilt the quartz container in the quartz tube in a certain orientation before initiating the growth and loading the solution and the substrate in the two ends of the container respectively. The vertical method is to place the solution in the graphite crucible before starting the growth and loading the substrate on the substrate holder above the solution. The sliding method refers to the epitaxial growth process that is carried out in the graphite boat with multiple solution tanks. In the epitaxial growth process, solution cooling can be carried out by four methods: equilibrium method, rapid cooling method, supercooling method and two-phase method.

Compared with other epitaxial methods; LPE has the following advantages: (1) the growth equipment is relatively simple; (2) higher growth rate; (3) a wide range of dopant selection; (4) good crystal integrity, The dislocation density of the epitaxial layer is lower than the substrate; (5) The crystal has high purity. There are no highly toxic and corrosive raw materials and products in the growth system, and the operation is safe and simple.

The disadvantage of LPE is that the growth process cannot be performed well if the difference in lattice constant between the epitaxial layer and the substrate is greater than 1%, Secondly, it is difficult to control the uniformity of doping and multi-component composition through the film thickness due to the difference in the segregation coefficient, except for the epitaxial layer which is very thin. Furthermore, the surface of the epitaxial layer grown by LPE is generally not as good as the epitaxial layer grown by other techniques.

4.3.2 MBE Method

MBE is an abbreviation for Molecular Beam Epitaxy, which is a technique for growth of high quality crystal films on a substrate. The material vapor under ultra-high vacuum conditions, which is generated by heating the boat containing various desired components and is collimated, become molecular beam or atomic beam. The beams are directly sprayed onto a substrate. Simultaneously controlling the molecular beam to scan the substrate, the molecules or atoms can be grown on the substrate to form a thin film. The advantages of this technology are: (1) the substrate temperature is low; (2) the growth rate of the film is low; (3) the beam intensity is easy to control, and (4) the film composition and doping concentration can be quickly adjusted as the source changes. With this technique, it is possible to prepare a single crystal film as thin as several tens of atomic layers. An ultrathin layer quantum well microstructure material can be grown by alternately growing different compositions and doped thin films.

The characteristics of MBE include: (1) The growth rate is extremely slow, about 1 m/hr, which is equivalent to grow a single atomic layer per second. This facilitates the precise control of thickness, structure and composition, and formation of steep heterostructures. In fact, MBE is an atomic processing technology. It is particularly suitable for growing superlattice materials. (2) The temperature of epitaxial growth is low. This thereby reduces the lattice mismatch effect introduced by thermal expansion at the interface and the effect of substrate impurities on the self-doping diffusion of the epitaxial layer. (3) Since the growth is carried out in an ultra-high vacuum, the surface of the substrate can be completely cleaned by treatment. The contamination can be avoided during the epitaxial process so that an excellent quality epitaxial layer can be grown. Generally, there is in situ instrument in molecular beam epitaxy for detecting surface structure, composition and vacuum residual gas. The composition and structural integrity of the epitaxial layer can be monitored at any time, which is beneficial to scientific research. (4) MBE is a dynamic process of growing incident neutral particles (atoms or molecules) one-by-one on a substrate rather than a thermodynamic process. It can therefor grow films that are difficult to grow according to ordinary thermal equilibrium growth methods. (5) MBE is an ultra-high vacuum physical deposition process that does not require intermediate chemical reactions and is not affected by mass transfer. The use of shutters can also instantaneously control and interrupt the growth process. Therefore, the composition and doping concentration of the film can be rapidly adjusted as the source changes.

4.3.3 MOCVD Method

Metal Organic Chemical Vapor Deposition (MOCVD) was developed by Amano et al. [3] in 1968 to prepare a compound semiconductor single crystal thin film material. It is a chemical vapor deposition technology widely used in compound optoelectronic devices and microelectronics device. MOCVD uses N_2 or H_2 as a carrier gas. The metal organic compounds (TMGa, TEGa, TMIn, Cp2Mg, TMAl, etc.) stored in the source cylinder are thoroughly mixed and transferred to the reaction chamber. They undergo a chemical reaction with the groups V source gases (NH_3, AsH_3, etc.) and simultaneously reach the surface of the substrate. The desired film is obtained on the surface of the substrate, and the side-reaction product is discharged to the exhaust treatment system with the carrier gas.

MOCVD equipment can have multiple source gas cylinders. There are many types of metal organic compounds that can be selected. It is suitable for material growth of compound semiconductors containing various metals (InGaN, AlGaAs, AlGaInP, etc.). In addition, the wafer carrier of MOCVD is far from the temperature balance region. Therefore, the process window is relatively wide. There is no strict process condition required by methods such as LPE. Compared to MBE, LPCVD (low pressure chemical vapor deposition), PECVD (plasma enhanced chemical vapor deposition) and other equipment, MOCVD is suitable for large-scale commercial production and can obtain a stable uniform film. This is unique in manufacturing devices. MOCVD has been widely used in various LED manufacturers. Amano et al. [4] first used a radio frequency induction heating atmospheric pressure MOCVD technology to produce single crystal GaN film. Nakamura et al. [5] developed a dual beam atmospheric pressure growth method in 1990. With the breakthrough of key technologies such as buffer layer technology [5] and annealing technology [6], the research on MOCVD in GaN materials and devices has been rapidly developed [7, 8].

MOCVD, which has been commercialized on a large scale in the world, is mainly produced by AIXTRON (Thomas swan of the United Kingdom has been acquired by Aixtron) and VEECO (Acquisition of turbodisc MOCVD division of Emcore Company in the United States). These two companies monopolize the world 90% market share [9]. Other manufacturers are mainly Japan's Taiyo Nippon Sanso and Nissin Electric, but their markets are basically limited in Japan. AIXTRON's main equipment, the Planetary G5 and the Close-Coupled Showerhead Crius II, has a 56 × 2-inch epitaxial wafer. VEECO's main Turbo-disk reactor MOCVD uses a new nozzle design to achieve better uniformity and crystal quality. The stand-alone capacity has reached 55 × 2-inch. AIXTRON equipment is more uniform and stable, and the source gas is also relatively economical. However, the maintenance of AIXTRON equipment is quite complex. It takes time to clean the Ceiling and Showerhead. Such steps directly affect the quality of the subsequent epitaxial LEDs. Although the VEECO K465i type MOCVD uses a relatively large amount of source gas, the intermediate cleaning and maintenance time are relatively short, which is suitable for large-scale commercial production.

4.3.3.1 Principles of MOCVD

The epitaxial process of MOCVD:

1. Metal-organic such as TMGa and TMIn and non-metal hydrides are transported into the reaction chamber through carrier gases such as N_2 and H_2;
2. The mixed source gas is transported on the surface of the heated substrate and thermally decomposed;
3. Atoms from the chemical reaction are adsorbed on the surface of the substrate;
4. The reacting atomic cluster migrates to the nucleation site to form a stable physical adsorption;
5. The by-product of the surface reaction leaves the surface of the substrate and is pumped to the exhaust gas treatment system;
6. The source gas, which is not participating in the reaction, is taken out together with the carrier gas to the exhaust gas treatment system.

Taking GaN material as an example, the chemical reactions of TMGa and NH_3 are as follows:

$$Ga(CH_3)_3 + NH_3 \rightarrow GaN + 3CH_4 \tag{4.2}$$

where GaN is adsorbed on the surface of the substrate, and CH_4 is discharged to the exhaust gas treatment system along with the carrier gas.

This reaction does not take into account the complex intermediate reaction process. Due to the complexity of the III–V nitride reaction, it is very difficult to obtain a detailed intermediate reaction process. Nishzawa and DenBaars and others reported several reactions that may be involved in the reaction process of materials such as GaAs and InP:

(1) Decomposition process of TMGa

$$Ga(CH_3)_3 \rightarrow Ga(CH_3)_2 + Ga(CH_3) + Ga + CH_3 \tag{4.3}$$

(2) Decomposition process of NH_3

$$NH_3 \rightarrow NH_2 + NH + N + H \tag{4.4}$$

(3) Synthetized process of GaN:

$$Ga + N \rightarrow GaN \tag{4.5}$$

(4) Synthetized process of CH_4, H_2, N_2

$$CH_3 + H \rightarrow CH_4$$
$$N + N \rightarrow N_2$$

$$H + H \rightarrow H_2 \tag{4.6}$$

(5) Side reaction process:

$$Ga(CH_3)_3 + NH_3 \rightarrow Ga(CH_3)_3 : NH_3 + \left[Ga(CH_3)_3\right]nNH_3 \tag{4.7}$$

The above is the simplest reaction process of GaN thin films. The reaction processes of AlGaN and InGaN are more complicated because the pre-reaction of TMAl, TMIn and NH_3 is more complex. The products of these side reactions are adsorbed on the wafer carrier or the ceiling, which is very difficult to clean. Baking is typically carried out multiple times after the epitaxial growth of III–V nitride.

According to the principle of chemical reaction equilibrium, in order to enable TMGa and NH_3 to proceed in the direction of positive reaction, we need a relatively large N partial pressure, that is, a large V/III ratio. The product of the reaction is also required to be able to discharge out of the reaction chamber in time.

4.3.3.2 Component of MOCVD Equipment

MOCVD is mainly composed of gas delivery system, reactor and heater system, control system, exhaust and security system, and in-situ monitor system.

(1) Gas delivery system

The gas delivery system includes various reaction sources such as metal organic sources, ammonia, pipes for delivery gases, valves, and various pressure gauges and flow meters for controlling gas flow pressure and flow. The metal organic sources are stored in the bubblers and dipped into the water bath with certain temperature. The gas delivers various reaction gases to the reaction chamber in a certain order and flow rate through carrier gases such as N_2 and H_2. In order to avoid pre-reaction between the Group III and Group V source gases, it is often necessary to transport the organic source and hydride separately into the reaction chamber through separate lines. In order to enable rapid switching of the reactant gases during the growth, a valve is required to control the switching of the run-vent gas path.

The metal organic source used in the III–V nitride are mainly trimethylgallium (TMGa), trimethylaluminum (TMAl), trimethylindium (TMIn), and ferrocenylmagnesium (Cp2Mg), etc. The group V gas is generally NH_3, and the carrier gas used is hydrogen and nitrogen (the epitaxial indium-containing nitride uses nitrogen as a carrier gas). The organic source gas is typically stored in a sealed steel cylinder and placed in a water bath at certain temperature to ensure a stable flow. Generally, a pressure gauge is used at the outlet of the source bubbler to ensure the stability of the vapor pressure of the organic source. The organic source gas in the bubbler can ensure the flow stability at different stages. The stainless-steel pipe is used to carry the source where a heating wire is generally wrapped around the pipe to ensure stability during gas transport. In order to grow a quantum well with a sharp interface, it is necessary to quickly switch the source gas during the growth process. Each corresponding organic source gas line has a vent line that allows the gas to stabilize to a suitable state before switching (Fig. 4.1).

(2) Reactor and heater system

The reaction chamber is the heart of the entire MOCVD equipment. According to the structure of the reaction chamber, it can be divided into two types, vertical and horizontal. The reactants in the vertical reaction chamber are injected vertically from the top injector onto the rotating substrate. The laminar flow of the substrate surface is achieved by varying the flow rate of the reactants, the rotational speed of the substrate, the pressure, the temperature, and the like. The direction of injection of the reactants in the horizontal chamber is parallel to the substrate which is placed

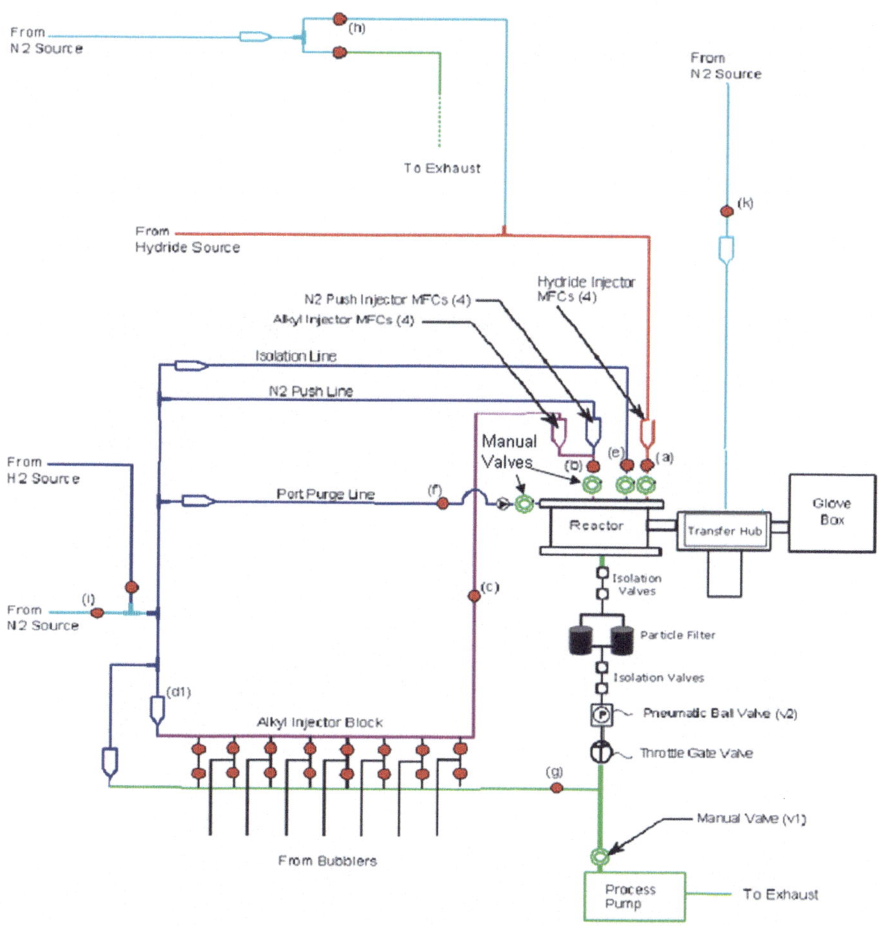

Fig. 4.1 Schematic diagram of the MOCVD source gas delivery system [11]

horizontally. Both chambers provide stable planar flow to improve the uniformity of the epitaxial layer. The MOCVD of AIXTRON in Germany uses "planetary rotation" technology, which uses low-speed substrate rotation and pedestal revolution to achieve airflow stability and uniformity. Various reaction source gases are injected through the nozzles at the center of the top of the reaction chamber, and become an advection mode above the substrate, which tends to cause uneven distribution of the source. The consistency of rotation of the rotating graphite is poor, and it requires frequent startup and maintenance, which will affect the consistency from run to run. The ceiling of the reaction chamber participates in the deposition and needs to be cleaned frequently, increasing downtime. Figure 4.2 is a schematic view of the AIXTRON G4 MOCVD reaction chamber, which can be loaded with 42 × 2-inch epitaxial wafers at a time. AIXTRON's latest G5 MOCVD has been able to load 56 × 2-inch epitaxial wafers in a single pass. Heating by means of radio frequency heating (RF) enables a fast and stable temperature rise and fall process. The use of quartz Ceiling can reduce the adhesion of solid particles generated by the reaction and reduce the frequency of cleaning. The maintenance time is greatly reduced.

The biggest feature of the MOCVD reaction chamber of the British THOMAS SWAN company acquired by AIXTRON is the use of the near-close coupled showerhead technology, which injects the organic source and the ammonia gas through the grid. The distance between the substrate and the nozzle in this case is very short. The organic source and the ammonia gas only mix at a short distance above the substrate, thereby greatly reducing the pre-reaction of the organic source with the ammonia.

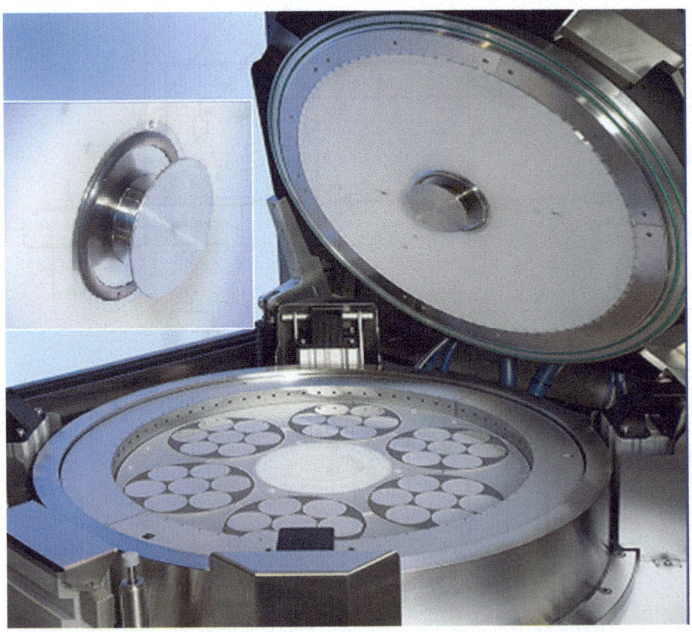

Fig. 4.2 Aixtron G4 MOCVD reaction chamber [12]

Fig. 4.3 Aixtron CRIUS I MOCVD reaction chamber [12]

The disadvantage of the reaction chamber is that the shower head is easy to block and difficult to clean (Fig. 4.3).

VEECO's MOCVD equipment in the United States uses the "TurboDisc" technology, which has a strong advantage in obtaining a uniform epitaxial layer and increasing the growth rate. The reaction chamber is simple in structure and easy to clean and maintain to reduce downtime and cost. The disadvantage is that the utilization rate of raw materials is low, and the pre-reaction in the reaction chamber is large. VEECO's new K465i MOCVD adopts a newly designed injector, which has greatly improved the uniformity of the epitaxy. Furthermore, the gas consumption is much less than that of the K465 [10].

VEECO's new MaxBright MOCVD uses four independent reaction chambers that can hold up to 216 × 2-inch epitaxial wafers at a single run. The loading of the pallet is operated by an automatic robotic arm, saving the cost of manual maintenance and operation (Fig. 4.4).

(3) Control System

MOCVD control system can be divided into two parts: the circuit control and computer control, mainly to ensure the controllability, the ease of operation and the security of the MOCVD system to facilitate the growth of various complex device structures. The circuit control part consists of four parts: digital input and output, analog input and output. The computer control section contains a main program and an interrupt program, in which the main program loops execute the digital input, output, analog input, output and refresh, and status display operations.

Fig. 4.4 Schematic diagram of Veeco Cluster4 MOCVD reaction chamber [13]

(4) Exhaust gas treatment and safety system

Since the metal organic source and NH_3 used in the MOCVD system are both toxic and flammable, an exhaust gas treatment system (scrubber) is required to prevent harmful gases from being directly discharged into the air and causing environmental pollution. Generally, a pyrolysis furnace or a cooling furnace is used to firstly break the harmful gas in the exhaust gas into a harmless gas, and then adsorbed by an activated carbon absorber or acid cleaning, and then discharged to the atmosphere.

(5) In situ monitor system

At present, MOCVD in situ monitor systems mostly use optical methods [14–17]. The in situ monitoring system can detect the temperature in the reaction chamber, the reflectivity and growth rate of the nitride film in real time. The deflectmeter can also detect the degree of curve of the epitaxial wafer in real time. The reflectometer usually uses a dual wavelength reflectometer and an emissivity corrected pyrometer to monitor real-time epitaxial growth. The fiber optic thermometer is based on the principle of black body radiation, simulating the gas atmosphere in the reaction chamber to monitor the temperature of the substrate surface, and determining the actual heating temperature by a coefficient with the actual temperature. Fiber optic thermometers generally have three test points inside and outside to ensure the uniformity of the temperature field in the reaction chamber.

4.3.4 HVPE Method

Hydride Vapor Phase Epitaxy (HVPE) HVPE is the earliest technology used to study GaN thin films and plays an important role in the early development of nitrides. In 1969, Maruska et al. modified the GaAs hydride equipment to form a GaN single crystal film by vapor deposition of GaCl and NH_3 on a sapphire substrate using a hot-wall reaction chamber [18]. However, due to the difficulty in reducing background carrier concentration and p-type doping, this technique was completely replaced by MOVCD and MBE technology in the late 1980s. In recent years, HVPE technology has once again attracted people's attention due to its own characteristics, and has become a new research topic.

HVPE method usually uses the reaction product among GaCl, metal Ga, and HCl gas as the group III source, and NH_3 as the group V source for epitaxial GaN material. The basic principle and process are as follows: in the low temperature region of the growth chamber, HCl gas flows through the quartz tube and the metal Ga in the boat reacts with it to form GaCl. GaCl and NH_3 flow to the substrate in the high temperature region driven by carrier gas, respectively. They mixed and reacted above the substrate, and a part of the generated GaN is deposited on the surface of the substrate. The remaining portion is deposited on the wall of the reaction tube. Unreacted NH_3 and HCl proceed and react at the outlet of the reaction tube to form solid NH_4Cl white powder.

The main reactions are as follows [19]:

$$\text{Low temperature zone}: \ \mathrm{Ga(l)} + \mathrm{HCl}(g) \rightarrow \mathrm{GaCl}(g) + \mathrm{H_2} \tag{4.8}$$

High temperature zone (1050 °C):

$$\mathrm{GaCl}(g) + \mathrm{NH_3} \rightarrow \mathrm{GaN} + \mathrm{HCl}(g) + \mathrm{H_2} \tag{4.9}$$

In addition, there are side reactions in the chamber:

$$\mathrm{NH_3}(g) + \mathrm{HCl}(g) \rightarrow \mathrm{NH_4Cl}(s) \tag{4.10}$$

$$\mathrm{GaCl}(g) + 2\mathrm{HCl}(g) \rightarrow \mathrm{GaCl_3}(g) + \mathrm{H_2}(g) \tag{4.11}$$

HVPE growth systems generally consist of furnace bodies and reactors, gallium boats and gas pipes, gas distribution systems, and exhaust.

The processing system consists of four parts. According to the structural characteristics of the reactor, the HVPE system can be divided into two types: horizontal and vertical. The horizontal HVPE systems are used by Karl-Marx University and AIXTRON, Germany, Wisconsin University, and Russian Soft-impact. The vertical type has been used by the University of Lincoping in Sweden, the University of Justus-Liiebig in Germany, and the MIT in the United States. Both methods have their own advantages and disadvantages. The horizontal structure is relatively mature.

However, the surface of the substrate is prone to large temperature gradients and growth concentration gradients, which tends to cause depletion of reactants downstream that is not favorable to the uniform epitaxial wafer. The development of the vertical growth chamber starts relatively late. Usually, the substrate is placed on the rotatable table below the growth chamber, which makes it easier to obtain high-speed rotation of the substrate. The uniformity of the grown GaN is better, but the substrate is more susceptible to the GaN particles and by-products from the top above the wafer. This can contaminate and affect the quality of the epitaxial wafer.

HVPE equipment is relatively simple and easy to maintain. It offers fast growth rate, where the growth rate can exceed 300 μm/h. It is also easy to obtain uniform and large size GaN thick film. High-quality GaN substrate can be obtained by laser stripping and other technologies. Moreover, as the thickness (or the reaction) increases, the quality of the material can be continuously improved. Combined with the lateral epitaxy technique, the dislocation density can be reduced to the order of 10^5 cm^{-2} [20]. Its disadvantage is that it is difficult to precisely control the film thickness and composition. It is also difficult to grow multi-nitride materials and heterostructures (such as quantum wells, superlattices). The reaction gas has a strong corrosive effect on the growth equipment, which can limit the improvement in material purity. In addition, the by-product NH_4Cl is easy to block the pipeline due to the serious parasitic reaction. For example, polycrystalline GaN can easily grow on the quartz tube and is difficult to remove, which reduces the lifetime of the Ga boat and the quartz tube.

4.4 Two-Step Growth Method for MOCVD Grown Nitride Materials

Due to the large lattice mismatch and thermal mismatch between GaN and sapphire, GaN grown directly on the sapphire substrate generally has a poor crystal quality and a rough surface. Special process must be employed in order to grow GaN material with a flat surface and high quality on a sapphire substrate. Similar to the growth of GaAs on a Si substrate, the two-step growth method is a typical process for GaN/sapphire MOCVD growth techniques. Since the growth process is accompanied by a change in growth mode, many phenomena can be explained by the surface dynamics of the film growth. Here, the principle is introduced, and the general procedure for growing GaN by the two-step method is introduced in combination with the kinetic principle.

4.4.1 Surface Dynamics for Film Growth

In epitaxial growth, since the vacuum deposition process is in a thermodynamic equilibrium state, nucleation and growth of the material is a dynamic process. Because of this, film growth leads to a series of rich surface topography in a non-equilibrium state, as well as the corresponding lattice relaxation problems.

In the epitaxial growth of a thin film, the deposited atoms fall on the substrate. Firstly, they meet and combine in a certain way to form an atomic group, and then new atoms are continuously added to these already formed atomic groups so that they stably grow into larger particles family (often referred to as the "island"). As the deposition continues, the atoms will fill the gaps between the islands, forming a continuous film. During film growth, the size and distribution of the nucleation of the deposited atoms, the islanding process, and the pattern of the initial growth will greatly affect the quality of the entire film to be formed.

4.4.1.1 Island Size Effect and Distribution

Critical Island Size: In the process of nucleation, when an island contains more atoms than the critical island size, the island is stable; when an island contains less atoms than the critical island size, the island is unstable and will continue to absorb new atoms for growth. The relationship between island density and critical island size is as follows:

$$N = \left(\frac{D}{F}\right)^{-\frac{i}{i+2}} \exp\left[\frac{Ei}{(i+2)K_B T}\right] \tag{4.12}$$

where N is the density of the total nucleation islands, D is the diffusion coefficient of the atoms on the surface of the substrate, E_i is the bond energy, and F is the deposition flow rate. According to the Formula (4.12), the critical island size can be derived by experimentally measuring the variation of the island density with respect to the deposition flow rate.

4.4.1.2 Coarsening Mechanism

The surface of a real material usually has irregular steps, deposited atoms, atomic groups, vacancies, and islands. Their existence indicates that the surface is in a thermodynamic non-equilibrium state. Therefore, as long as conditions permit, such surfaces will relax and go back to equilibrium. In many systems, surface relaxation is achieved by atomic diffusion among islands. Such a process is commonly referred to as Ostward Ripening, and it is also referred to as Coarsening due to the increase in surface roughness accompanied with the process.

Ostwald Ripening exists in many physical and chemical systems. Under the framework of Ostwald Ripening, the atoms in the smaller islands have higher activity, so the equilibrium vapor pressure is higher and the chemical potential is higher. Therefore, when two islands with different sizes are next to each other, there is a tendency to evaporate atoms because the smaller-size island has a higher chemical potential.

On the other hand, the larger island has a tendency to absorb atoms due to its lower chemical potential. As a result, the larger islands grow up by absorbing atoms, while smaller islands disappear by losing atoms. Therefore, there is a net atomic flow from the small island to the big island. The small island gradually shrinks and becomes smaller, and eventually degrades and disappears; while the big island grows up. This process seems like that the small island is being "eaten" gradually by the big island.

4.4.1.3 Growth Mode

The growth process of the film can be divided into three modes: Frank-vander Merve or layer-by-layer (LBL, layered) growth, Volmer-Weber (island) growth, and Stranski-Krastanov (mixed) growth as shown in Fig. 4.5.

(1) Frank-van der Merwe (LBL) mode: When the wettability between the deposited material and the substrate is good, the atoms of the deposited material tend to bond with the substrate atoms. Therefore, the film adopts a two-dimensional growth mode from the nucleation stage.

(2) Volmer-Weber (3D) mode: This growth mode indicates that the atoms or molecules of the deposited material tend to bond with each other. Their wettability with the substrate is not good. Therefore, they avoid to be bonded with substrate atoms. This mode will form a plurality of islands, results in a rough surface.

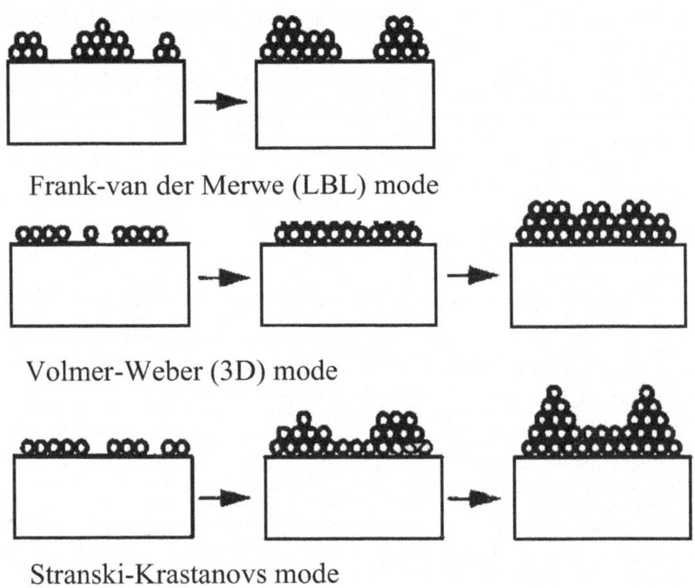

Fig. 4.5 Schematic diagram of the growth pattern of thin films

(3) Stranski-Krastanovs mode: It begins with layered growth in one or two atomic layer thicknesses, and then is converted into island-like growth. That is, firstly the layered growth mode and then converted to the island growth mode.

Of the above three growth modes, the latter two modes are three-dimensional growth. At the atomic scale, producing a smooth and uniform film means layered growth or two-dimensional growth, instead of three-dimensional growth.

The MOCVD growth process of GaN materials also follows surface dynamics of thin film growth. Therefore, many experimental phenomena observed during the growth of GaN can be explained by this theory. The following is a brief introduction to the common MOCVD two-step growth procedure of GaN on sapphire.

4.4.2 Two-Step Growth Program for GaN/Sapphire by MOCVD

Figure 4.6 shows the in situ monitoring reflectance curve of MOCVD growth of GaN materials. Generally speaking, the two-step growth of intrinsic GaN materials requires the following steps:

(1) First, high temperature treatment at 1100 °C is carried out for 15 min under H_2 atmosphere. This procedure will eliminate surface contamination of the substrate.

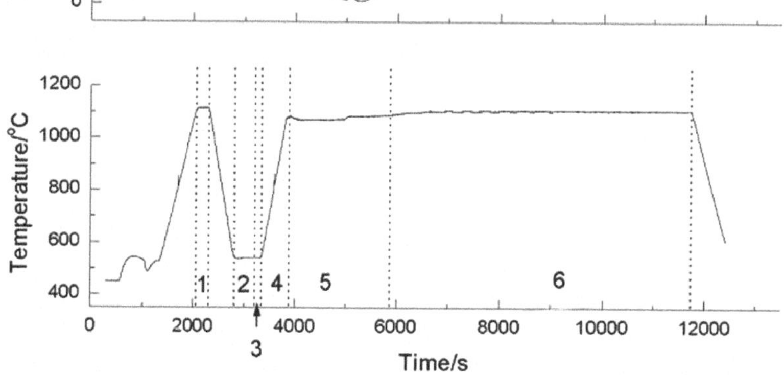

Fig. 4.6 In-situ monitoring curve of GaN growth

(2) Stabilizing the temperature after cooling to about 535 °C, and nitridating the substrate surface using NH_3. In other words, heat treatment of the substrate is done in the environment of NH_3. Sapphire and NH_3 will react under heating to form AlN_xO_{1-x}. The formation of such compound can reduce the lattice mismatch between the substrate and the epitaxial layer.

(3) Low-temperature buffer layer growth is performed at about 535 °C. Since the temperature is low, the surface diffusion coefficient of the atomic group is small. This makes it easier to adhere to the surface of the nearby substrate, and thus the density of the nucleation island is high and the reflectance increases. The growth conditions of the buffer layer, such as thickness, temperature, pressure, etc. greatly affect the size and distribution of the nucleation islands, and thereby affect the growth mode of the subsequent high temperature epitaxial layer. Therefore, precise control and optimization of the growth conditions are required. Its AFM image is shown in Fig. 4.7a. Low-temperature buffer layer growth is performed at about 535 °C. Since the temperature is low, the surface diffusion coefficient of the atomic group is small. This makes it easier to adhere to the surface of the nearby substrate, and thus the density of the nucleation island is high and the reflectance increases. The growth conditions of the buffer layer, such as thickness, temperature, pressure, etc. greatly affect the size and distribution of the nucleation islands, and thereby affect the growth mode of the subsequent high temperature epitaxial layer. Therefore, precise control and optimization of the growth conditions are required. Its AFM image is shown in Fig. 4.7a.

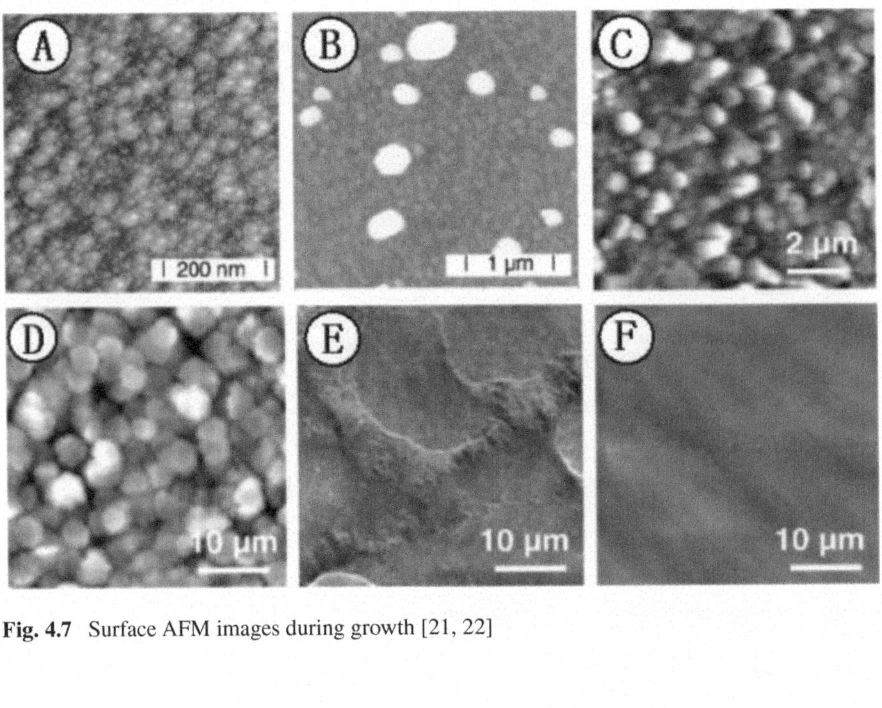

Fig. 4.7 Surface AFM images during growth [21, 22]

(4) After the buffer layer is grown, the temperature is raised from a low temperature of 535 °C to a high temperature of 1080 °C. This section is generally referred to as an annealing stage of the buffer layer. During this procedure, the amorphous buffer layer is recrystallized so that the reflectance is slightly increased at the beginning stage. However, when the temperature is raised above about 850 °C, the GaN nucleation island begins to coarsen, resulting in a rapid decline in reflectance. Its AFM image is shown in Fig. 4.7b.

(5) The roughening layer, which is the growth stage of the GaN nucleation island, is a three-dimensional growth. At high temperatures, high-density nucleation islands gradually grow. As large nucleation islands gradually absorb nearby atoms, such a process makes the islands larger and sparse distribution. It is therefore that the reflectivity of this segment is very low. The large and sparse nucleation island can reduce the dislocation density. The roughening speed can be promoted by elevating the pressure and lowering the V/III ratio, making the roughened island larger and sparse. The AFM image of the island's merger process is shown in Fig. 4.7c, d.

(6) The recovery layer and the high temperature layer are the lateral and vertical growth of the island at the nucleation sites. The lateral growth causes the nucleation islands to gradually merge into a quasi-two-dimensional plane. Then, high-temperature growth of the GaN epitaxial layer can be performed on this basis with a two-dimensional growth mode to obtain a high-quality GaN material. The AFM image of its surface topography is shown in Fig. 4.7e, f.

Among them, steps 1–5 are the initial stages of GaN two-step epitaxial growth, providing a basis for high temperature epitaxial growth listed in step 6).

Figure 4.7 shows the surface morphology of the buffer layer obtained by the method of growth interruption at different growth stages. The surface topography corresponds to the reflectance curve of Fig. 4.6 respectively. Observation of the surface topography at each stage helps us to better understand the changing process of buffer layer.

The initial growth conditions have a great impact on the growth mode of the subsequent high temperature epitaxial layer. If the initial growth conditions are not suitable, the surface of the material is rough and the crystal quality deteriorates sharply. Therefore, it is quite necessary to optimize the growth conditions of the initial stage for the GaN epitaxy.

4.5 Influence of Growth Conditions on Epitaxial Layer Quality of Group III Nitride Materials

The initial growth stage includes high-temperature pretreatment, surface nitridation, buffer layer growth, buffer layer recrystallization, and roughening layer growth during the two-step GaN growth process, which is the basis for the growth of high-temperature GaN materials. In particular, the growth parameters of the buffer layer and the roughening layer have a great impact on the material properties of the high temperature intrinsic GaN. The properties of GaN materials grown under different initial growth conditions were characterized. The influence of the growth parameters of GaN on the properties of high temperature GaN materials was obtained. The material properties of intrinsic GaN were improved by optimizing the initial growth parameters.

4.5.1 *Effect of Buffer Layer Growth Conditions on Material Quality*

4.5.1.1 Effect of Buffer Layer Thickness

The structural and electrical properties of intrinsic GaN materials grown with buffer layers of different thicknesses (other growth conditions are the same) were investigated. It was found that when the thickness of the buffer layer is increased from 18 to 28 nm, the FWHM of the (002) and (102) peaks of the X-ray double crystal diffraction of the material decreases gradually, the background carrier concentration decreases, and the mobility increases.

According to Heinke et al. [23], Beaumont et al. [24], Ayers [25], the dislocation density is proportional to the full width at half maximum of the X-ray diffraction curve, as shown in Eq. (4.13):

$$\rho = \frac{FWHM(\mathrm{hkl})}{4.36b^2} \tag{4.13}$$

where b is the Burgers Vector of the dislocation, *FWHM* (hkl) is the full width at half maximum of the measured (hkl) plane, and ρ is the dislocation density. Thus, a relatively large value FWHM from X-ray diffraction of the thin buffer layer indicates an increase in dislocation density, which in turn leads to the degradation in the crystal quality.

From the electrical properties of GaN with different buffer layers shown in Table 4.2, the background carrier concentration of the sample increases and the

Table 4.2 Structure and electrical properties of intrinsic GaN grown by buffer layers of different thickness

Sample	Buffer layer thickness (nm)	Structural property		Electrical property	
		(002) FWHM (arcsec)	(102) FWHM (arcsec)	Carrier concentration (10^{16} cm^{-3})	Mobility (cm^2/Vs)
A	28	284	316	1.2	430
B	18	302	388	1.82	320

mobility decreases as the thickness of the buffer layer decreases. This is consistent with the XRD results, which also indicates the epitaxial material with thin buffer layer has poor crystal quality. Poor quality of high temperature epitaxial layer by using thin buffer layer can be attributed to a loose distribution of nucleation islands that are not favorable seed crystals for the lateral growth. In other words, they cannot provide a good template for the growth of the subsequent high temperature GaN film, making the high-temperature growth mode away from the two-dimensional growth and therefore a high dislocation density and poor crystal quality [26–28]. According to the results of Kuznia et al. [29], the optimized buffer layer thickness should be between 20 and 30 nm. Further increasing the thickness of the buffer layer also causes the deterioration of crystal quality. Therefore, much thicker buffer layer is not widely investigated.

PL characterization was performed on intrinsic GaN samples with different buffer layer thicknesses. The PL results are shown in Fig. 4.8. It is found that the intensity of the band edge peaks of samples A and B with different buffer layer thickness is similar, but the intensity of the yellow light peak of the sample with thin buffer layer is obviously larger. The intensity ratio of the yellow peak/band edge peak of the two samples A and B are 1.14 and 2.19 respectively. At present, researchers generally believe that yellow band is caused by gallium vacancy V_{Ga}. Therefore, it is generally true that the intrinsic GaN material with thin buffer layer has more V_{Ga} [30]. This can be considered as a result of the change in growth mode due to the thin buffer layer, which in turn leads to the introduction of excessive internal defects V_{Ga} during growth.

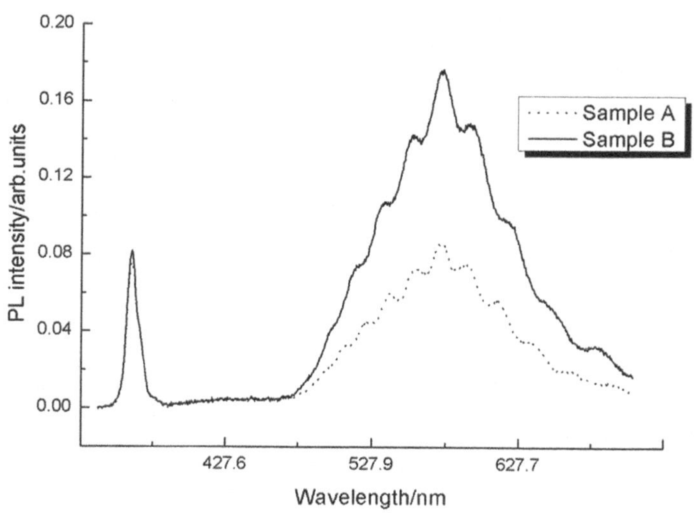

Fig. 4.8 The intrinsic GaN sample PL diagram of different buffer layer thickness

4.5.1.2 Effect of the Growth Temperature of Buffer Layer

It can be seen from Table 4.3 that different temperature to grow buffer layer has a great influence on the electrical properties of the material. As the growth temperature of the buffer layer increases (515–535 °C), the carrier mobility of the intrinsic GaN material increases significantly, and the background carrier concentration decreases. Nevertheless, higher temperature (550 °C, data is not listed) to grow buffer layer can result in a decreased carrier mobility. It indicates that a suitable growth temperature for buffer layer can improve the crystal quality of the intrinsic GaN material.

With proper temperature, the nucleation island of the buffer layer can exhibit a large and sparse distribution. As shown in Fig. 4.9b, the subsequent high temperature growth tends to be more lateral growth. The growth mode of high-temperature GaN is two-dimensional growth under such processing conditions, and can result in reduced dislocation density and improved crystal quality. For the buffer layer grown under lower temperature, the migration ability of the atomic group on the surface is lower. It causes smaller-size and higher-density nucleation islands, which accelerates the islands merging process and causes the high-temperature GaN growth mode to deviate from the two-dimensional growth. This results in more dislocations and defect densities. Besides, the low growth temperature of the buffer layer also deteriorates the surface topography, resulting in a serious reduction in reflectance. While excessive high temperature will cause the kinetic energy of the reactants to increase, making the distribution of the island smaller and denser as shown in Fig. 4.9c. This will interrupt and/or destroy the continuous two-dimensional growth.

The decrease of growth temperature will lead to decreased migration ability for the atomic group on the surface, which will cause it to grow in the nearby nucleation island and form high density of the nucleation islands. Therefore, this will accelerate

Table 4.3 Electrical properties of intrinsic GaN grown at different buffer layers

Sample	Buffer layer growth temperature (°C)	Resistivity (Ωcm)	Mobility (cm²/Vs)	Electron concentration (10^{16} cm^{-3})
C	535	0.32	455	4.2
D	515	0.28	340	6.1

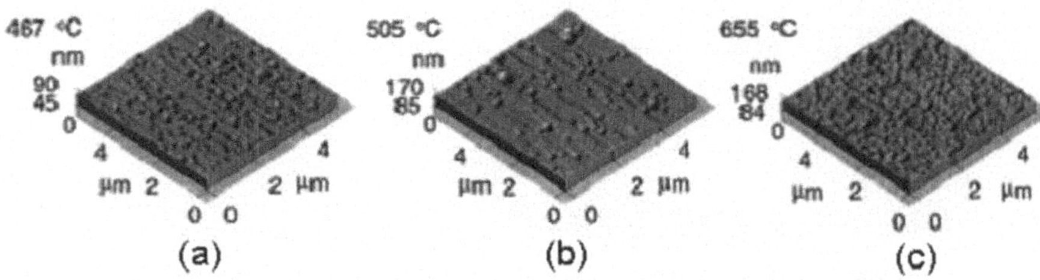

Fig. 4.9 AFM results for buffer layer at different growth temperatures [31]

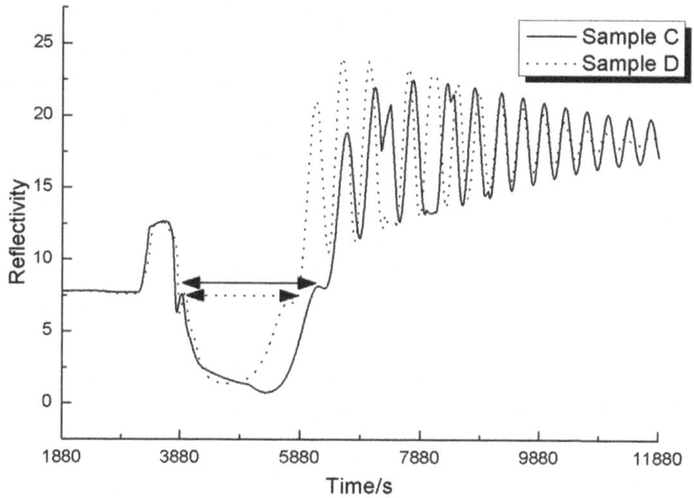

Fig. 4.10 In-situ monitoring curves of intrinsic GaN grown at different layer temperature

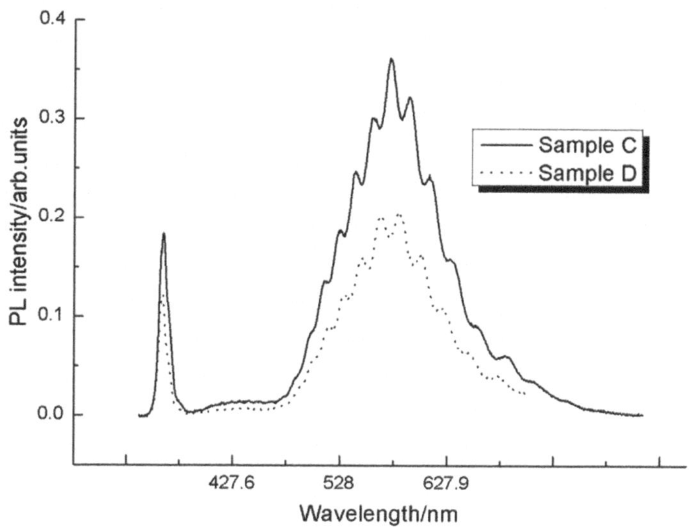

Fig. 4.11 PL results of intrinsic GaN samples grown with different buffer layer temperature

the merger process of the island during the post-high temperature growth process. The reflectance curve shows a faster recovery as shown in Fig. 4.10. The severely reduced reflectance of the high-temperature growth layer indicates that the increase of surface roughness and the deterioration of the surface morphology due to the low-temperature buffer layer.

Figure 4.11 shows the PL spectra of the intrinsic GaN sample grown with different buffer layer temperature. The intensity of the band edge peak of the intrinsic GaN sample with a suitable-temperature for growth of buffer layer is significantly higher, which is consistent with the results reported by Yi et al. [31]. It is believed that this is resulted from a suitable buffer layer temperature which improves the quality of post-high temperature growth GaN materials and reduces non-radiative recombination. The optimized temperature for the buffer layer was 535 °C from our experiments. There is 30 °C difference from the 505 °C reported by Yi et al. [31]. The difference could be caused by the variation in growth equipment or temperature monitoring.

4.5.1.3 Effect of TMGa Flow for Buffer Layer

In experiments of different TMGa flow rates for the buffer layer, we found that the slope of the buffer layer reflectance curve is larger for high TMGa flow. This indicates that the Ga flow rate determines the growth rate in the buffer layer, and the buffer layer is grown under N-rich condition. That is to say, the dissociation of NH_3 is sufficient even at such a low temperature owing to the large V/III ratio (V/III ratio is 10,000). This also indicates that the growth rate of the buffer layer does not increase with temperature, which is consistent with the results of Fig. 4.12 (the growth rate of the buffer layers with different growth temperature is the same). In addition, when the buffer layer is grown with a small Ga flow rate, the reflectance of the sample recovers more rapidly during the roughening stage (with the same growth conditions in the roughening stage). This can be explained by the faster island consolidation during high temperature growth. Such a phenomenon is caused by the insufficient growth of the nuclear islands and the resulted small size, high density, and small spacing

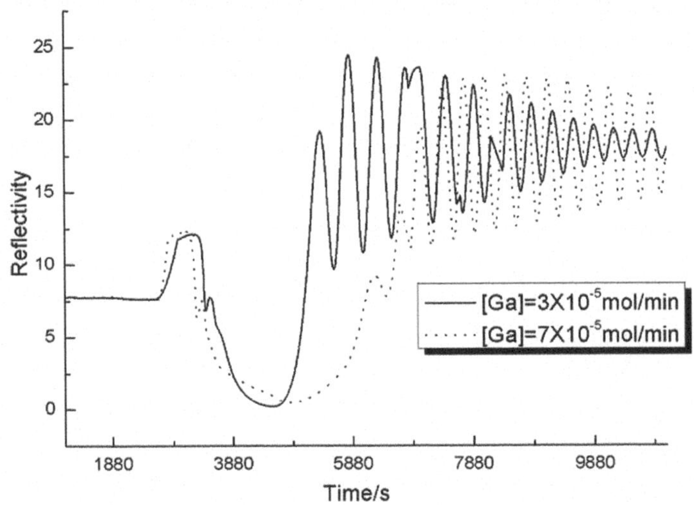

Fig. 4.12 In-situ monitoring curve of different Ga flow growth buffer layers

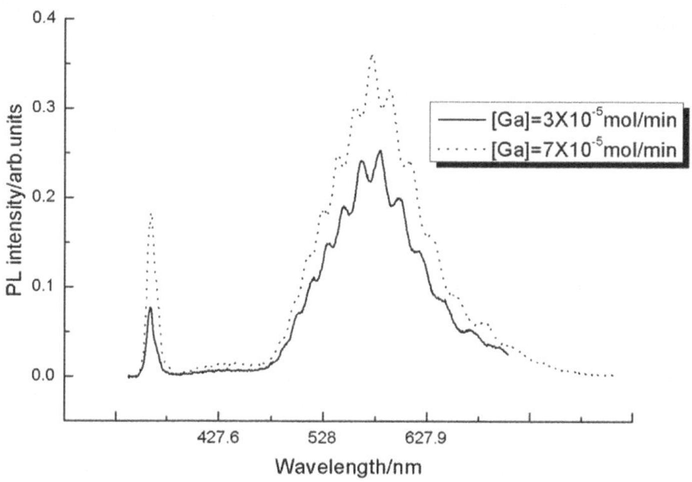

Fig. 4.13 PL result of intrinsic GaN for different Ga flow growth buffer layers

islands due to the small Ga flow rate in a limited time. The post-high temperature growth deviates from the two-dimensional growth mode due to the unsatisfactory distribution of nucleation islands. Therefore, the surface becomes rough, which leads to the sharp reduction in the amplitude of the reflectance curve at high temperature stage for the sample with small Ga flow buffer layer as shown in Fig. 4.12.

The PL characterization was carried out on intrinsic GaN grown under different Ga flow rate as shown in Fig. 4.13. It was found that the band edge peak intensity of the sample with a small Ga flow buffer layer is only about 50% of the sample with a high Ga flow buffer layer. This can be explained by the high dislocation or defect density since the low-Ga flow rate buffer layer may cause a large islands density and a quick merge and hence a growth mode deviated from the lateral growth. The band edge peak intensity should be depressed due to its competition with defect-related luminescence. Cho et al. [32] analyzed the intrinsic GaN of different buffer layer growth rates using a deep-level transient energy spectrometer (DLTS) and found that the edge dislocations and deep-level defect densities increase as growth rate decreases. This result also supports our conclusion.

Hall test is also carried out on the samples with different Ga flow buffer layers as shown in Table 4.4. The decrease in buffer layer growth rate has small effect on the carrier mobility of high-temperature GaN, but the background carrier concentration

Table 4.4 Electrical properties of intrinsic GaN of different Ga flow growth buffer layers

Sample	TMGaflow (μmol/min)	Growth rate (nm/min)	Mobility (cm^2/Vs)	Electron concentration (10^{16} cm^{-3})
E	30	5	370	9.5
F	70	12	360	2.4

is increased by a factor of four. This can be explained by the high defect density caused by the growth mode deviated from lateral growth. The experimental results are basically consistent with that reported by Lee et al. [33] and Kim et al. [34], but different from that reported by Juang et al. [35]. The reason for the difference could be that the Ga flow taken in Juang's experiment is too small (Ga flow is 4–6 sccm, corresponding to 20–30 μmol/min).

4.5.1.4 Effect of Buffer Layer Growth Pressure

We conducted a buffer layer experiment with different pressures under the same Ga flow. The experimental results show that the decrease of pressure will reduce the migration ability of the reactant atomic groups, and lead to a small and dense distribution of nucleation islands on the surface of the substrate. Therefore, the reflectivity increases rapidly in the initial stage of buffer layer growth. However, after high temperature annealing, the reflectivity of the subsequent coarse layer GaN growth does not decrease as shown in Fig. 4.14. This is because the nucleation islands are not fully grown. The nucleation islands are too dense and they merge quickly. It has a high reflectivity during the annealing phase, which indicates that a large and sparse nucleation island distribution is not formed. This is consistent with the results reported by Chen et al. [36]. Cao et al. [37] also achieved ultra-low pressure (20 Torr) growth of GaN buffer layer under optimization of growth conditions in virtue of in situ monitoring, but the crystal quality of the material was not the best. These results also support our conclusion from certain aspects (Figs. 4.15, 4.16 and 4.17).

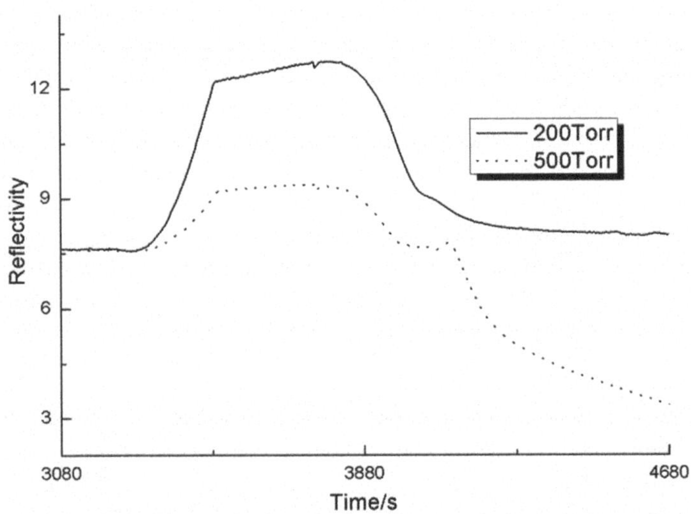

Fig. 4.14 In-place monitoring curve for different pressure of buffer layers

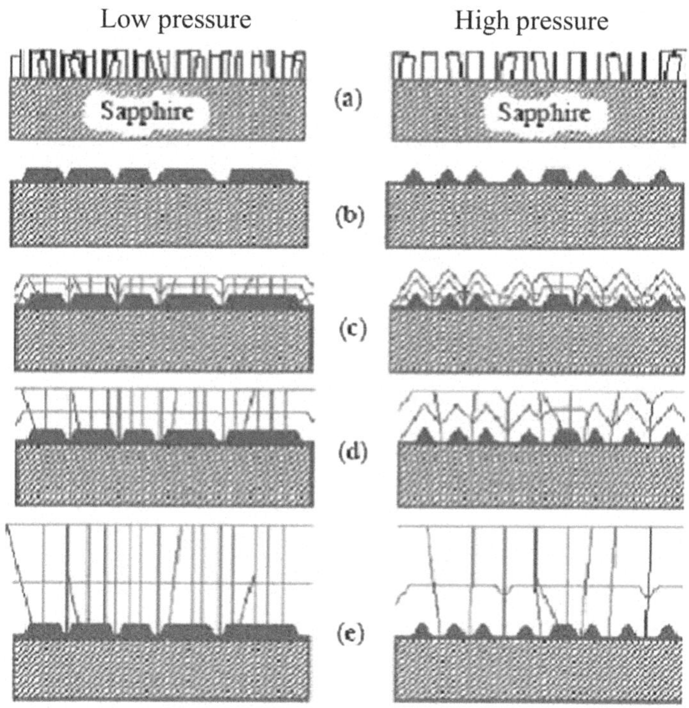

Fig. 4.15 Schematic diagram of the change of the buffer layer under different pressure

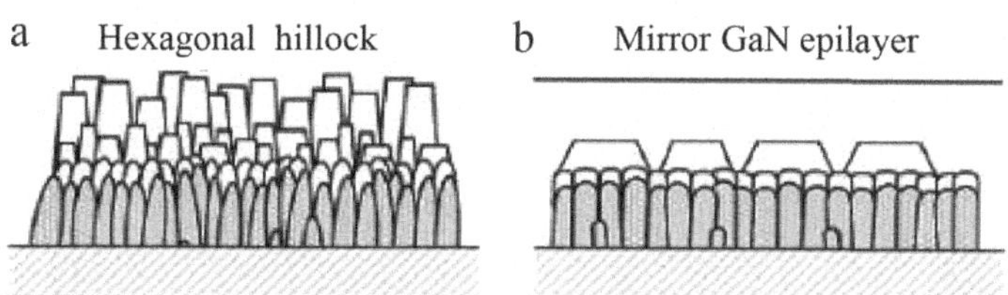

Fig. 4.16 Schematic diagram of different growth pressure of the coarsening layer. **a** Under low pressure and **b** under high pressure

The buffer layer grown under low pressure also causes an increase in the dislocation density and a decrease in crystal quality. Together with the buffer layer variation diagram shown in Fig. 4.18, we can explain why a low-pressure during film growth causes the deterioration of crystal quality. The vertical line in the figure represents the edge dislocation, and the oblique line represents the mixed type threading dislocation. In the actual physical process, new edge dislocations will be formed when the islands merge. Most of these dislocations will penetrate the thickness of GaN epitaxial layer. Due to the presence of mixed dislocations, interaction occurs when

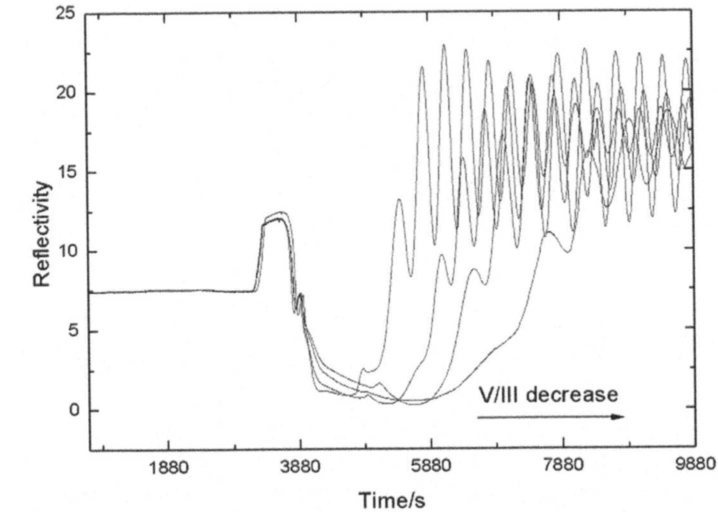

Fig. 4.17 In-situ monitoring curve under different V/III of coarsening layer

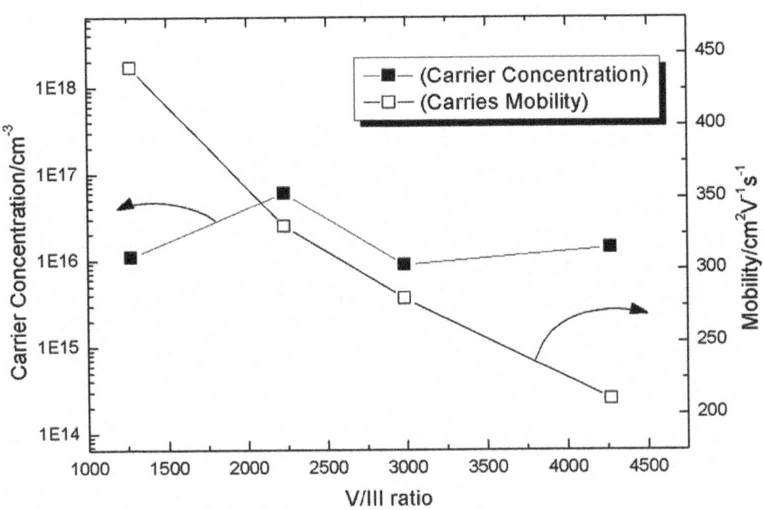

Fig. 4.18 Electrical properties of intrinsic GaN under different V/III ratios of coarsening layer

mixed dislocations meet edge dislocations. If there are two dislocations with respective Burgers vector of **b1** and **b2**, they can merge into a new dislocation with Burgers vector b3, i.e., $\mathbf{b3} = \mathbf{b1} + \mathbf{b2b3} = \mathbf{b1} + \mathbf{b2}$. If $\mathbf{b1} = -\mathbf{b2}$, then $\mathbf{b3} = \mathbf{0}$. This indicates that two dislocations can annihilate at the same time. Whether the dislocations are merged or annihilated, the dislocation density penetrating into the epitaxial layer can be reduced as shown in Fig. 4.15d, e.

When the buffer layer is grown under a low pressure, the island has a small size and a high density in the initial stage of nucleation. Therefore, the islands are rapidly merged, and a large number of edge dislocations are generated during the merging process. The dislocations extend along the normal direction of the substrate into the epitaxial layer. The process is shown on the left in Fig. 4.15. While under high pressure conditions during film growth, the island has a large size and a small density. This makes the distance between the islands become larger, and the merging process becomes slower. The edge dislocations generated during the merging process will be much affected by the mixed dislocations. The density of dislocations extended to the epitaxial layer is normally reduced due to the interaction between the edge dislocations and mixed dislocations.

4.5.2 *Effect of Rough Layer Growth Conditions*

After undergoing buffer layer growth and recrystallization, the surface has formed a nucleation islands, which provides seed crystals for post-high temperature GaN material growth. With the growth of high-temperature GaN epitaxial materials, the growth mode changes: First, at the GaN nucleation island, the island grows horizontally and vertically. In the lateral direction, the large islands gradually become larger by absorbing small islands, while in the longitudinal direction, the growth rate of the islands gradually decreases. This stage is three-dimensional growth. Subsequently, the adjacent islands are merged to form a quasi-two-dimensional plane, which is due to the competition between the coarsening caused by island growth and the flattening brought about by the island merge. The growth mode at this time should be the mixed growth mode. Finally, when a two-dimensional plane is formed, the mismatch stress will be moderated, which is beneficial for the film to tend to grow two-dimensionally, making the surface smooth as a mirror. The growth conditions in the roughening stage also have a great impact on the epitaxial growth mode, which in turn affects the material quality, so it requires to be investigated.

4.5.2.1 Influence of Rough Layer Pressure

The growth pressure of the roughened layer can also affect the size and distribution of the island. When the pressure is small, it can only grow at the nearby nucleation island due to the small activity of the reactants, which reduces the lateral growth effect and thus speeds up the combination of the islands. During the formation of islands and merging islands process, it can result in even higher density of stacking faults, micro twin dislocations, and defects as shown in Fig. 4.16.

4.5.2.2 Influence of Roughing Layer V/III Ratio

In addition to increasing the pressure of the roughened layer, it is also possible to reduce the merge speed of the island by reducing the V/III ratio of the roughened layer. This is because the low V/III ratio makes the growth of nucleation islands more inclined to 3D growth. While for high V/III ratio, there will be more NH_3. This will increase the viscosity of the airflow and make the MO source difficult to pass through the boundary layer. It therefore leads to small and dense nucleation islands, makes the high temperature growth island merge faster, reduces the lateral growth during the merger process, and increases the dislocation density and deteriorates the crystal quality. In comparison with the in situ monitoring curves under different roughing layer V/III conditions shown in Fig. 4.17, it can be seen that the reflectivity recovery becomes slower as the V/III ratio decreases. This will allow the nucleation islands to grow fully. Such a process is beneficial to enhance the lateral growth mode in the post-high temperature growth [38, 39].

Hall measurement is carried out for the intrinsic GaN samples grown under different V/III ratio conditions for the roughened layer as shown in Fig. 4.18. It is found that the background carrier concentration of the samples with different V/III ratios is similar, but the mobility increases rapidly with the decrease of V/III ratio. It can be seen that the low V/III ratio for the roughening layer can reduce the island merging speed, make the island growth fully, enhance the lateral growth of the post-high temperature GaN, reduce the dislocation density and the amount of implanted impurities, and hence improve the carrier mobility.

The surface morphology of the intrinsic GaN samples with roughened layer of different V/III ratios is also studied. It can be seen that the surface morphology of the sample with a low V/III ratio is better. This is because the low roughening layer V/III ratio can reduce the merging speed of the island, which results in a two-dimensionally growth mode for the subsequent high-temperature layer deposition and therefore a better surface topography.

4.6 Epitaxial Technology of High Quality GaN on SiC Substrate

4.6.1 Basic Properties of SiC

SiC is a Group IV–IV binary compound semiconductor and is the only solid compound comprised Group IV elements from the Periodic Table of the Elements. SiC is consisting of two elements Si and C. Each atom is surrounded by four heterogeneous atoms, which forms tetrahedral units by oriented strong tetrahedral SP_3 bonds (Fig. 4.19) [40]. The bond length between Si–C atoms is 1.89 Å, and the bond length between Si–Si or C–C is 3.08 Å. The SiC crystals are formed by interconnecting these tetrahedrons at the corners. The SiC crystal has particular strong ionic covalent

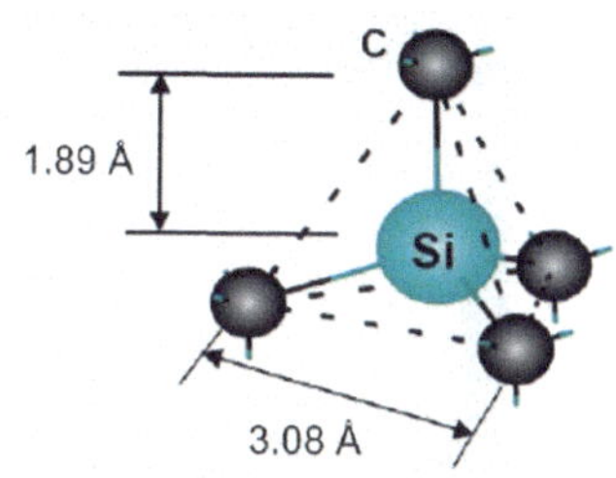

Fig. 4.19 The regular tetrahedron unit of SiC [40]

bonds, indicating that SiC has a structure with stabilized combination energy. This is reflected by a high atomic energy value of 1250 kJ/mol. SiC has a high Debye temperature of 1200–1400 K. Therefore, SiC has excellent mechanical and chemical properties such as good stability, high hardness, wear resistance, and high melting temperature under normal pressure.

The Si atomic layer and the C atomic layer in the single crystal SiC constitute a basic diatomic layer. Si–C diatomic layers are alternately stacked in a certain sequence to form SiC crystals. The difference in the stacking sequence of the diatomic layers leads to different crystal structures, i.e., the formation of polyisomers of SiC, or homogenous polymorphisms. More than 250 crystal structural morphologies have been discovered [41]. The most important one are 3C, 2H, 4H, 6H, 8H, 9R, 10H, 14H, 15R, 20H, 21H, 24R, etc., belonging to three basic crystallography types: cubic (C), hexagonal (H), diamond (R). Generally, the polytypes of hexagonal and rhomboid structures are collectively referred to as α-SiC; the SiC of cubic structure is collectively referred to as β-SiC. Due to the different butting of the SiC tetrahedron, the SiC material shows a difference in cubic structure or hexagonal structure. As shown in Fig. 4.20, the different butts of the SiC tetrahedron form the cubic and hexagonal phases respectively [42].

The crystal form of hexagonal phase SiC substrate is the same as the group III nitride. Thus, the hexagonal Si surface of SiC substrate is selected for epitaxial growth of the nitride. Due to the polytype characteristics of SiC materials, the crystal structure is more complex than the elemental semiconductors Si and Ge. The defects in SiC can be very complex as well. So far, researchers have measured the distribution

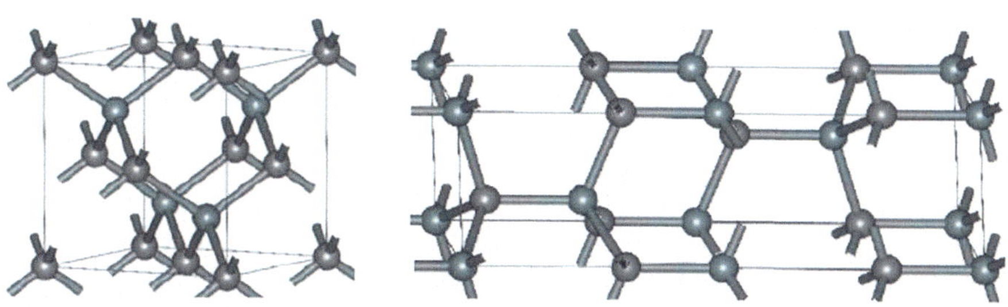

Fig. 4.20 The cubic and hexagonal phases are respectively formed due to the different joint of the SiC tetrahedrons [42]

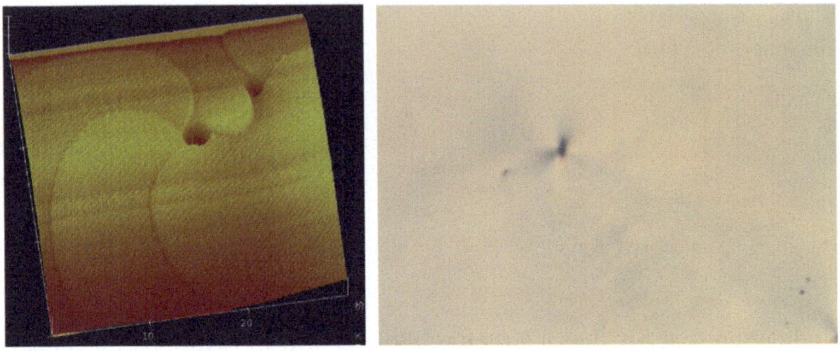

(a) (b)

Fig. 4.21 **a** AFM image of microtubule in SiC. **b** Morphology of microtubule

and density of defects in SiC crystals such as dislocations, ridges, stacking faults, and micropipes using chemical etching, X-ray diffraction, and TEM techniques. Among them, microtubules are the common defect in SiC crystals which will have a serious impact on subsequent epitaxial growth and device performance. Microtubule defects (also known as voids) generally propagate from the roots of the seed crystals or propagate through the SiC re-epitaxial to the deep crystals as shown in Fig. 4.21a. Studies have shown that a spiral dislocation with a large Burgers vector should have an empty core, or open core, at equilibrium. When nucleated by seed crystals, the microtubes have a hexagonal cross section with a diameter between 0.1 and 0.4 microns. The formation mechanism of microtubules involves various factors such as thermodynamics, growth kinetics and control techniques. The microtubules exhibit a butterfly-like stripe under the microscope as shown in Fig. 4.21b.

4.6.2 Nucleation and Growth of GaN on SiC Substrate

Due to the lack of single crystal GaN materials in nature and the extreme difficulty in growth of bulk single crystal GaN, GaN materials are mainly obtained by heteroepitaxial growth. Sapphire (Al_2O_3), silicon (Si) or silicon carbide (SiC) are usually used as the substrates for heteroepitaxial growth. SiC substrate is more suitable as a substrate material for nitride because of its smaller lattice mismatch with GaN material, higher thermal conductivity, easy cleavage and ability to fabricate conductive substrate [43, 44]. In recent years, Cree, Eudyna, NEC, RFMD and other companies in US and Japanese have launched commercial 4–6 inch large-size SiC substrates. However, the substrate is very expensive, which seriously restricts the development of Group III nitrides on SiC substrates. SiC substrate development in China is relatively late. There is still a certain gap in comparison with the imported substrate. Institutes in China engaged with SiC substrate research and development include the Institute of Physics of the Chinese Academy of Sciences, the Shanghai Institute of Ceramics of the Chinese Academy of Sciences, Shandong University, Shandong Tianyue and Hebei Tongguang and other research institutes and enterprises. Mass production of 3–4 inch SiC substrates has been achieved. The quality and yield of 6 inch SiC substrates are being solved. The research and development of SiC substrates in China provide convenient conditions for the epitaxial growth of SiC-based GaN and SiC. Table 4.5 shows the characteristics comparison between SiC substrate and nitride epitaxial material:

Although the lattice parameters of SiC substrate are similar to GaN, there are still many difficulties in epitaxial growth of GaN on SiC substrate, mainly in the following aspects: (1) difficulty in GaN nucleation on the surface of SiC substrate: due to the large chemical difference between GaN and SiC, the wettability of Ga atoms on the surface of SiC substrate is poor. This makes GaN nucleation difficult on SiC surface, which directly influence the growth rate and difficulty to obtain high quality epitaxial layer [45]; (2) Thermal mismatch: The thermal expansion coefficient of SiC is smaller than that of GaN, with a difference of 33.1%. For GaN materials grown at around 1000 °C, the strain in the epitaxial film layer is compressive with a strain of about 0.1% [46]. Due to the difference in thermal expansion coefficient, the stress in GaN epitaxial layer changes from compressive to tensile strain during the cooling process. If the tensile stress is accumulated to a certain extent, it will directly cause cracking of the GaN epitaxial layer, thus affecting the performance and reliability of the device. (3) Lattice mismatch: There is a 3.5% lattice mismatch between the SiC substrate and the GaN material, which also affects the crystal quality of the GaN epitaxial layer.

Therefore, in order to solve the poor wettability, reduce the mismatch stress, and grow a high quality GaN epitaxial layer, it is necessary to use a buffer layer to reduce the tensile stress in the GaN epitaxial layer on the SiC substrate. At present, the common buffer layers for GaN MOCVD epitaxial growth on SiC substrates are mainly AlN and AlGaN. There are also a few reports using low temperature GaN as a nucleation layer. Among them, the AlGaN buffer layer is mainly used for vertical structure optoelectronic devices such as LEDs. The AlN buffer layer is mainly used for microelectronic devices [47–51]. The use of AlN or AlGaN as a buffer layer has the following advantages: (1) AlN has better wettability on the surface of SiC substrate. In comparison with GaN, AlN is easier to nucleate and grow on the surface of SiC substrate; (2) The lattice mismatch between AlN and SiC is only 1%, which alleviates the lattice mismatch between GaN and SiC. (3) The thermal expansion coefficient of AlN is similar to that of SiC. Therefore, AlN is less prone to cracking

Table 4.5 Comparison of physical properties of nitride materials and substrate materials

Parameter	Unit	H-InN	H-GaN	H-AlN	Al_2O_3	SiC
Lattice constant a	Å	3.548	3.189	3.112	4.748	3.080
Coefficient of thermal expansion Δa/a	1/K	3.8×10^{-6}	5.59×10^{-6}	4.2×10^{-6}	7.3×10^{-6}	4.2×10^{-6}

due to the smaller tensile stress during the cooling process. If AlGaN is used as the buffer layer, the content of Al component should be greater than 6% to reduce the tensile strain in the epitaxial layer [45]; (4) For microelectronic materials, semi-insulating AlN can improve the high-voltage breakdown capability of the buffer layer.

Studies by Davis et al. [52] show that AlN grows mostly in the polycrystalline state between 500 and 1050 °C. At temperature above 1100 °C, it is possible for AlN to form single crystal. Since the saturated vapor pressure of N in AlN is lower than that of GaN [53], the high temperature annealing process has little effect on AlN and does not form a three-dimensional growth process similar to that of GaN buffer layer. High temperature growth condition is mostly used for AlN on SiC substrate. Studies have shown that [54] nucleation of AlN or GaN directly on SiC substrate will exist as Al polar plane or Ga polar plane. Nitride grown by nitridation will appear as N polar plane. Since the polarity of AlN directly affects the polarity of the subsequent GaN material, we can determine the surface polarity characteristics of the GaN epitaxial layer according to the nitridation condition. Even if the lattice mismatch of AlN and SiC substrates is only 1%, the epitaxial growth of AlN on SiC substrate is still heteroepitaxy, which is manifested by the presence of the stacking faults of Si/C and Al/N diatomic layers at the interface between AlN epitaxial layer and SiC substrate [55]. Coupled with the weak migration of Al atoms on the surface of the substrate, the growth of the AlN epitaxial layer on SiC substrate is in the three-dimensional island mode at the initial stage as shown in Fig. 4.22 [56]. Since the lattice mismatch between SiC and AlN is small, the mismatch stress during epitaxial growth is small. Meanwhile, the growth of AlN is typically done at high temperature, it is therefore that the growth of AlN tends to be two-dimensional layered growth after several atomic layers. In summary, since the growth mode of AlN on the SiC substrate have a transition from a three-dimensional island growth mode to a two-dimensional growth mode [57], it is necessary to grow a certain thickness of the AlN layer to obtain a flat surface. In addition, the AlN buffer layer not only provides nucleation centres for GaN growth, but more importantly provides a compressive

Fig. 4.22 The 3D mode of the AlN buffer layer grown on the SiC substrate [56]

strain for the subsequent GaN epitaxial layer that can shield the tensile stress caused by the different thermal expansion coefficients. Therefore, the quality and thickness of the AlN buffer layer directly determine the quality and cracking characteristics of the GaN epitaxial layer.

The AlN buffer layer provides a nucleation center for GaN growth, and GaN can be directly epitaxially grown on AlN nucleation sites. Therefore, considering the nucleation alone, the thickness of the AlN buffer layer has little effect on the epitaxial growth of GaN. Researchers like Waltereit in Germany [50] have grown high-quality GaN using only a 5 nm AlN buffer layer. However, considering the factors such as cracking, the AlN buffer layer must have a certain thickness. The growth of GaN on AlN buffer is heteroepitaxial. GaN on SiC is quite different from sapphire. Sapphire-based GaN can be grown using low-temperature GaN as a buffer layer, for which GaN is homoepitaxially grown on the GaN buffer layer. Therefore, the surface energy of epitaxial growth is not much different. The epitaxial growth window of GaN on sapphire is wide. Good material quality materials can be obtained within a certain range. For GaN on SiC, however, the epitaxial growth is more difficult although SiC and GaN is very close in terms of lattice parameters. On the one hand, GaN is difficult to nucleate on the surface of SiC substrate due to the large difference in chemical properties between GaN and SiC. On the other hand, with the introduction of AlN or AlGaN buffer layer, AlN on SiC as well as GaN growth on the AlN buffer layer is heteroepitaxial. This is why it is difficult to obtain high quality epitaxial GaN layers on SiC substrate.

Since the lattice mismatch of AlN and GaN is 2.38%, the critical thickness of epitaxially grown dislocation-free GaN on AlN is only 4 nm [56]. Beyond this critical thickness, a large number of dislocations will be introduced in the GaN epitaxial layer due to stress accumulation. Studies have shown that [58] for GaN grown on the AlN buffer layer it still exhibits a three-dimensional growth mode at the initial stage, except for that the thickness of GaN grown by the three-dimensional mode is very thin (about 5 nm) as shown in Fig. 4.23 [58], and subsequently changes into the two-dimensional growth mode quickly.

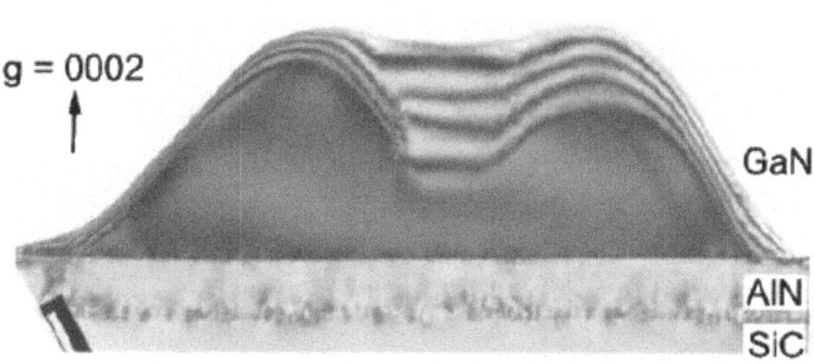

Fig. 4.23 GaN island particles after the AlN buffer layer [58]

4.6.3 Roots of GaN Stress on SiC Substrates

Epitaxial GaN layer on SiC substrate will subject to tensile strain during cooling due to the inherent defects of SiC substrate and the difference of thermal expansion coefficient of 33.1%. If the stress accumulation reaches a certain level, it will cause the GaN epitaxial layer to crack. This will seriously affect the performance and reliability of the device.

In order to solve the tensile strain and cracking problem, AlN or AlGaN is commonly used as a buffer layer to modulate the stress in the GaN epitaxial layer. Since the lattice parameter of AlN is smaller than that of GaN, the GaN epitaxial layer on AlN is inevitably subjected to compressive stress. The tensile stress between the SiC substrate and the GaN epitaxial layer is shielded by the compressive stress provided by AlN.

We discuss the stress in the GaN epitaxial layer in two cases. One case is that the GaN material is directly epitaxially grown on the SiC substrate or a very thin AlN buffer layer. Since the lattice mismatch between GaN and SiC substrates is only 3.4%, the critical thickness of GaN on SiC substrates is only a few tens of nanometers thick. At the growth temperature of GaN, the GaN epitaxial layer is mainly subjected to compressive stress from the SiC substrate. As the thickness of GaN increases, the compressive stress introduced by lattice mismatch also accumulates. When the thickness of GaN exceeds its critical thickness on the SiC substrate, the stress in the GaN epitaxial layer will be released by the formation of dislocations. Heying' studies have shown that stresses due to lattice mismatch during growth will be relaxed in the manner of producing edge dislocations [59]. This is also the reason why the (002) dichroic diffraction of GaN or AlN epitaxial layer grown on the SiC substrate has a small full width at half maximum (FWHM), while the (102) diffraction has a large FWHM. During the cooling process, a difference in thermal expansion coefficient of 33.1% between GaN and SiC substrate will cause a large tensile stress in the GaN epitaxial layer. However, since the compressive stress generated by the lattice mismatch during growth has been relaxed by dislocations and the like, the tensile stress generated during the cooling process cannot be shielded, and thus cracking will be difficult to avoid. Moreover, the more dislocations in the GaN epitaxial layer are, the larger the compressive stress through dislocation relaxation is. Also, the more the tensile stress introduced during the cooling process will increase, the more severe the cracking will be.

Another case is to use an AlN or AlGaN buffer layer [60]. When employed as a buffer layer, the mismatch between SiC and GaN will relax through the AlN layer. During epitaxial growth, compressive stress is generated between SiC and AlN due to lattice mismatch. At the interface, the mismatch stress is released by the three-dimensional growth mode of the AlN buffer layer. Therefore, the compressive stress in the GaN epitaxial layer will mainly come from the AlN buffer layer [50]. The compressive stress generated by epitaxial growth of GaN on AlN will be divided into two steps for relaxation [51]: 70% relaxation at the AlN/GaN interface; the remaining stress decreases exponentially with increasing GaN thickness [61]. If the residual compressive stress in the GaN epitaxial layer is greater than the tensile stress introduced by GaN during thermal cooling due to thermal mismatch, GaN will be subjected to compressive stress at normal temperature and is less prone to cracking. However, GaN grown on the surface of AlN is heteroepitaxial and tends to nucleate where the surface of the AlN is undulating and defective. The GaN growth involves both two-dimensional and three-dimensional mode [58, 62]. Therefore, the residual compressive stress is largely released at the AlN/GaN interface. Since the residual compressive stress is smaller than the tensile stress introduced by the difference in thermal expansion coefficient, cracking occurs due to tensile stress enrichment in the GaN epitaxial material. In summary, the stress in the SiC-based GaN epitaxial layer ultimately depends on the superposition of the tensile stress caused by the difference in compressive stress and thermal expansion coefficient caused by lattice mismatch.

References

1. www.bridgelux.com
2. www.cree.com
3. H. Amano, N. Sawaki, I. Akasaki et al., Metalorganic vapor-phase epitaxial-growth of a high-quality GaN film using an AlN buffer layer. Appl. Phys. Lett. **48**, 353 (1986)
4. H. Amano, M. Kito, K. Hiramatsu et al., p-type conduction in Mg-doped GaN treated with low-energy electron-beam irradiation (LEEBI). Jpn. J. Appl. Phys. **28**, L2112 (1989)
5. S. Nakamura, M. Senoh, T. Mukai, High-power GaN p-n-junction blue light emitting diodes. Jpn. J. Appl. Phys. **30**, L1708 (1991)
6. Z. Yu, M. Johnson, J. Brown et al., Study of the epitaxial lateral overgrowth process for GaN on sapphire. J. Cryst. Growth **195**, 333 (1998)
7. R. Davis, T. Gehrke, K. Linthicum et al., Conventional and pendeo-epitaxial growth of GaN(0001) thin films on Si (111) substrates. J. Cryst. Growth **231**, 335 (2001)
8. K. Hiramatsu, K. Nishiyama, M. Onishi et al., Fabrication and characterization of low defect density GaN using facet controlled epitaxial lateral overgrowth. J. Cryst. Growth **221**, 316 (2000)
9. http://lights.ofweek.com/2015
10. L. Wu, Proceedings of China SSL, Guangzhou, China (2014)
11. VEECO K465 MOCVD manual
12. M. Heuken, *Latest MOCVD Production Technology for Solid-State Lighting* (SSL, China, 2009)
13. www.veeco.com
14. K. Schmidegg, G. Neuwirt, D. Stifter et al., In situ optical analysis of low temperature MOCVD GaN nucleation layer formation via multiple wavelength ellipsometry. J. Cryst. Growth **272**, 106 (2004)
15. A. Springthorpe, T. Humphreys, A. Majeed et al., In situ growth rate measurements during molecular beam epitaxy using an optical pyrometer. Appl. Phys. Lett. **55**, 2138 (1989)
16. H. Grothe, F. Boebel, In situ control of Ga(Al)As MBE layers by pyrometric interferometry. J. Cryst. Growth **127**, 1010 (1993)
17. F.G. Bobel, H. Moller, A. Wowchak, et al., Pyrometric interferometry for real time molecular beam epitaxy process monitoring. J. Vacuum Sci. Technol. B. **12**, 1207 (1994)
18. H. Maruska, J. Tietjen, Preparation and properties of vapor-deposited single-crystalline GaN. Appl. Phys. Lett. **15**, 327 (1969)
19. V. Ban, Mass spectrometric studies of vapor-phase crystal growth 2 GaN. J. Electrochem. Soc. **119**, 761 (1972)

20. A. Sakai, H. Sunakawa, A. Usui, Transmission electron microscopy of defects in GaN films formed by epitaxial lateral overgrowth. Appl. Phys. Lett. **73**, 481 (1998)
21. S. Figge, T. Bottcher, S. Einfeldt et al., In-situ and ex-situ evaluation of the film coalescence for GaN growth on GaN nucleation layers. J. Cryst. Growth **221**, 262 (2000)
22. S. Einfeldt, T. Bottcher, S. Figge et al., Thermally induced stress in GaN layers with regard to film coalescence. J. Cryst. Growth **230**, 357 (2001)
23. H. Heinke, V. Kirchner, S. Einfeldt et al., X-ray diffraction analysis of the defect structure in epitaxial GaN. Appl. Phys. Lett. **77**, 2145 (2000)
24. B. Beaumont, M. Vaille, T. Boufaden et al., High quality GaN grown by MOVPE. J. Cryst. Growth **170**, 316 (1997)
25. J.E. Ayers, The measurement of threading dislocation densities in semiconductor crystals by X-ray diffraction. J. Cryst. Growth **135**, 71 (1994)
26. S. Nakamura, T. Mukai, M. Senoh, In situ monitoring and Hall measurements of GaN grown with GaN buffer layers. J. Appl. Phys. **71**, 5543 (1991)
27. J. Wu, X. Han, J. Li, et al., Influence of buffer layer thickness on the properties of an undoped GaN layer grown on sapphire substrate by metalorganic chemical vapor deposition. J. Syn. Cryst. **34**, 466 (2005) (in Chinese)
28. T. Wang, T. Shirahama, H. Sun et al., Influence of buffer layer and growth temperature on the properties of an undoped GaN layer grown on sapphire substrate by metalorganic chemical vapor deposition. Appl. Phys. Lett. **76**, 2220 (2000)
29. J. Kuznia, M. Khan, D. Olson et al., Influence of buffer layers on the deposition of high quality single crystal GaN over sapphire substrates. J. Appl. Phys. **73**, 4700 (1993)
30. G. Li, S.J. Chua, S.J. Xu et al., Nature and elimination of yellow-band luminescence and donor–acceptor emission of undoped GaN. Appl. Phys. Lett. **74**, 2821 (1999)
31. M. Yi, H. Lee, D. Kim et al., Effects of growth temperature on GaN nucleation layers. Appl. Phys. Lett. **75**, 2187 (1999)
32. H. Cho, K. Kim, C. Hong et al., Electron traps and growth rate of buffer layers in unintentionally doped GaN. J. Cryst. Growth **223**, 38 (2001)
33. C. Lee, S. Son, I. Lee et al., Growth and characterization of GaN epilayers grown at various flow rates of trimetylgallium during growth of nucleation layers. J. Cryst. Growth **171**, 27 (1996)
34. K. Kim, C. Oh, K. Lee et al., Effects of growth rate of a GaN buffer layer on the properties of GaN on a sapphire substrate. J. Appl. Phys. **85**, 8441 (1999)
35. F. Juang, Y. Su, S. Chang, et al., Effects of buffer layer growth conditions on GaN epilayer quality by MOCVD, in *Proceedings of 4th Conference on Light Emitting Diodes*, 26–27 January 2000, San Jose, CA
36. J. Chen, S. Zhang, B. Zhang et al., Influences of reactor pressure of GaN buffer layers on morphological evolution of GaN grown by MOCVD. J. Cryst. Growth **256**, 248 (2003)
37. B. Cao, K. Xu, Y. Ishitani et al., In-situ spectroscopic ellipsometry investigation and control of GaN growth mode in metal-organic vapor phase epitaxy at low pressures of 20 Torr. Thin Solid Films **455**, 661 (2004)
38. S. Kim, J. Oha, J. Kang et al., Two-step growth of high quality GaN using V/III ratio variation in the initial growth stage. J. Cryst. Growth **262**, 7 (2004)
39. T. Yang, K. Uchida, T. Mishima et al., Control of initial nucleation by reducing the V/III ratio during the early stages of GaN growth. Physica Status Solidi A-Appl. Mater. Sci. **180**, 45 (2000)
40. L.S. Ramsdell, Studies on Silicon Carbide. Am. Miner. **32**, 64 (1947)
41. G.R. Fisber, P. Barns, Phi. Mag. B. **61**, 217 (1990)
42. Wikimedia Foundation, Inc., Silicon carbide [EB/OL]. http://en.wikipedia.org/wiki/Silicn_carbide
43. T. Weeks, M. Bremser, K. Ailey et al., GaN thin films deposited via organometallic vapor phase epitaxy on alpha(6H)-SiC(0001) using high temperature monocrystalline AlN buffer Layers. Appl. Phys. Lett. **67**, 401 (1995)
44. B. Moran, F. Wu, A. Romanvo et al., Structure and morphological evolution of GaN grown by metalorganic chemical vapor deposition on SiC substrates using an AlN initial layer. J. Cryst. Growth **73**, 38 (2004)

45. H. Lahreche, P. Vennegues, M. Vaillc et al., Comparative study of GaN layers grown on insulating AlN and conductive AlGaN buffer layers. Semicond. Sci. Technol. **14**, L33 (1999)
46. S. Einfeldt, A. Roskowski, E. Preble et al., Strain and crystallographic tilt in uncoalesced GaN layers grown by maskless pendeoepitaxy. Appl. Phys. Lett. **80**, 953 (2002)
47. V. Davydov, N. Averkiev, I. Gonecheruk et al., Raman and photoluminescence studies of biaxial strain in GaN epitaxial layers grown on 6H-SiC. J. Appl. Phys. **82**, 5097 (1997)
48. B. Skromme, H. Zhao, D. Wang et al., Strain determination in heteroepitaxial GaN. Appl. Phys. Lett. **71**, 829 (1997)
49. W. Perry, T. Zheleva, M. Bremser et al., Correlation of biaxial strains, bound exciton energies, and defect microstructures in GaN films grown on AlN/6H-SiC (0001) substrates. J. Electron. Mater. **26**, 224 (1997)
50. P. Waltereit, O. Brandt, A. Trampert et al., Influence of AlN nucleation layers on growth mode and strain relief of GaN grown on 6H-SiC (0001). Appl. Phys. Lett. **74**, 3660 (1999)
51. T. Weeks, M. Bremser, K. Ailey et al., Undoped and doped GaN thin films deposited on high-temperature monocrystalline AlN buffer layers on vicinal and on-axis alpha (6H)-SiC (0001) substrates via organometallic vapor phase epitaxy. J. Mater. Res. **11**, 1101 (1996)
52. R. Davis, T. Weeks, M. Bremser et al., Growth of AlN and GaN thin films via OMVPE and gas source MBE and their characterization. Solid-State Electr. **41**, 129 (1997)
53. K. Yasuyuki, A. Tetsuya, K. Naoki, Thermal stability of low-temperature GaN and AlN buffer layers during metalorganic vapor phase epitaxy monitored by in situ shallow-angle reflectance using ultraviolet light. Jpn. J. Appl. Phys. **37**, L1208 (1998)
54. Y. Wu, A. Hanlon, J. Kaeding et al., Effect of nitridation on polarity, microstructure, and morphology of AlN films. Appl. Phys. Lett. **84**, 912 (2004)
55. S. Tnaka, R. Kern, R. Davis, Initial stage of AlN film growth on 6H-SiC by plasma assisted, gas source MBE. Appl. Phys. Lett. **66**, 37 (1995)
56. F. Ponce, B. Krusor, J. Major et al., Microstructure of GaN epitaxy on SiC using AlN buffer layer. Appl. Phys. Lett. **67**, 410 (1995)
57. Z. Chen, S. Newman, D. Brown et al., High quality AlN grown on SiC by metal organic chemical vapor deposition. Appl. Phys. Lett. **93**, 191906 (2008)
58. B. Moran, F. Wu, A. Romanov et al., Structural and morphological evolution of GaN grown by metalorganic chemical vapor deposition on SiC substrates using an AlN initial layer. J. Cryst. Growth **273**, 38 (2004)
59. B. Heying, X. Wu, S. Keller et al., Role of threading dislocation structure on the X-ray diffraction peak widths in epitaxial GaN films. Appl. Phys. Lett. **68**, 643 (1996)
60. S. Einfeldt, Z. Reitmeier, R. Davis, Surface morphology and strain of GaN layers grown using 6H-SiC (0001) substrates with different buffer layers. J. Cryst. Growth **253**, 129 (2003)
61. S. Heikman, S. Keller, S. DenBaars et al., Growth of Fe doped semi-insulating GaN by metalorganic chemical vapor deposition. Appl. Phys. Lett. **81**, 439 (2002)
62. T. Nishida, N. Kobayashi, Nucleation control in MOVPE of group III-nitrides on SiC substrates. J. Cryst. Growth **221**, 297 (2000)

Chapter 5
InGaN/GaN Multiple Quantum Wells Materials as Well as Blue and Green LEDs

InGaN/GaN quantum wells are commonly used in III-nitride based blue or green LED chips. Back in the earlier days, it was difficult for GaN materials to be doped for the formation of high crystal quality p-type GaN. The first GaN-based LED introduced in 1971 could only adopt MIS (metal-insulator-semiconductor) structure. In 1986, Hiroshi Amano and Akasaki et al. [1] demonstrated GaN films with a good surface structure and high crystal quality by MOCVD through low-temperature AlN buffer epitaxial layer. In 1989, Amano et al. [2] used low-energy electron irradiation (LEEBI) technology and successfully achieved p-type doping of GaN materials. These breakthrough technologies accelerated the development of LED. This led to the rapid progress in terms of output power, quantum efficiency, spectral quality, etc. However, InGaN/GaN multiple quantum wells LEDs also face many problems. Currently, InGaN-based LEDs are grown on the c-plane sapphire along the [0001] polarization direction. The InGaN quantum well is in a compressed elastic deformation, and a piezoelectric polarization field is produced along the growth direction up to the order of MV/cm. The polarized electric field will aggravate the quantum confined Stark effect (QCSE) in the InGaN quantum well, which causes the separation of electron and hole wave functions in the quantum well and reduces the probability of radiation recombination. At the same time, the polarized electric field will also cause carriers to leak from the well, which reduces the carrier concentration involved in radiation recombination and aggravates the droop effect. These will be discussed in subsequent chapters.

In addition, the dislocation density in InGaN/GaN multiple quantum wells grown on c-plane sapphire is very high. On the other hand, the internal quantum efficiency of InGaN LEDs is actually not low: the internal quantum efficiency of blue LEDs exceeds 80% and about 60% for green LEDs. This phenomenon has a lot to do with the localization of carriers in InGaN/GaN multiple quantum wells. Both the localization effect and the quantum confined Stark effect affect the spontaneous emission of InGaN/GaN multiple quantum wells.

J. Li et al., *III-Nitrides Light Emitting Diodes: Technology and Applications*, Springer Series in Materials Science 306,
https://doi.org/10.1007/978-981-15-7949-3_5

Polarization effect and quantum confined Stark effect in InGaN/GaN multiple quantum wells materials become even more obvious with the increase of In incorporation. The development of semi-polar and non-polar LEDs can effectively alleviate these effects on nitride LEDs performance.

This chapter mainly introduces the polarization effects, quantum confined Stark effect, and carrier localization effect closely related to the InGaN/GaN multiple quantum wells materials. The effects of these properties on blue and green LEDs will be also discussed. Finally, a brief overview of the research on semi-polar and non-polar LEDs is provided.

5.1 Introduction to InGaN Material System

The InGaN material is a ternary alloy composed of GaN and InN. The energy gap of this material is tunable from 0.7 to 3.4 eV with a very wide spectral range as shown in Fig. 5.1. As can be seen from the figure, the spectra cover the green, blue and ultraviolet. InGaN material has high breakdown voltage and stable chemical properties. These properties make it possible to achieve high efficiency blue and green LEDs.

The emission wavelength of InGaN/GaN multi-quantum wells LEDs can adjusted in two ways: one is to change the well width and the number of wells in the quantum well or the thickness of the barrier layer, and the other is to adjust the composition of In in the active layer of InGaN alloy. It is easier to achieve and control different luminescence wavelengths by adjusting the composition of In in the InGaN alloy. The In composition can be adjusted by controlling the flow rate of the carrier gas ratio of TMIn and TMGa during the MOCVD growth process. Theoretical analysis shows that the emission wavelength of InGaN LED belongs to the blue light category when the composition of In is 15–20%. When the composition of In is 25–30%, the emission wavelength of InGaN LED belongs to the green light category. Generally, in order to extend the emission wavelength to the rest of the visible light wavelength, it is necessary to configure the In composition to a large extent to change the bandgap. However, it causes an excessive lattice mismatch, accompanied by a higher order

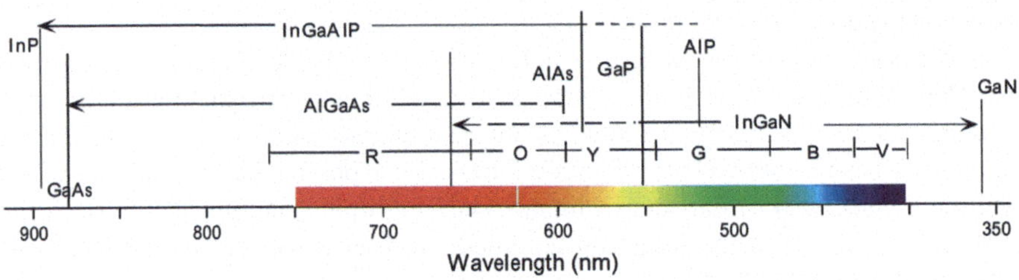

Fig. 5.1 Wavelength range corresponding to LEDs of different materials [3]

of dislocation densities. Therefore, InGaN material is not suitable for other spectral regions than the visible spectrum.

So far, the growth of InGaN/GaN multiple quantum wells structure is mainly through MOCVD technology [4, 5]. In the growth process of multiple quantum wells, the large change of temperature is very important, because the GaN barrier layer is generally required to grow at a relatively high temperature (1050 °C), which is most advantageous for growing high-quality GaN crystal. However, in order to keep stability during growth, the InGaN layer must grow at a lower temperature (800 °C). This requires the growth system to have a capacity to change the substrate temperature by 100 °C per minute. For large-scale manufacturing, therefore, only MOCVD can provide such capability at present.

5.2 Polarization Effects in InGaN/GaN Multiple Quantum Wells Materials

5.2.1 Polarity of GaN-Based Materials

In sphalerite and wurtzite structures, cations and anions are produced due to the charge transfer between atoms. For the wurtzite structure, the material exhibits polarization macroscopically due to the deviation of the positive and negative charging centers of the unit cell. This material is called a polar material. GaN is a polar material. Generally, if the epitaxial layer has a Ga terminated surface, the epitaxial material has a Ga plane polarity. Conversely, if the epitaxial layer has an N plane as the termination surface, the epitaxial material has an N plane polarity. It is worth noting that polarity is an integral property of a material, not a surface property. For example, covering a single atomic layer of N atoms on a Ga-polar material does not change the properties of the Ga-plane polarity of the material. In addition, there are cases where the termination surface of the epitaxial layer can include both Ga atoms and N atoms. In this case there is generally a dominant polarity. In the III-V nitride materials, the conventional epitaxial layer grown by MOCVD is along the (0001) direction of sapphire. Generally, a high-quality GaN epitaxial layer grown by a low-temperature buffer layer on a sapphire substrate by MOCVD has a Ga surface polarity.

The polarity direction of GaN is the (0001) direction, and it can also be represented by a GaN plane, a + c, a gallium polarity, or a Ga polar plane. The opposite direction is the N polar plane. For the Ga polar plane, the Ga atoms exist with one bond facing up and the three bonds facing downward. The N atoms are the opposite with three bonds facing upward and one bond facing downward. The N polar plane is opposite to Ga polar plane.

5.2.2 Spontaneous Polarization and Piezoelectric Polarization

Polarization is an important property of nitride materials. Polarization effects exist in both Ga-polar and N-polar materials. Polarization effects include spontaneous polarization and piezoelectric polarization. Spontaneous polarization is caused by the asymmetry of the wurtzite structure in the (0001) direction. Piezoelectric polarization is caused by the arrangement shift of anion and cations accompanied by the lattice distortion. The polarity of the III-V material determines the direction of spontaneous polarization and piezoelectric polarization while affecting the structure and electrical properties of the material.

As early as 1997, Bernardini et al. [6] predicted that there is a strong spontaneous polarization effect in the III-nitride materials. In the materials of the Ga face polarity (cationic polarity), the material lattice constant c/a is smaller than the ideal factor of 1.633, and the electronegativity between the group III cation and the N atom is largely different. Therefore, the polarization vector P_1 is larger than the polarization vector P_2 in the material structural unit, so there is a top-down spontaneous polarization P_{sp} throughout the material which is opposite to the material growth direction [0001]. In the case of the N-face polarity, the polarity is opposite to that of the Ga surface, and there is an upward spontaneous polarization in the material that is the same as the material growth direction [000-1].

Figure 5.2 shows the spontaneous polarization coefficients of GaN, InN and AlN [6]. It can be seen from the figure that the spontaneous polarization coefficient of GaN is the smallest. The difference between InN and GaN is not large. The spontaneous polarization of AlN material is the largest.

The spontaneous polarization intensity of the nitride material is related to the composition. Referring to the theoretical calculations by Fiorentini et al. [7], the spontaneous polarization of the ternary nitride can be expressed by the following formula:

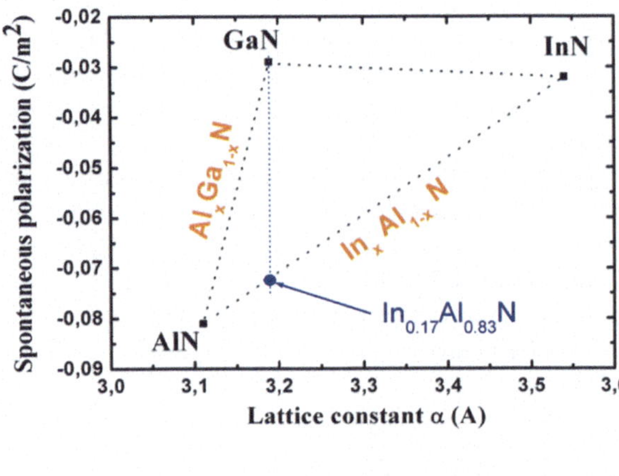

Fig. 5.2 Spontaneous polarization coefficient of group III nitride

$$P^{sp}_{In_xGa_{1-x}N} = -0.042x - 0.034(1-x) + 0.037x(1-x) \tag{5.1}$$

$$P^{SP}_{Al_xGa_{1-x}N} = 0.09x - 0.034(1-x) + 0.019x(1-x) \tag{5.2}$$

When the Group III nitride undergoes lattice distortion from biaxial stress from the outside, piezoelectric polarization effects are also produced in the material [8–11]. In the nitride crystal of the Ga polar plane, when the epitaxial material is subjected to biaxial tensile stress, the bond angle θ between the metal cation and the three nitrogen atoms becomes large in the structural unit cell of the material, in which $P_1 > P_2$ exists in the entire structural units because the resultant polarization vector P_2 between the metal cation and the three nitrogen atoms gets smaller, thereby generating a piezoelectric polarization along [000-1] direction in the material. When the material is subjected to biaxial compressive stress, the bond angle θ becomes smaller so that the resultant polarization vector P_2 increases. In this case, $P_1 < P_2$, a piezoelectric polarization along the [0001] direction is generated in the material. The magnitude of the piezoelectric polarization in the nitride material can be calculated from the piezoelectric coefficient and the strain of the material. In nitrides, the heterojunction materials composed of GaN, InN, and AlN or their alloy materials have a certain stress due to their different lattice constants, and piezoelectric polarization occurs in the material [12].

In 2002, Bernardini et al. [7] proposed a nonlinear polarization model of III-nitride materials through experimental tests and theoretical simulations. In the nonlinear polarization model, the polarization surface charge density at the heterojunction interface is a second order function of the material composition. Later, Yu et al. [13] studied the correctness and validity of the nonlinear polarization model. In InGaN/GaN quantum wells, the polarization effect causes spatial separation of the wave functions of electrons and holes in the quantum well, reducing the luminous efficiency of the quantum well. The strongly polarized electric field also causes the LED emission wavelength to red-shift.

The macroscopic polarization is strong in the wurtzite nitride heterostructure. The polarization includes (1) piezoelectric polarization caused by lattice mismatch strain and (2) spontaneous polarization caused by charge accumulation at the heterojunction interface. In a typical nitride heterostructure grown along (0001) plane, there is a large piezoelectric polarization due to the large piezoelectric constant (see Table 5.1). Piezoelectric polarization causes an internal electric field to be generated in the [0001] growth direction.

Piezoelectric polarization is caused by lattice mismatch between heterojunctions. For nitride materials, when the material is grown along the [0001] direction and the material is subjected to biaxial stress, its piezoelectric polarization can be expressed as:

$$P_{pz} = 2\left(e_{33}\varepsilon_z + e_{31}\left(\varepsilon_x + \varepsilon_y\right)\right) \tag{5.3}$$

Table 5.1 Fundamental constants of nitride materials [6]

Parameters	Material type		
	GaN	InN	AlN
P_{sp} (c/m^2)	−0.034	−0.042	−0.090
e_{33} (cm^2)	0.73	0.97	1.46
e_{13} (cm^2)	−0.49	−0.57	−0.60
c_{13} (GPa)	10.8	92	12
c_{33} (GPa)	39.9	224	39.5
a_0 (nm)	3.189	3.545	3.11

where e_{31} and e_{33} are piezoelectric constants and ε_z is the in-plane strain which can be calculated by using formula

$$\varepsilon_z = (c - c_0)/c_0 \tag{5.4}$$

where c is the in-plane lattice constant of nitride, and c_0 is the lattice constant after relaxation.

Lattice constants of wurtzite group III-V nitrides have the following relationship:

$$\frac{c - c_0}{c_0} = -2\frac{c_{13}}{c_{33}}\frac{a - a_0}{a_0} \tag{5.5}$$

where c_{13} and c_{33} are elastic coefficients.

$$P_{pe} = 2\frac{a - a_0}{a_0}\left(e_{31} - e_{33}\frac{c_{13}}{c_{33}}\right) \tag{5.6}$$

According to the theoretical calculation results of Fiorentini et al. [7], the relationship between piezoelectric polarization intensity and components of multicomponent nitrides can be expressed as follows:

$$\begin{aligned} P_{pz}\left(Al_x In_y Ga_{1-x-y} N\right) = {} & P_{pz}(AlN)x + P_{pz}(AlN)x + P_{pz}(InN)y \\ & + P_{pz}(GaN)(1 - x - y) \end{aligned} \tag{5.7}$$

where

$$P_{pz}(AlN) = -1.808\varepsilon + 5.624\varepsilon^2, \quad \text{when } \varepsilon < 0, \tag{5.8}$$

$$P_{pz}(AlN) = -1.808\varepsilon - 7.888\varepsilon^2, \quad \text{when } \varepsilon > 0, \tag{5.9}$$

$$P_{pz}(GaN) = -0.918\varepsilon + 9.541\varepsilon^2 \tag{5.10}$$

$$P_{pz}(InN) = -1.373\varepsilon + 7.559\varepsilon^2 \tag{5.11}$$

where ε is the in-plane strain of c plane:

$$\varepsilon(x) = [a_{subs} - a(x)]/a(x) \tag{5.12}$$

where a(x) is the lattice parameter in c plane when the component is x, and a_{subs} is the lattice constant of adjacent materials.

Spontaneous polarization of ternary nitrides can be given by the following formula:

$$P_{sp\_In_xGa_{1-x}N} = 0.418x - 0.0424x(1-x) \tag{5.13}$$

$$P_{sp\_Al_xGa_{1-x}N} = -0.525x - 0.0282x(1-x) \tag{5.14}$$

The total polarization intensity (expressed as P) is the sum of piezoelectric polarization and spontaneous polarization,

$$P = P_{sp} + P_{pz} \tag{5.15}$$

At the sharp interface of the multiple quantum wells structure (InGaN/GaN or InGaN/AlInGaN), the polarization intensity can be increased or decreased in the bilayer to generate the surface charge density of the polarization:

$$\sigma = P^w - P^b = \left\{P^w_{sp} + P^w_{pe}\right\} - \left\{P^b_{sp} - P^b_{pe}\right\} \tag{5.16}$$

For a multiple quantum wells structure composed of only two materials, namely a well and a barrier, with no free charges and no voltage drop along the multiple quantum wells, the electric field caused by polarization of the well layer can be calculated by the following equation,

$$F_P = L_b\left(P^b - P^w\right)/[\varepsilon_0(L_w\varepsilon_b + L_b\varepsilon_w)] \tag{5.17}$$

where L_b and L_w are the thickness of the barrier layer and the well layer, respectively, the ε_0 is the vacuum dielectric constant, the ε_b and the ε_w are the relative dielectric constant of the barrier layer and the well layer, respectively. When there is polarization surface charge at the interface, background carriers or injected carriers will gather at the interface and partially shield the polarization electric field. Considering the free carrier shielding, the well layer electric field can be calculated by the following equation.

$$F_P = L_b\left(\sigma + P^b - P^w\right)/[\varepsilon_0(L_w\varepsilon_b + L_b\varepsilon_w)] \tag{5.18}$$

When a polarization electric field exists in the well layer, the rectangular quantum well needs to be replaced by a triangular situation well. In this case, the radiation combination energy of electrons and the holes at ground state can be expressed by the following formula.

$$E_{eh} = E^{g}_{InGaN} - eF_P L_w + \left(\frac{9\pi heF_P}{8\sqrt{2}}\right)^{2/3}\left(\frac{1}{m^*_{InGaN,e}} + \frac{1}{m^*_{InGaN,h}}\right)^{1/3} \tag{5.19}$$

where $m^*_{InGaN,e}$, $m^*_{InGaN,h}$ are the electron and hole effective mass of InGaN, and

$$E_{eh} = E^{g}_{InGaN} - eF_p L_w + 0.287e\left(F_p\right)^{2/3} \tag{5.20}$$

5.3 Quantum-Confined Stark Effect

Polarization generates an electric field in the InGaN/GaN multiple quantum wells, which tilts the conduction band and the valence band edge and changes the subband energy level and the bound state wave function, resulting in a change in transition energy and intensity. The electric field separates the space between electrons and holes, reducing the spatial overlap of the carrier wave function, thereby reducing the probability of radiation recombination and increasing the radiative recombination lifetime [14–17]. Since the electrons and holes in the quantum well are separated into different sides of the well, the combination energy is different from the band edge energy difference. This is the quantum confined Stark effect in nitride LEDs, where the amount of red-shift increases as the well width increases. When the well width is small, the quantum confined effect causes the quantum well luminescence energy to be blue-shifted compared to the bulk material. As the well width increases, the quantum confined Stark red-shift is dominant. At high photoexcitation density or electrical injection, high-density electrons and holes move to different directions of the well, shielding part of the polarization-induced surface charges and generating an electric field opposite to the direction of the polarized electric field. Such an effect can partially attenuate the polarized electric field. Adding a voltage opposite to the polarized electric field can also eliminate the polarization effect [18].

The effect of the quantum confined Stark caused by the polarization effect on InGaN light-emitting diodes is mainly concentrated in two aspects: first, the effect on the transition level in the InGaN quantum well; second, the effect on the combination probability in the quantum well.

5.3.1 Effect on Transition Energy Levels

The electron and hole transition energy in a quantum well is a function of several parameters such as band gap, biaxial strain, sub-band energy level caused by quantum confined effect, electric field strength in the well, and sample temperature, etc., which can be expressed by the following formula:

$$hv(L, T, a_{well}) = E_0 + \Delta E_{QW}(L) + \Delta E_{strain}(a_{well}) + \Delta E_{E-field}(L, a_{well}) + \Delta E_{temp}(T) \tag{5.21}$$

where L is the quantum well width, T is the temperature, a_{well} is the actual in-plane lattice constant of the well layer, and E_0 is the band gap under ideal conditions. Below we discuss the impact of each terms of the formula:

ΔE_{QW} is the sub-band levels movement generated by quantum confined effect. Using a simple infinite deep potential well model, this movement can be expressed as

$$\Delta E_{QW} = \frac{h^2}{2m}\frac{n^2\pi^2}{L^2} \quad n = 1, 2, 3 \tag{5.22}$$

where L is the well width and m is the sum of the effective mass of electrons and holes.

Regarding the third term in the equation, a lot of work about the effect of strain caused by lattice mismatch on the band gap of InGaN have been studied [19–21]. The contribution of strain to the change of the band gap can be expressed by the following formula.

$$\Delta E_{strain} = -14.755\varepsilon - 51.148\varepsilon^2 \tag{5.23}$$

where ε is the biaxial strain.

The last term in the equation is the dependence of transition energy on temperature. From this, we can study the effect of localization on the optical properties of InGaN quantum wells, which will be detailed in the next section.

For the InGaN/GaN MQWs structure, since the spontaneous polarization constant difference of GaN and InN is small and the difference in lattice constant is large, the contribution of piezoelectric polarization is dominant. The dependence of the transition energy level and the polarization field in a quantum well is generally estimated by the following equation.

$$E_{eh} = E^g_{InGaN} - eF_pL_w + 0.287e\left(F_p\right)^{2/3} \tag{5.24}$$

where E^g_{InGaN} is the band gap of InGaN, and F_p is the total piezoelectric polarization intensities in the quantum well. When the lattice mismatch between the well and barrier increases, the intensity of the polarized electric field in the quantum well increases, resulting in a decrease in the electron and hole transition level. Then $\Delta E_{E-field}(L, a_{well})$ changes to negative, as a result of transition energy level caused by the polarization field.

5.3.2 *Effect on Luminous Intensity*

The polarized electric field bends the energy band in the active region, and the electrons and holes are spatially separated. This reduces the wave function overlap of the carriers, and finally reduces the radiation recombination probability. Due to the spatial separation of the carrier wave function caused by the polarization field, the increase of the thickness of the well layer further promotes the decrease of the electron and hole wave function overlap probability. The radiation combination intensity of the InGaN quantum well will decrease. The electroluminescence intensity of the InGaN/GaN multiple quantum wells LED epitaxial wafer varies with the width of the well. It can be seen that as the well width increases, the peak wavelength has a blue-shift and the electroluminescence intensity decreases.

5.4 Carrier Localization in InGaN/GaN Multiple Quantum Wells

Due to lattice mismatch, InGaN/GaN multiple quantum wells generate large stresses during growth, resulting in high screw dislocation density. Even so, the internal quantum efficiency of InGaN LEDs can be very high (the internal quantum efficiency of blue LEDs exceeds 80%, and that of green LEDs can reach about 60%), which is in contrast to carriers in InGaN/GaN multiple quantum wells that has a lot to do with carriers localization in InGaN/GaN multiple quantum wells [22]. The ideal band structure should be flat, but the presence of point defects and changes in growth parameters [23] can cause potential fluctuations, which in turn make carriers easily confined to the minimum value of potential energy. The phenomenon is called the localization effect [24–26]. The reason for the formation of localization effects vary widely, but it can be summarized into three types: quantum well width fluctuations, alloy fluctuations [27], and In clustering (indium clustering). The effect of localization of carriers on the luminescence properties of InGaN/GaN multiple quantum wells can be analyzed by temperature-dependent photoluminescence [28].

As shown in Fig. 5.3, when the temperature is 10 K, the carriers are randomly distributed at the minimum point of potential energy. As the temperature increases from 10 to 70 K, the carriers in the weakly localized states are activated by heat, and then they are relaxed to strongly localized states by a Hopping mechanism. The energy (E_p) corresponding to the luminescence peak appears red-shift, and the carriers are released to the lower-energy tail state which are concentrated in the

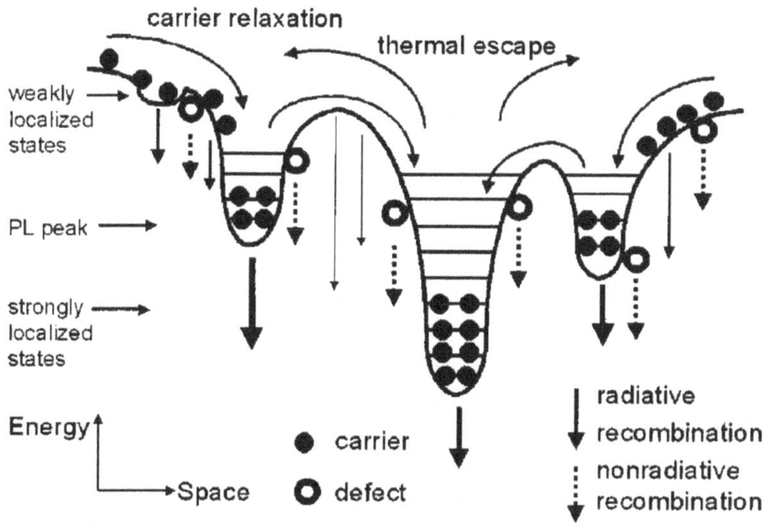

Fig. 5.3 Schematic diagram of carrier localization model

strong localized states for combination luminescence. In addition, the corresponding FWHM is reduced. After 70 K, as the temperature rise, carriers begin to occupy the higher energy level of the local state, and the delocalized as well as non-radiative combination increase. Under such circumstances, the peak position of the corresponding luminescence peak has a blue-shift. As the carriers occupy more high energy levels, the corresponding FWHM increases. When the temperature rises to 150 K, the carriers confined in the local states will start to become free exciton states. After 150 K, the common temperature thermal effect will become more and more significant, and the energy of the luminescence peak has a red-shift due to the shrink of forbidden band. Then, the corresponding FWHM gradually increases [28].

For most semiconductors (including GaN), in general, an increase in temperature leads to a decrease in band edge transition energy and the band gap narrows. The relationship between the band gap of the semiconductor and the temperature is in accordance with the Varshni formula, which is the shrinkage effect of the band gap with increasing temperature [29].

$$E_g(T) = E_g(0) - \frac{\alpha T^2}{\beta + T} \tag{5.25}$$

where $E_g(T)$ is the band gap at temperature T, $E_g(0)$ is the band gap at 0 K, and constants α and β are Varshni thermal coefficients. Due to the fluctuation of the In composition, the local state exists in the quantum well. Therefore, we use the band-tail model to modify the Varshni formula to study the variation of the peak position with temperature.

$$E_g(T) = E_g(0) - \frac{\alpha T^2}{\beta + T} - \frac{\sigma^2}{kT} \quad ((5.26))$$

where k is the Boltzmann constant and σ is the amount that characterizes the degree of localization.

This formula can be used to characterize the energy of the local state.

5.5 Green LED and Non-polar, Semi-polar LED

At present, the external quantum efficiency of red LEDs represented by AlGaInP and blue LEDs represented by InGaN can reach more than 80%, indicating that the internal quantum efficiency and light extraction efficiency of red and blue LEDs can exceed 90% by using appropriate processing methods. However, the luminescence efficiency of InGaN-based LEDs decreases sharply with increasing wavelength of light. The internal quantum efficiency drops below 30% when the wavelength is 520 nm. In addition, the decreased efficiency of green LEDs is also more serious than blue and violet LEDs. As the injection current increases, the external quantum efficiency of green LEDs decreases faster than red and blue LEDs, resulting in low output power of green LEDs under high current. This can affect the light mixing efficiency of high power white light. Therefore, the low-efficiency green LED has become the bottleneck of the current development of RGB three-primary LEDs.

At present, there are two main ways to improve the internal quantum efficiency of InGaN based green LEDs. The first method is to continuously optimize the epitaxial growth process and structural design of the InGaN quantum wells region to minimize the occurrence of defects in the active region. For example, growth interruption method is used to improve the quantum well interface [30]; InGaN pre-strain layer is inserted below the multiple quantum well layer to block dislocation propagation and reduce the stress in the quantum well [31]; the sapphire is replaced with GaN homoepitaxial substrate for heterogeneous growth [32] to reducing quantum well penetration dislocation density and so on. The second method is to epitaxially grow green LEDs in the non-polar and semi-polar directions from the crystal structure so that the grown material does not spontaneously polarize and grow in the direction perpendicular to the polar axis [0001]. The strain can avoid the piezoelectric polarization problem, thereby eliminating the influence of the polarized electric field and increasing the probability of radiation recombination. At present, most researchers focus on the second approach. Depolarization around non-polar and semi-polar GaN has become a new research topic.

5.5.1 Polar Surface High in Composition Green LEDs

Since 1995, Nichia Chemical Co., Ltd. has launched commercial InGaN-based green LEDs. The United States, Japan, South Korea, the mainland China and Taiwan have all explored the development of nitride green LED structures and optimized their growth parameters. They also studied the effect of material quality on quantum efficiency in the region. In view of the key scientific and technical issues such as material quality and piezoelectric polarization field in high indium composition quantum wells materials, researchers at the Institute of Semiconductors of the Chinese Academy of Sciences have further studied the optimization of the growth conditions of high-indium composition and designed new active regions structures to reduce the influence of the piezoelectric polarization electric field on the radiation recombination process. As shown in Fig. 5.4, the use of the traditional electron blocking layer to affect the piezoelectric polarization electric field in high-indium composition quantum wells is proposed. The electron blocking layer of the gradient composition is also used to replace the traditional electron blocking layer. The result shows that the polarization field in the active region of green LEDs is improved. The peak wavelength shift of the green light decreases with the change of the current, which maintains the high color stability of the device. Under the current condition of 20 mA, the optical output power is increased by 163%. In order to further weaken the polarized electric field and the droop effect in the green LED quantum well, the researchers used the Mg doping method in the quantum barrier to study the shielding effect of the activation of Mg impurity in the active region on the polarized electric field in the multiple quantum wells. The piezoelectric polarization electric field in the multiple quantum wells is suppressed. Furthermore, the blue-shift of the illuminating wavelength in a large current range is improved, and the droop effect is successfully suppressed.

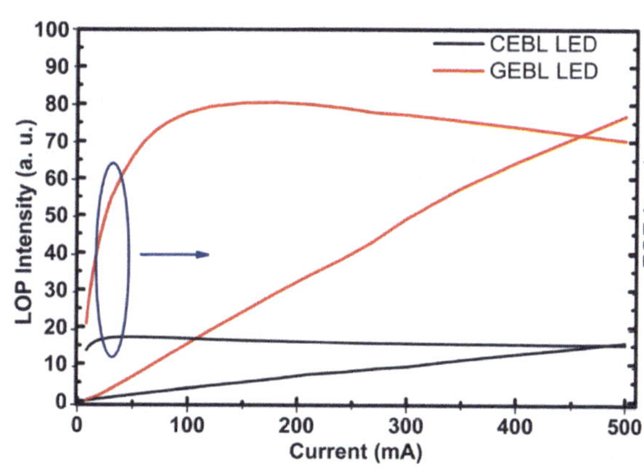

Fig. 5.4 Optical output power and external quantum efficiency of samples with different current densities

5.5.2 Semi-polar and Non-polar Materials

As shown in Fig. 5.5, both the (0001) Ga plane and the (000-1) N plane are c-plane in the nitride semiconductors. The c-plane nitride material is a kind of polar material, in which the surface of epitaxial material thin film is a polar surface. If the surface of the film material is neither parallel nor perpendicular to the c-axis but an angle between 0° and 90°, the film is a semi-polar material. If the surface of the film material is the m-plane or the a-plane parallel to the polar axis (i.e., the c-axis), the material has no polarization effect along the growth direction, and the material is a nonpolar material [33].

As shown in Fig. 5.6, non-polar and semi-polar GaN materials can be achieved by epitaxy on a heterogeneous substrate with a specific crystal plane, e.g., a-plane GaN grown on a r-plane sapphire substrate, a-plane and m-plane GaN grown on a-plane

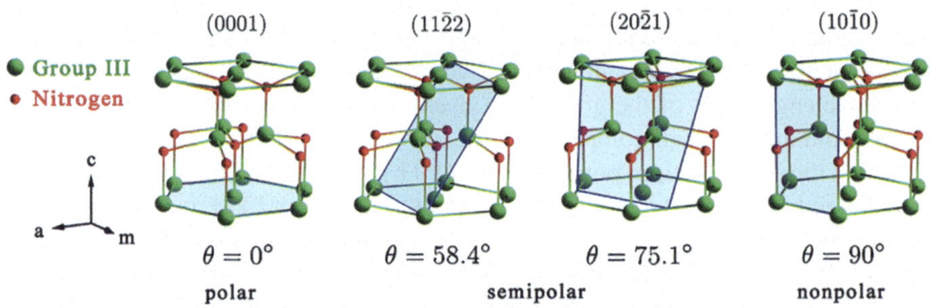

Fig. 5.5 Polar, semi-polar and non-polar plane of wurtzite structure GaN

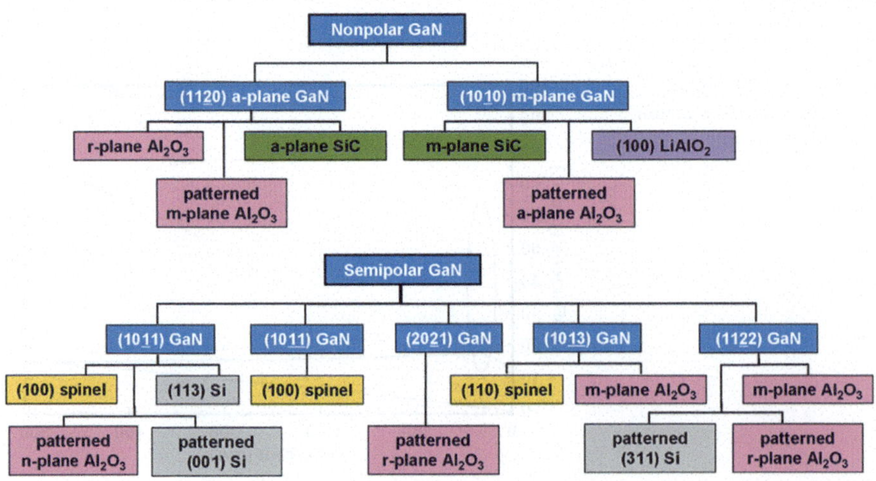

Fig. 5.6 Non-polar and semi-polar GaN materials grown on different epitaxial substrates

and m-plane SiC [33]. Due to the obvious anisotropy of material properties in all in-plane directions, stacking faults are also an important type of extensional defects in non-polar and semi-polar materials in addition to dislocations [34].

5.5.3 Research Progress on Semi-polar and Non-polar LEDs

In 2000, Waltereit et al. [35] first reported their non-polar GaN results on Nature, and obtained a non-polar m-plane AlGaN/GaN MQWs structure on a $LiAlO_2$ substrate, showing the removal of polarization effects and thus opening up the new study of non-polar GaN. In the following years, the internationally renowned GaN research institutions, including the University of California at Santa Barbara (UCSB), the University of South Carolina, the University of Nagoya, Japan, the Japanese Roma, and the US Cree are actively engaged in non-polar research on GaN. They have grown (11-20) a-plane GaN on r-plane sapphire, or (1-100) m-plane GaN on $LiAlO_2$ and m-SiC substrates. However, in the experiment, it was found that the growth window of non-polar GaN is narrow, and the In is difficult to be incorporated in the non-polar InGaN materials. In 2005, Sharma et al. [36] proposed semi-polar GaN which can reduce the polarization electric field, and obtained the first semi-polar green LED with crystal orientation along the (10-13) direction. Semi-polar GaN is a non-polar surface with a growth plane between the c-axis and the non-polar plane, which is oblique to the c-axis at an angle, including (11-22), (10-13), (1-101) and (20-21), etc. Semi-polar materials can greatly reduce the polarization electric field. In 2008, Gühne et al. [37] prepared (11-22) green LEDs on m-plane sapphire. As the current increased, the EL peak wavelength did not move, which proved that the effect of polarization effect was basically eliminated. Since the In doping in the InGaN material grown along the semi-polar direction is easier, the growth window is also wider than the non-polarity. Thus, the semi-polar InGaN material can be extended to emit longer wavelengths. However, semi-polar GaN grown directly on m-plane, r-plane sapphire, $MgAl_2O_4$, etc. by MOCVD technology has not yet achieved the same results as traditional common growth process such as GaN because of the late start, although it has certain advantages over non-polar GaN growth. There are still a series of problems such as surface roughness, stacking faults and dislocations, etc. [38, 39], which causes the subsequent growth of semi-polar green LED structures to be greatly affected.

At present, the research on semi-polar LEDs in the world is studied mainly by UCSB in US, Kyoto University of Japan, and the famous city university. Better device results are obtained on the GaN substrate. Among them, the research team led by "Father of Blue LED" Nakamura from UCSB has been the leader in the field of non-polar and semi-polar GaN. They have effectively combined homoepitaxy and semi-polar growth to reduce polarization effects on the basis of previous work. In addition, a GaN substrate grown by HVPE (Mitsubishi Chemical) was used for beveling to obtain semi-polar GaN materials with a variety of crystal orientations, followed by homoepitaxial semi-polar LEDs. Breakthroughs have been made along this direction. At present, the level of semi-polar long wavelengths yellow-green LEDs has surpassed the best level of traditional c-plane yellow-green LEDs, letting people see the hope. At the end of 2010, UCSB released a blue semi-polar LED on the (10-11) GaN substrate. The output power was 31.1 mW at 20 mA and the external quantum efficiency was 54.7% [40]. At the same time, the droop effect of the LED was greatly reduced compared with conventional c-plane devices. UCSB immediately reported the green LED results obtained on the (11-22) GaN substrate, with an output power of 9.9 mW at 20 mA and the external quantum efficiency of up to 20.4% [41], showing the superiority of semi-polarity in the growth of high In composition InGaN. In 2011, green LEDs were homoepitaxially grown on (20-21) GaN substrates through doping Mg [42] in quantum barriers, which increases the hole recombination concentration in semi-polar quantum wells and further improves the performance of green LEDs.

References

1. H. Amano, N. Sawaki, I. Akasaki, Y. Toyoda, Metalorganic vapor-phase epitaxial-growth of a high-quality GaN film using an AlN buffer layer. Appl. Phys. Lett. **48**, 353–355 (1986)
2. H. Amano, M. Kito, K. Hiramatsu, I. Akasaki, p-type conduction in Mg-doped GaN treated with low-energy electron-beam irradiation (LEEBI). Jpn. J. Appl. Phys. Part 2 Lett. **28**, L2112–L2114 (1989)
3. H. Morkoç, *Light-Emitting Diodes and Lighting. Nitride Semiconductor Devices* (Wiley-VCH Verlag GmbH & Co., KGaA, 2013), pp. 209–266
4. R.D. Dupuis, Epitaxial growth of III-V nitride semiconductors by metalorganic chemical vapor deposition. J. Cryst. Growth **178**, 56–73 (1997)
5. O. Ambacher, Growth and applications of group III nitrides. J. Phys. D-Appl. Phys. **31**, 2653–2710 (1998)
6. F. Bernardini, V. Fiorentini, D. Vanderbilt, Spontaneous polarization and piezoelectric constants of III-V nitrides. Phys. Rev. B. **56**, 10024–10027 (1997)
7. F. Bernardini, V. Fiorentini, Nonlinear behavior of spontaneous and piezoelectric polarization in III-V nitride alloys. Phys. Status Solidi A-Appl. Mater. Sci. **190**, 65–73 (2002)
8. B. Doshi, K.F. Brennan, R. Bicknell-Tassius, F. Grunthaner, The effect of strain-induced polarization fields on impact ionization in a multi-quantum-well structure. Appl. Phys. Lett. **73**, 2784–2786 (1998)
9. Y.Z. Deng, Y.C. Kong et al., Study on strain and piezoelectric polarization of AlN thin films grown on Si. J. Vac. Sci. Technol. A **23**, 628–630 (2005)
10. A.E. Romanov, T.J. Baker, S. Nakamura, J.S. Speck, E.J.U. Grp, Strain-induced polarization in wurtzite III-nitride semipolar layers. J. Appl. Phys. **100**(2), 023522 (2006)
11. B. Gil, P. Bigenwald, O. Briota, Polarization fields in wurtzite strained layers grown on (hkl) planes. Superlattices Microstruct. **44**, 291–301 (2008)
12. N. Zhang, polarization engineering and approaches for high-performance GaN-based green light-emitting diodes. Ph.D. thesis, University of Chinese Academy of Sciences (2013) (in Chinese)
13. E.T. Yu, X.Z. Dang, P.M. Asbeck, S.S. Lau, G.J. Sullivan, Spontaneous and piezoelectric polarization effects in III-V nitride heterostructures. J. Vac. Sci. Technol. B **17**, 1742–1749 (1999)
14. L.Y. Chen, C.Y. Wang, C.P. Chen, Y.W. Cheng, M.Y. Ke, J.J. Huang, Quantum confine stark effect suppressed nanorod structure light emitting diode, in *2008 IEEE Photonics-global@Singapore (Ipgc)*, vols. 1 and 2 (2008), pp. 129–132

15. T. Wang, J. Bai, S. Sakai, The investigation on the emission mechanism of InGaN/GaN quantum well structure, in *Proceedings of International Workshop on Nitride Semiconductors*, vol. 524–527 (2000), p. 1002
16. T. Wang, J. Bai, S. Sakai, I.I. Ipap, The investigation on the emission mechanism of InGaN/GaN quantum well structure, in *Proceedings of the International Workshop on Nitride Semiconductors* (Inst Pure Applied Physics, Tokyo, 2000), pp. 524–527
17. C.C. Lin, C.H. Chiu, H.W. Huang, S.P. Chang, H.C. Kuo, C.Y. Chang, Highly efficient InGaN-based light emitting devices grown on nanoscale patterned substrates by MOCVD, in *Proceedings Paper, Display, Solid-State Lighting, Photovoltaics, and Optoelectronics in Energy III*, vol. 8312 (2011)
18. F. Della Sala, A. Di Carlo et al., Free-carrier screening of polarization fields in wurtzite GaN/InGaN laser structures. Appl. Phys. Lett. **74**, 2002–2004 (1999)
19. M.D. McCluskey, C.G. Van de Walle, C.P. Master, L.T. Romano, N.M. Johnson, Large band gap bowing of $In_xGa_{1-x}N$ alloys. Appl. Phys. Lett. **72**, 2725–2726 (1998)
20. P.R.C. Kent, G.L.W. Hart, A. Zunger, Biaxial strain-modified valence and conduction band offsets of zinc-blende GaN, GaP, GaAs, InN, InP, and InAs, and optical bowing of strained epitaxial InGaN alloys. Appl. Phys. Lett. **81**, 4377–4379 (2002)
21. W. Lu, C.R. Li, Z. Zhang, Study of band structure $In_xGa_{1-x}N$/GaN multiple quantum wells by high-resolution electron microscopy and electron holography. Appl. Phys. Lett. **86** (2005)
22. S.W. Feng, Y.C. Cheng et al., Impact of localized states on the recombination dynamics in InGaN/GaN quantum well structures. J. Appl. Phys. **92**, 4441–4448 (2002)
23. M.A. Semina, R.A. Sergeev, R.A. Suris, Electron-hole complexes localized on the quantum well interface roughnesses. Phys. Status Solidi C Curr. Top. Solid State Phys. **4**(2), 363–365 (2007)
24. L.J. Brillson, T.M. Levin, G.H. Jessen, F.A. Ponce, Localized states at InGaN/GaN quantum well interfaces. Appl. Phys. Lett. **75**, 3835–3837 (1999)
25. M.Y. Ryu, J.H. Song et al., Localized excitons in $In_xGa_{1-x}N$/GaN quantum well structure. J. Korean Phys. Soc. **33**, S316–S318 (1998)
26. L.E. Golub, S.V. Ivanov et al., Low-temperature kinetics of localized excitons in quantum-well structures. Phys. Status Solidi B-Basic Solid State Phys. **205**, 203–208 (1998)
27. J.H. Na, R.A. Taylor, et al., Dependence of carrier localization in InGaN/GaN multiple-quantum wells on well thickness. Appl. Phys. Lett. **89** (2006)
28. Y.H. Cho, G.H. Gainer et al., "S-shaped" temperature-dependent emission shift and carrier dynamics in InGaN/GaN multiple quantum wells. Appl. Phys. Lett. **73**, 1370–1372 (1998)
29. G. Sun, G.B. Xu, et al., Investigation of fast and slow decays in InGaN/GaN quantum wells. Appl. Phys. Lett. **99** (2011)
30. K. Jacobs, B. Van Daele, M.R. Leys, I. Moerman, G. Van Tendeloo, Effect of growth interrupt and growth rate on MOVPE-grown InGaN/GaN MQW structures. J. Cryst. Growth **248**, 498–502 (2003)
31. C.F. Huang, T.Y. Tang, et al., Prestrained effect on the emission properties of InGaN/GaN quantum-well structures. Appl. Phys. Lett. **89** (2006)
32. Y. Yang, X.A. Cao, C.H. Yan, Rapid efficiency roll-off in high-quality green light-emitting diodes on freestanding GaN substrates. Appl. Phys. Lett. **94** (2009)
33. M. Kneiss, J. Rass, L. Schade, U. Schwarz, Growth and optical properties of GaN-based non- and semipolar LEDs, in *III-Nitride Based Light Emitting Diodes and Applications*, ed. by T.-Y. Seong, J. Han, H. Amano, H. Morkoc (Springer Netherlands, 2013), pp. 83–119
34. Y. Hao, J.F. Zhang, J.C. Zhang, *Nitride Based Wide Bandgap Semiconductor Materials and Electronic Devices*. China Science Publishing Ltd., ISBN: 9787030367174 (2013) (in Chinese)
35. P. Waltereit, O. Brandt et al., Nitride semiconductors free of electrostatic fields for efficient white light-emitting diodes. Nature **406**, 865–868 (2000)
36. R. Sharma, P.M. Pattison, et al., Demonstration of a semipolar (10(1)over-bar1(3)over-bar) InGaN/GaN green light emitting diode. Appl. Phys. Lett. **87** (2005)
37. T. Guhne, P. DeMierry, M. Nemoz, E. Beraudo, S. Chenot, G. Nataf, Demonstration of semipolar (11-22) InGaN/GaN blue-green light emitting diode. Electron. Lett. **44**, 231–U221 (2008)

38. J.S. Speck, S.F. Chichibu, Nonpolar and semipolar group III nitride-based materials. MRS Bull. **34**, 304–312 (2009)
39. A.E. Romanov, T.J. Baker, S. Nakamura, J.S. Speck, Strain-induced polarization in wurtzite III-nitride semipolar layers. J. Appl. Phys. **100** (2006)
40. Y.J. Zhao, J. Sonoda, et al., 30-mW-class high-power and high-efficiency blue semipolar (10(1)over-bar(1)over-bar) InGaN/GaN light-emitting diodes obtained by backside roughening technique. Appl. Phys. Express **3** (2010)
41. S. Yamamoto, Y.J. Zhao, et al., High-efficiency single-quantum-well green and yellow-green light-emitting diodes on semipolar (20(2)over-bar1) GaN substrates. Appl. Phys. Express **3** (2010)
42. C.Y. Huang, Q.M. Yan, et al., Influence of Mg-doped barriers on semipolar (20(2)over-bar1) multiple-quantum-well green light-emitting diodes. Appl. Phys. Lett. **99** (2011)

Chapter 6
AlGaN-Based Multiple-Quantum-Well Materials and UV LEDs

As illustrated in Fig. 6.1, ultraviolet (UV) light, which is shorter than the wavelength of visible light, is attractive in various application scenarios. Among them, UV-C light (250–280 nm) can destroy nucleic acids, the genetic material of microorganisms, through photochemical action. UV treatment can prevent the replication of microorganisms and cause the death of microorganisms, hereby sterilizing water, air and the surfaces of object. The band gap of AlGaN is 3.43–6.04 eV. The corresponding band edge radiation wavelength is 365–210 nm, which makes AlGaN-based UV LEDs attractive for a broad range of applications in water/air purification, sterilization, bio-probe, medical, and non-field-of-sight communication. UV LEDs have attracted extensive international research interest [1]. In the field of disinfection, AlGaN-based UV LEDs have the following advantages: First, small size has great potential in portable, highly integrated products. Second, AlGaN-based UV LEDs are rugged and durable, more resistant than the mercury lamp with the quartz glass. Third, AlGaN-based UV LEDs have higher energy efficiency. Compared with mercury lamps, the energy consumption of UV LEDs can be as low as 70%. Fourth, UV LEDs are eco-friendly. UV LEDs do not contain harmful substances such as mercury, which are certified by RoHS (Restriction of Hazardous Substances). Fifth, UV LEDs have lower working voltage, only about 3.3–4.5 V. Compared with the high-voltage mercury lamp, the safety is much improved and the cost of the driving circuit is reduced. Sixth, the power of UV LEDs is easier to adjust. Seventh, the cooling system for UV LEDs is much simpler, further reducing system cost. Eighth, the optical system is simple and more suitable for practical applications. UV LEDs have a compact beam angle and a uniform beam pattern without an external lens, thereby reducing the cost and enhancing the system reliability. Because of these advantages, AlGaN-based UV LED technology is becoming an attractive option for the UV sterilization industry.

However, achieving high efficiency AlGaN-based UV LEDs are still facing many technical difficulties and obstacles such as low internal quantum efficiency, low current injection efficiency, and low light extraction efficiency. This chapter will

J. Li et al., *III-Nitrides Light Emitting Diodes: Technology and Applications*,
Springer Series in Materials Science 306,
https://doi.org/10.1007/978-981-15-7949-3_6

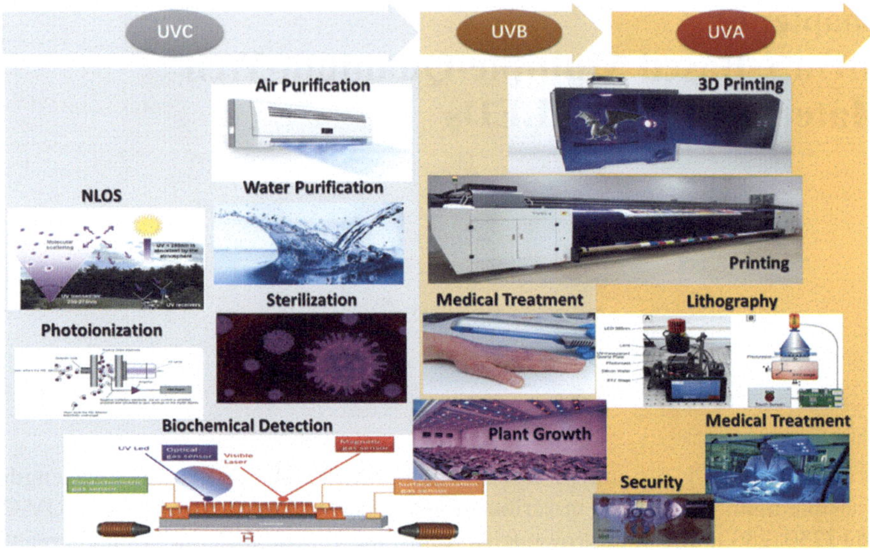

Fig. 6.1 Potential applications of DUV LEDs

introduce the AlGaN material, including its optical and electrical properties, and elaborate its epitaxial growth and doping techniques. The structure design and fabrication techniques of AlGaN-based UV LED are also introduced.

6.1 Introduction of AlGaN Material System

There are three crystal structures of AlGaN material: wurtzite, sphalerite and rock salt. Under atmospheric conditions, wurtzite is a thermodynamically stable structure of AlN and GaN with the space group is C_{v4}^{6} and P6mc. Figure 6.2 shows that the unit cell of its hexagonal wurtzite is a hexagonal prism structure consisting of two Group-III atoms and two N atoms. One Group-III (or N) atoms in the center and four surrounding nearest neighboring N (or Group-III) atoms in the apex position form a tetrahedron. Figure 6.2b shows the schematic diagram of a hexagonal wurtzite structure of III-nitride, involving two lattice constants **a**, **c** and a parameter **u**. The lattice constant **a** determines the distance between two identical hexagonal planes, and the lattice constant **c** determines the distance of atoms in the crystal plane. The parameter **u** is defined as the bond length of anion and cation, that is, the ratio of the nearest neighbor atom spacing and the lattice constant **c**, which indicates the deformation of the crystal lattice. In the ideal wurtzite structure, $\mathbf{c}/\mathbf{a} = \sqrt{8/3} \approx 1.633$, $u = 3/8 = 0.375$ [2]. As shown in Fig. 6.2a, the hexagonal wurtzite structure has two sets of hexagonal close-packed structures composed of group-III atoms and N atoms respectively, and is formed by moving 3**c**/8 in the c-axis direction. The stacked

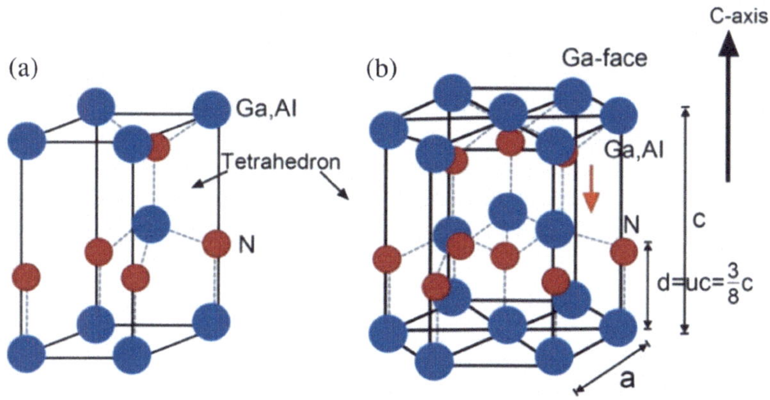

Fig. 6.2 The schematic diagram of **a** the unit cell and **b** wurtzite structure of group-III nitride

order is…AaBb-AaBb… in the [0001] direction. The sphalerite structure is a cubic structure, and the two sets of face-centered cubic structures are formed by translating 1/4 diagonal lines in the [111] direction. The stacked order is… AaBbCc–AaBbCc….

Since the III-nitride with the wurtzite structure lacks the inversion symmetry center in the [0001] direction, the III-nitride has two different polarity in the [0001] direction, Group-III atoms (Al, Ga or In) polarity or [0001] polarity and N polarity or [000–1] polarity. If the vertical bond is from group-III atom to N atom, or the group-III atom is above the (0001) bilayer, this is Ga polarity; if the vertical bond is from N atom to group-III atom, or the N atom is located above the (0001) bilayer, this is N polarity [3]. The polarity of nitrides has an important effect on the doping and piezoelectricity during epitaxial growth [4]. In general, Ga polar face and N polar face depends on the substrate and growth process, especially the initial stage of growth [5].

The growth method of the AlGaN-based multiple quantum well structure is mainly MOCVD technology. As an example of III-nitride growth by MOCVD, without considering the intermediate reaction and the product, the simplified reaction equation can be written as:

$$\mathrm{M}(g) + \mathrm{NH_3}(g) \rightarrow \mathrm{MH}(s) + 3\mathrm{RH}(g) \tag{6.1}$$

where R is an organic compound group such as trimethyl or triethyl, M is a trivalent element of the deposit, such as Al and Ga. Therefore, the reaction equations for MOCVD growth of AlN, GaN, and $\mathrm{Al}_x\mathrm{Ga}_{1-x}\mathrm{N}$ can be abbreviated as follows:

$$(\mathrm{CH_3})_3\mathrm{Ca} + \mathrm{NH_3} \rightarrow \mathrm{GaN} + 3\mathrm{CH_4} \tag{6.2}$$

$$(\mathrm{CH_3})_3\mathrm{Al} + \mathrm{NH_3} \rightarrow \mathrm{AlN} + 3\mathrm{CH_4} \tag{6.3}$$

$$(CH_3)_3Al + (1 - x)(CH_3)_3Ca + NH_3 \rightarrow Al_xGa_{1-x}N + 3CH_4 \tag{6.4}$$

In fact, MOCVD growth of III-nitrides is a very complex process, accompanied by multiple parallel and series of reaction steps, including complex decomposition reactions, pre-reactions, parasitic reactions, isomorphisms and isomerization reactions. Figure 6.3a, b show the theoretical kinetic models of gas phase and surface chemistry in the epitaxial growth of AlN and GaN by MOCVD [6, 7].

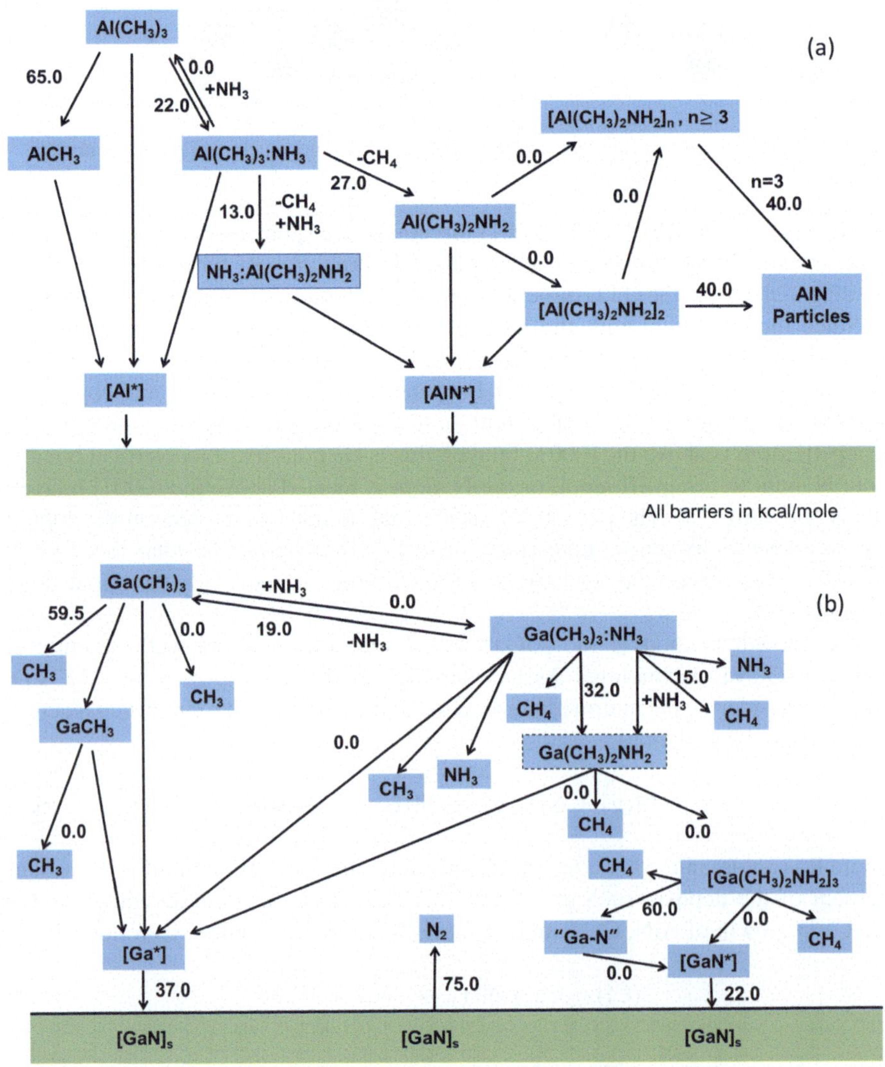

Fig. 6.3 The schematic diagram of AlN **a** and GaN **b** growth process in MOCVD

6.2 Optical and Electrical Properties of AlGaN Materials

The forbidden bandwidths of AlN and GaN at room temperature are 6.03 and 3.42 eV, respectively [8]. The forbidden bandwidth of the ternary alloy does not change linearly with the change of the Al composition, but a bending coefficient **b** is required. Thus the forbidden bandwidth of $Al_xGa_{1-x}N$ is expressed as:

$$E_g(Al_xGa_{1-x}N) = x \cdot E_g(AlN) + (1-x) \cdot E_g(GaN) - b \cdot x(1-x) \quad (6.5)$$

The bending coefficient b is usually related to strain, buffer layer properties and growth conditions [9]. For $Al_xGa_{1-x}N$, b is usually taken as 0.7. The lattice constant can be calculated according to the Vegard theorem as follows:

$$a(Al_xGa_{1-x}N) = x \cdot a(AlN) + (1-x) \cdot a(GaN) \quad (6.6)$$

The AlGaN material has a strong spontaneous polarization effect (its spontaneous polarization direction is [0001]), which is toward the substrate. Its spontaneous polarization effect increases with increasing the Al composition. When the Al mole fraction of the well and the barrier of AlGaN/AlGaN MQWs are much different [10], a highly polarized electric field (built-in electric field) is generated at the heterojunction interface or quantum well region along the c-axis, resulting in band-bending, spatial separation of electron and hole wave functions, so-called Quantum Conned Stark Effect (QCSE) as shown in Fig. 6.4. QCSE increases carrier lifetime, reduces radiation recombination efficiency, and causes red shift of the emission wavelength. Even without external bias, the spontaneous polarization and piezoelectric polarization electric field are also as high as MV/cm. The formation of two-dimensional electron gas is advantageous for field effect devices such as HEMT. However, for optoelectronic devices, the internal quantum efficiency will be largely reduced.

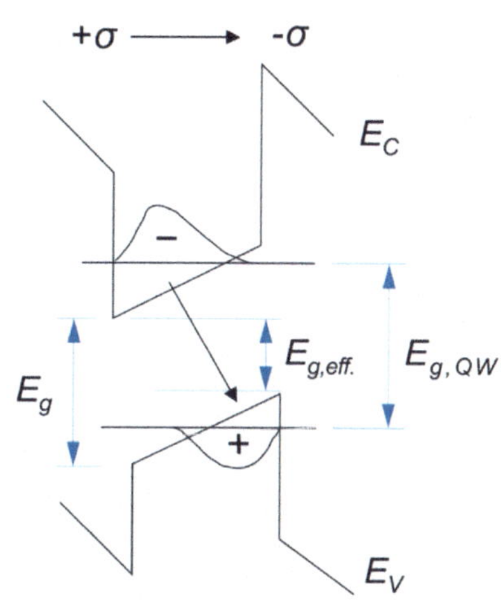

Fig. 6.4 The schematic diagram of the effect of QCSE on the energy band of MQW

6.3 Epitaxial Growth and Doping Techniques for AlGaN Materials

For epitaxial growth of high-quality III-nitride thin films, the choice of substrate is particularly important. Homogeneous substrate is theoretically the best substrate choice for III- nitride growth. However, commercial production of high quality, large size, low cost III-nitride homogenous substrates is still difficult [11], especially for AlN single crystal substrates. Therefore, commercially available III-nitride devices mainly use heterogeneous substrates such as (0001) sapphire, (111) Si, and 6H–SiC substrates, where the (0001) sapphire is the most common substrate. For the heteroepitaxial growth of III-nitrides, the lattice constant and thermal expansion coefficient are two critical parameters. Table 6.1 lists the lattice constants and thermal expansion coefficients and mismatch of AlN and different substrates. On one hand, due to the existence of lattice mismatch, the strain accumulated in the epitaxial layer relaxes by generating dislocations at the interface between substrate and epitaxial layer, thus causes a large number of dislocations in the epitaxial layer. On the other hand, due to the existence of thermal mismatch, cracking and dislocations of the epitaxial layer are caused by the mismatch in lattice deformation between substrate and epitaxial layer during heating or cooling process. To alleviate this problem, a low-temperature AlN or GaN buffer layer is introduced as an initial nucleation layer to reduce dislocation density and compensates for strain in the epitaxial layer caused by thermal mismatch.

Figure 6.5 shows the lattice arrangement of (0001) AlN on the c-plane sapphire substrate. The lattice point of AlN is rotated by 30° with respect to the lattice point of sapphire [12]. Therefore, the actual lattice mismatch between AlN and sapphire (13.3%) is calculated as follows:

Table 6.1 The lattice mismatch and thermal mismatch between AlN and different substrate

	Lattice constant (Å)	Thermal expansivity (10^{-6}/K)	Lattice mismatch (%)	Thermal expansion mismatch (%)
AlN	$a = 3.112$ $c = 4.982$	$\Delta a/a = 4.2$ $\Delta c/c = 5.3$		
GaN	$a = 3.189$ $c = 5.186$	$\Delta a/a = 5.59$ $\Delta c/c = 3.17$	−0.24	−0.25
Sapphire	$a = 4.758$ $c = 12.991$ $a_{eff.} = 2.747$	$\Delta a/a = 7.5$ $\Delta c/c = 8.57$	−34.6 13.3	−44
6H-SiC	$a = 3.0817$ $c = 15.1123$	$\Delta a/a = 4.2$ $\Delta c/c = 4.68$	0.98	0
Si (111)	$a = 5.4301$ $a_{eff.} = 3.843$	$\Delta a/a = 3.59$	42.7 −19%	17

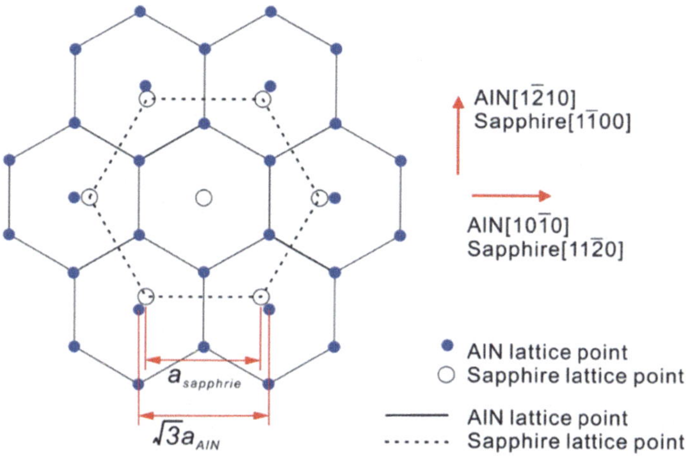

Fig. 6.5 The schematic diagram of lattice arrangement of (0001) AlN on c-plane sapphire

$$\Delta a = \frac{a\,\text{AlN}^{-a}\text{sapphrie}/\sqrt{3}}{a\,\text{sapphrie}/\sqrt{3}} = \frac{3.112 - 2.747}{2.747} = 13.3\% \tag{6.7}$$

Compared with sapphire substrates, hexagonal SiC substrates with high thermal conductivity have very close lattice constants and thermal expansion coefficients with AlN and GaN. But their UV absorption and relatively high price limit their applications in the UV light emitting devices. At present, (111) Si substrates have attracted international research interest in the field of GaN-based optoelectronic devices and power electronic devices due to their high thermal conductivity, low price and integration advantages [13, 14]. The research of AlN growth and UV light-emitting device on Si substrate have also been reported [15, 16]. Reducing the dislocation density of AlN on Si due to the large lattice mismatch between AlN and Si substrate is still a big challenge in improving the crystal quality. The high-quality AlN substrate is the best choice of substrate for growing low dislocation density and high crystal quality AlN materials. It is also an important way to further improve the performance of deep ultraviolet (DUV) LEDs and achieve DUV LED [17]. However, AlN substrate cannot be mass produced currently and its high price limits its wide application. In response to these problems, various methods have been proposed to reduce the dislocation density in the epitaxial layer.

Pulse Atomic Layer Epitaxy (PALE): In 1992, Asif Khan in the University of South Carolina used SALE (switched atomic layer epitaxy) technology to increase the surface mobility of Group-III atoms by modulating the Group-III and N source pulses. They successfully grew GaN, AlN and GaN/AlN short-period superlattices with high crystal quality and high surface morphology on sapphire substrate [18]. In 2001, Asif Khan's research team developed Pulsed Atomic Layer Epitaxy (PALE) technology to grow AlInGaN and achieve 305 nm UV-B LED [19, 20]. In 2002, they used PALE technology to grow ultra-high quality AlGaN and achieve 228 nm

UV-C light emission [21]. In 2002, Asif Khan's research group used the AlN/AlGaN superlattice structure grown by PALE technology as the stress release and dislocation filter layer, which effectively reduced the stress and threading dislocation density in the epitaxial layer and reduced the cracking probability of the epitaxial layer [22, 23]. Based on these, they reported sub-mW-level 250, 269, 285 and 324 nm AlGaN-based DUV LEDs [24–26] and mW-level 278 and 280 nm AlGaN-based DUV LED [27].

Migration enhanced epitaxial technology: In 2005, the research groups of SETI and Asif Khan in the United States developed migration enhanced MOCVD (MEMOCVD) epitaxial technology to grow high-quality AlN templates and AlN/AlGaN superlattices and improve the quality of AlGaN [28]. DUV LED was achieved at the peak wavelength of 280, 295 nm under mW level, and 265, 270, and 275 nm under sub-mW level [29]. Since 2008, Ryan G. Banal et al. in Kyoto University have optimized the AlN nucleation process by using the Migration Enhanced Epitaxy (MEE) and modified MEE (Modified-MEE) technology and achieved a 600 nm high-quality AlN on sapphire with the full width at half maximum (FWHM) of the XRD rocking curves of (002) and (102) as 45 and 250 arcsec, respectively. High-quality AlGaN/AlN MQWs were grown by using this technique as well [28, 29]. The quantum efficiency was as high as 36%.

Variable V/III ratio epitaxial growth: In 2006, Masataka Imura et al. in Meijo University used the method of multiple V/III ratio modulation (sequential reduction) during high temperature to grow 9 μm AlN by MOCVD. At the same time of preventing cracking, the dislocations are continuously annihilated in the transition process of growth mode. The dislocation density was reduced to 3×10^8 cm^{-2} [30, 31].

NH_3 pulse multilayer growth and multilayer AlN buffer layer technology: In 2007–2009, Hideki Hirayama in the Institute of Physical and Chemical Research reported the high-quality AlN and AlGaN epitaxial layers with low defect density on sapphire by NH_3 pulse multilayer growth and multilayer AlN buffer layer technology, respectively. They also reported 222–273 nm AlGaN-based DUV LEDs and 282 nm InAlGaN-based DUV LEDs [32–34].

Pulse lateral overgrowth (PLOG): Since 2006, the Asif Khan's team in the University of South Carolina has used the NH_3 pulsed lateral overgrowth to grow AlN with flat surface on a micron-sized trench-type AlN/sapphire template (PLOG-AlN), where the threading dislocation density is reduced by 10^2–10^3 cm^{-2} [35]. They further achieved 214 nm laser emission of AlN by optical pump at room temperature [36] and significantly improved the luminous efficiency, saturation characteristics and heat dissipation performance of 290 nm UV LED [37]. The lifetime was over 5,000 h. The 276 nm vertical UV-C LED was achieved by wet stripping the sapphire substrate [38]. In 2009, Hideki Hirayama et al. in the Institute of Physical and Chemical Research used the same method to grow AlN on the micron-sized grooved AlN/sapphire template to reduce the threading dislocation density to 10^8 cm^{-2} and achieved 270 nm UV-C LEDs with a maximum light output of 2.7 mW [39].

Migration enhanced lateral overgrowth: In 2008, R. Jain in SETI Corporation and M. S. Shur in Rensselaer Polytechnic Institute used MEMOCVD technology to laterally overgrow 20 μm AlN on a micro-groove-type AlGaN template to obtain a fully merged smooth surface AlN with low dislocation density [40, 41]. The FWHM of the XRD rocking curves of (002) and (102) are 157 and 291 arcsec, respectively. The threading dislocation density was reduced to 2.8×10^8 cm^{-2}.

High-temperature lateral epitaxy: Since 2006, the Isamu Akasaki's team in Meijo University has combined ELOG with high-temperature growth to grow AlN and AlGaN on micro-scale trench-type AlN/sapphire template [42, 43]. The threading dislocation density of AlN and AlGaN is reduced to 10^7 cm^{-2}. Based on this method, 345 nm UV-A LEDs with EQE > 6.7% [44, 45] and 356 nm UV-A LDs with electric pump [46] were achieved. The FWHM of (0002) and (10–10) XRD rocking curves of ELO-AlN on the grooved sapphire substrate were only 148 and 385 arcsec respectively. The LOP of 266 nm LED reached 5.3mW at 60 mA, and the EQE reached 1.9%. The LOP of 278 nm LED reached 8.4 mW, and the EQE reached 3.4%. High temperature growth can increase the surface mobility of Al atoms. However, it also causes greater thermal mismatch due to the large difference in thermal expansion coefficient between AlN and sapphire substrates. The pre-reaction of TMAl and NH_3 in the gas phase is also introduced simultaneously.

AlN bulk substrate: The homogeneous AlN bulk substrate is a better alternative to improve the crystal quality of Al(Ga)N materials and the performance of UV LEDs. At present, Crystal IS (acquired by Asahi Kasei from Japan in 2011), HexaTech and Nitride Solutions from USA, and CrystAL-N from Germany have made great progress in the growth of AlN substrates and high-performance UV LEDs. Crystal IS reported that the dislocation density of $Al_{0.5}Ga_{0.5}N$ was reduced to 7.5×10^5 cm^{-2} using PVT-AlN substrate [47]. In 2012, they achieved 260-nm UVC LEDs with IQE as high as 70% and output power of 53 mW at a WPE of 3.6% at 200 mA in CW operation [48]. In 2013, they presented 271 nm UVC LEDs with output power of 66.8 mW at a WPE of 2.5% at 300 mA by AlN substrate thinning and encapsulation [49]. HexaTech collaborating with Japan's HVPE developer Tokuyama successfully reduced the dislocation density of MQWs to less than 10^6 cm^{-2} on HVPE AlN substrates, and the 268 nm UVC LEDs exhibited output power of 28 mW at an EQE of 2.4% at 250 mA in 2012 [50]. The 261 nm DUV LEDs fabricated on HVPE-AlN substrates exhibited LOP of 10.8 mW at 150 mA, and an estimated lifetime over 5000 h at a high driven current of 150 mA in 2013 [51]. However, only 2" AlN wafers are available on the market at present and AlN substrates cannot be extensively used due to their much higher price than that of sapphire substrates.

For AlGaN and other III-nitrides, Si and Mg are the most common n-type and p-type doping elements, respectively. With the increase of Al composition, it's difficult to obtain n-type AlGaN with high conductivity and high carrier concentration. As the Si donor energy level becomes deeper, the activation energy increases continuously. The Si donor can be compensated by acceptor-type defects (such as group III cation vacancies and relevant complexes, impurities, dislocations) [52]. The p-type doping of AlGaN is even more difficult to realize. Due to the deep acceptor feature of Mg in AlGaN, the activation energy increases almost linearly from 160 meV in GaN to 510–600 meV in AlN as the Al composition increases. The acceptor activation efficiency is low, leading to low hole concentration and poor conductivity of p-type

AlGaN. It's should be noted that the hole concentration is much lower than the electron concentration [53]. For AlGaN-based optoelectronic devices, the asymmetric electron and hole concentration distribution can result in the electron injected in the active region leaking into the p-type region. As a result, the radiative recombination efficiency of quantum wells is reduced, accompanied by a long-wavelength parasitic luminescence in the p-type region. Besides, the bad current spreading capacity in the n-type and p-type regions brings in the current crowding effect and the heat accumulation, which can degrade the device luminous efficiency and reliability.

To improve the n-type doping efficiency, studies from Kansas State University, University of California, Santa Barbara, University of South Carolina, and NTT in Japan have shown that the n-type doping efficiency of AlGaN could be significantly improved by suppressing the formation of self-compensation defects, and even n-type conductive Si-doped AlN can be obtained [54–58]. Asif Khan's research group used the MEMOCVD technology to grow high-quality AlN and AlN/AlGaN SLs to reduce the defects, the compensation and scattering centers in the AlGaN layer. P. Cantu et al. found that the Si doping efficiency is strongly dependent on the Al composition in AlGaN. The free carrier concentration decreases as the Al composition increases. By utilizing Si-In co-doping, the formation of the deep acceptor-type centers could be suppressed and the free carrier concentration in $Al_{0.65}Ga_{0.35}N$ was increased by two orders of magnitude. Yoshitaka Taniyasu et al. from NTT optimized the growth process of Si-doped AlN, specifically suppressed the parasitic reaction of the source gas and accurately controlled the Si doping level. The dislocation density was reduced to 10^9 cm^{-2}. The electron concentration and the carrier mobility reached 1.75×10^{15} cm^{-3} and 125 cm^2/(V s) at 300 K, respectively. H. X. Jiang's research group employed the δ-doping technique of Si and Mg in the n-AlGaN layer and the p-GaN layer, respectively, to improve the material quality and reduce self-compensation defects. The free carrier concentration and the conductivity of n-type and p-type layers were both improved. The luminous efficiency of 340-nm UV LEDs was also enhanced.

To improve the p-type doping efficiency, researchers from Boston University, University of California at Santa Barbara, Carnegie Mellon University, and the US Air Force Laboratory reported theoretically and experimentally that the Mg-doped AlGaN/GaN superlattices, which were used to replace the conventional Mg-doped AlGaN layer, could reduce the acceptor ionization energy and increase the Mg doping efficiency, thereby increase the hole concentration and reduce the resistivity. The optical absorption edge also moved to a short wavelength [59–63]. M. S. Shur et al. and M. Shatalov et al. separately proposed to use p-GaN/p-AlGaN single heterojunction to obtain mass accumulation of holes near the heterojunction interface. Due to the existence of only one barrier and the assistance of carrier transition by tunneling and thermal emission, the vertical conductivity was improved. Currently, this method has been widely used in DUV LEDs to improve hole injection efficiency and luminescence efficiency. H.X. Jiang's research group adopted theδ-doping technique of Mg dopant [64]. The dislocation density of p-GaN and p-AlGaN layers was reduced by about one order of magnitude. Thanks to the improved material quality and reduced self-compensation defects, the lateral and vertical conductivities presented twofold and fivefold enhancements, respectively. John Simon et al. proposed a polarization doping method to induce Mg acceptor ionization. They achieved three-dimensional mobile hole gas with high density of 2×10^{18} cm^{-3} using N-face AlGaN with Al composition grading from 0 to 0.3 [65]. Lian Zhang et al. from our research group also reported three-dimensional mobile hole gas induced by polarization in metal-face Al-composition graded AlGaN [66].

To improve the current injection efficiency, Fujioka [67] and Sumiya [68] reported that inserting a 1-nm thin AlN layer between AlGaN-MQWs and p-AlGaN electron blocking layer could suppress carrier overflow and the 320-nm parasitic luminescence. The diffusion of Mg towards MQWs was also effectively prevented using this method. Hideki Hirayama et al. proposed a multi-quantum-barrier electron blocking layer to increase the effective barrier height to suppress electron leakage and increase the carrier injection efficiency [69]. The EQE of 250-nm DUV LEDs was improved by 2.7 times.

6.4 Structure Design and Fabrication of UV LEDs

High-efficiency DUV LEDs require high crystal quality AlGaN epitaxial layer, high carrier-confinement multiple-quantum-well structure, and high carrier-injection-capability electron and hole injection layers. Therefore, the improvement of the photoelectric performance of DUV LEDs depends on the crystal quality of AlGaN materials, the optimization of doping and quantum structure, and the improvement of the device fabrication process. At present, AlGaN-based UV LEDs still encounter significant obstacles such as low internal quantum efficiency, low current injection efficiency, and low light extraction efficiency.

Low internal quantum efficiency: High-efficiency AlGaN-based DUV LEDs require quantum structures with high carrier confinement capability. AlGaN material has a strong spontaneous polarization effect along the [000–1] direction. The spontaneous polarization is pointing to the substrate direction, and its intensity enhances as the Al composition increases. In addition, the significant Al composition difference between AlGaN quantum wells and barriers induces a strongly polarization electrical field (built-in electrical field) along the c-axis. In consequence, the energy bands bend and the electron and hole wave functions are separated spatially, which brings in internal quantum efficiency issue.

Low current injection efficiency: Si and Mg are widely used n-type and p-type doping elements in III-nitride materials. With increasing Al composition, it becomes challenge to obtain n-type AlGaN with high conductivity and high carrier concentration. This can be ascribed to the fact that Si donor energy level becomes deeper. The Si donor is compensated by acceptor-type defects (such as group III cation vacancies and relevant complexes, impurities, dislocations) [52]. Unfortunately, p-type doping of AlGaN is even more challenging. Activation energy of the acceptor increases from 160 meV in GaN to 510–600 meV in AlN. Such phenomenon leads to a fairly low hole concentration and poor conductivity of p-type AlGaN. It needs to be pointed out that the hole concentration is much lower than the electron concentration [53]. For AlGaN-based optoelectronic devices, the asymmetric electron and hole concentration distribution would cause electron leakage. In other words, injected electrons in the active region is likely to penetrate into the p-type region. In such case, radiative recombination efficiency of the quantum wells would be affected and reduced, accompanied by a long-wavelength parasitic luminescence in the p-type region. Meanwhile, relatively low current spreading capacity of the n-type and p-type regions would cause current crowding effect and heat accumulation. The undesired effects mentioned above will cause device degradation, which will reduce its luminous efficiency and reliability.

Low Light Extraction Efficiency: Another major factor limiting the EQE of AlGaN-based UV LEDs is the lower light extraction efficiency. It can be mainly attributed to three reasons. (1) The light emitted from the LED front side is nearly completely absorbed due to the strong absorption of UV light by p-GaN. That's why UV LEDs generally adopt a flip-chip structure. There are also some reports of UV LEDs with vertical structures [70, 71]. (2) The planar sapphire substrates cannot effectively reduce the internal total reflection. There lacks a simple and effective surface roughening method to process hard sapphire substrates. N. Lobo's simulation showed that flip-chip UV LEDs with no mirrors and encapsulation had a light extraction efficiency of only 7–9% [59]. (3) In AlGaN-based UV LEDs, the dominant emission mode converts from the TE mode to the TM mode as the Al composition increases and the wavelength decreases. For LEDs on the c-plane sapphire, the TE and the TM polarized light propagates in the vertical and horizontal directions, respectively. Thus, TM polarized light is more difficult to be extracted from the vertical direction than TE polarized light. Han-Youl Ryu's simulation showed that the light extraction efficiency of the TM mode in DUV LEDs is less than one tenth of that of the TE mode [72].

Figure 6.6 shows the typical epitaxial structure of an AlGaN-based DUV LED emitting at 280 nm [73]. It mainly includes 1 μm thick high-temperature AlN template layer grown on the sapphire substrate, 10 to 20 pairs of AlGaN/AlN superlattice structure for strain management and dislocation blocking, 2~3 μm thick Si-doped $Al_{0.55}Ga_{0.55}N$ n-type contact layer, 5 pairs of $Al_{0.5}Ga_{0.5}N/Al_{0.4}Ga_{0.6}N$ MQW active region, 20 nm thick Mg-doped $Al_{0.6}Ga_{0.4}N$ electron barrier layer (EBL), and a p^+-GaN (100 nm)/p-AlGaN hole injection layer. Generally speaking, DUV LEDs are fabricated in flip-chip structure to avoid optical absorption. Brief fabrication process of such structure is shown in Fig. 6.7.

(1) **Wafer cleaning**: The cleaning process has an important influence on the process and device performance, and is a critical step during the semiconductor process. It starts with organic cleaning to remove various hydrocarbons such as stains and oils. The wafer can be ultrasonically cleaned with hot acetone and isopropanol or ethanol, then rinsed in deionized water and dried with N_2. If there are residual metal indium spots on the wafer, they can be dissolved in aqua regia.
(2) **Meas etching**: The mesa shape is determined by photolithography and ICP etching. A mixed gas of $BCl_3/Cl_2/Ar$ is usually used to form low-damage mesas.

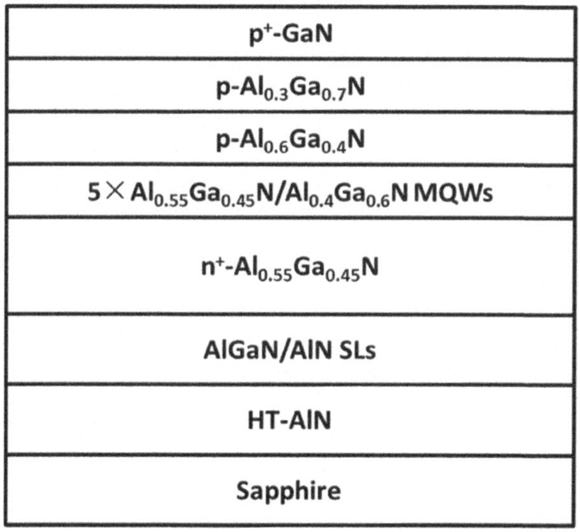

Fig. 6.6 Schematic diagram of the typical epitaxial structure of 280-nm AlGaN-based DUV LEDs

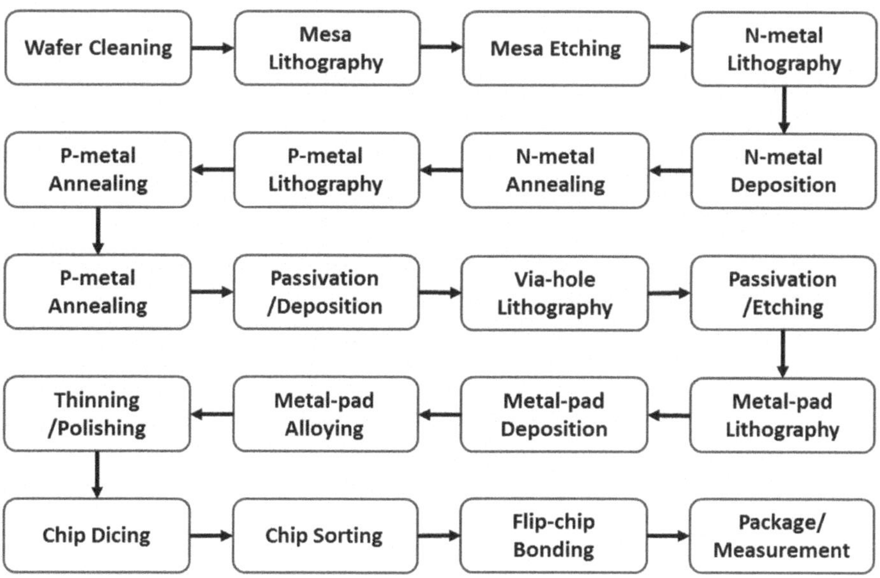

Fig. 6.7 The brief fabrication process flow of flip-chip DUV LEDs

(3) **N and p electrode deposition**: In order to obtain good ohmic contact performance on the n-AlGaN and p-GaN layers, it is necessary to separately fabricate n-type and p-type electrodes. The n-electrode usually adopts Ti/Al metal system and is annealed at 850 °C for 30 s in N_2 atmosphere in a rapid annealing furnace. The p-electrode usually adopts Ni/Au metal system, and is annealed at 550 °C for 1 min in air atmosphere. Before evaporating the p and n electrodes, it is necessary to remove the surface oxidation of n-AlGaN and p-GaN by immersing in acid solutions.
(4) **Passivation isolation**: In order to reduce the leakage current, avoid short circuit and suppress the possible surface recombination of exposed quantum wells after mesa etching, we use a SiO_2 or SiN layer deposited by plasma enhanced chemical vapor deposition (PECVD) as a passivation layer to protect the quantum wells and isolate the n and p electrodes.
(5) **Metal-pad deposition**: via-holes in the passivation layer are formed by photolithography and etching. Thick metal electrodes such as Cr/Pt/Au or Al/Ti/Au or Ti/Au are then filled in the via-holes with designed shapes.
(6) **Flip-chip bonding**: The DUV LED chips are flip-chip mounted on high thermal conductivity substrates or sub-mounts with metal bump electrodes. The substrate has an important influence on the electrical and thermal properties of DUV LEDs. Silicon-based or ceramic (AlN, etc.)-based substrates are usually used.
(7) **Package**: The package has an important influence on the light extraction, light field distribution, optical output power and reliability of UV LEDs. The traditional epoxy resin is suitable for UV LEDs with wavelength longer than 380 nm due to its disadvantages on its high temperature heat-resistance and UV transmittance.

Figure 6.8a–e show micrographs of the DUV LED chip after dicing, the silicon sub-mount, the flip-chip bonded DUV LED chip on the silicon sub-mount, the encapsulated DUV LED chip and the spectrometer-integrating sphere system for output power testing, respectively.

The LOP-I-EQE curves (light output power-current-external quantum efficiency curves) of a bare 293-nm DUV LED and the EL spectrum at 20 mA are shown in Fig. 6.9a, b, respectively. At an injection current of 20 mA, the DUV LED has an output power of 1.97mW, a corresponding EQE of 2.32%, and an EL spectrum with a full width at half maximum of 9.7 nm. At 30 mA, the LOP reaches 2.47mW and begins to saturate. When the current increases to 40 mA, the LOP saturates at a value of 2.59mW and a corresponding EQE of 1.53%. The saturation of LOP comes from the aggravation of thermal effect with the increasing current density and measurement time.

Figure 6.10a, b show the I-V curves of the DUV LED. The forward voltage is reduced to 5.1 V at 20 mA after optimization. The leakage current of -10 V is as low as 10^{-4} mA.

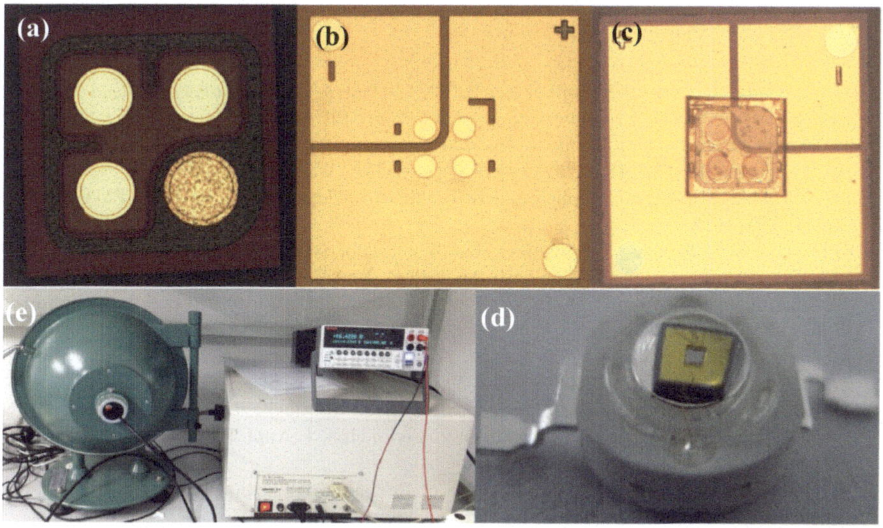

Fig. 6.8 Micrographs of **a** the DUV LED chip after dicing; **b** the silicon sub-mount; **c** the flip-chip bonded DUV LED chip on the silicon sub-mount; **d** the encapsulated DUV LED chip and **e** the spectrometer-integrating sphere system

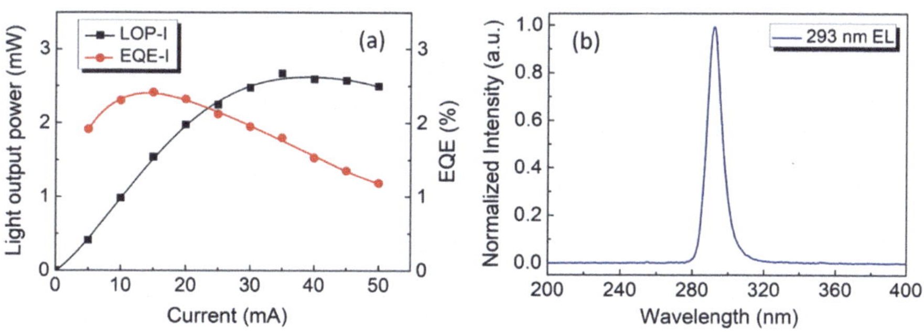

Fig. 6.9 **a** LOP and EQE as functions of the injection current of a bare 293-nm DUV LED; **b** the electroluminescence (EL) spectrum at 20 mA

Fig. 6.10 The forward and reversed I-V curves of the DUV LED after ohmic contact optimization in: **a** linear-linear scale and **b** linear-log scale

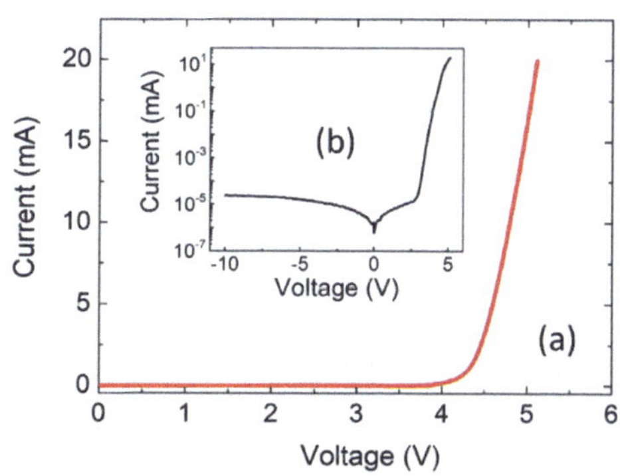

References

1. H. Hirayama, N. Noguchi, S. Fujikawa et al., 222–282 nm AlGaN and InAlGaN based deep-UV LEDs fabricated on high-quality AlN template. Proc. SPIE **7216**, 721621 (2009)
2. H. Morkoç, *Handbook of Nitride semiconductors and devices: materials properties, physics and growth* (WILEY-VCH Verlag GmbH & Co. KGaA, Weinheim, 2008)
3. O. Ambacher, Growth and applications of group III-nitrides. J. Phys. D Appl. Phys. **31**(20), 2653–2710 (1998)
4. O.H. Nam, M.D. Bremser, T.S. Zheleva et al., Lateral epitaxy of low defect density GaN layers via organometallic vapor phase epitaxy. Appl. Phys. Lett. **71**(18), 2638–2640 (1997)
5. M. Stutzmann, O. Ambacher, M. Eickhoff et al., Playing with polarity. Physica Status Solidi(b) **228**(2), 505–512 (2001)
6. T.G. Mihopoulos, V. Gupta, K.F. Jensen, A reaction-transport model for AlGaN MOVPE growth. J. Cryst. Growth **195**(1–4), 733–739 (1998)
7. K. Harafuji, Transport of gas-phase species stored in stagnant volumes under a GaN metalorganic vapor phase epitaxy horizontal reactor. Jpn. J. Appl. Phys. **40**(11), 6263–6283 (2001)
8. B. Heying, X.H. Wu, S. Keller et al., Role of threading dislocation structure on the X-ray diffraction peak widths in epitaxial GaN films. Appl. Phys. Lett. **68**(5), 643–645 (1996)
9. S.R. Lee, A.F. Wright, M.H. Crawford et al., The band-gap bowing of $Al_xGa_{1-x}N$ alloys. Appl. Phys. Lett. **74**(22), 3344–3346 (1999)
10. O. Ambacher, J. Majewski, C. Miskys et al., Pyroelectric properties of Al(In)GaN/GaN hetero- and quantum well structures. J. Phys.: Condens. Matter **14**(13), 3399–3434 (2002)
11. A. Khan, K. Balakrishnan, T. Katona et al., Ultraviolet light-emitting diodes based on group three nitrides. Nat. Photonics **2**(2), 77–84 (2008)
12. C.J. Sun, P. Kung, A. Saxler et al., A crystallographic model of (001) aluminum nitride epitaxial thin film growth on (001) sapphire substrate. J. Appl. Phys. **75**(8), 3964–3967 (1994)
13. S. Guha, N.A. Bojarczuk, Ultraviolet and violet GaN light emitting diodes on silicon. Appl. Phys. Lett. **72**(4), 415–417 (1998)
14. N. Ikeda, Y. Niiyama, H. Kambayashi et al., GaN power transistors on Si substrates for switching applications. Proc. IEEE **98**(7), 1151–1161 (2010)
15. T. Mino, H. Hirayama, T. Takano et al., Development of 260 nm band deep-ultraviolet light emitting diodes on Si substrates. Proc. SPIE **8625**, 86251Q (2013)
16. Y. Zhang, S. Gautier, C.-Y. Cho et al., Near milliwatt power AlGaN-based ultraviolet light emitting diodes based on lateral epitaxial overgrowth of AlN on Si(111). Appl. Phys. Lett. **102**(1), 011106 (2013)
17. http://www.compoundsemiconductor.net/csc/features-details.php?id=19732954
18. M.A. Khan, J.N. Kuznia, D.T. Olson et al., GaN/AlN digital alloy short-period superlattices by switched atomic layer metal organic chemical vapor deposition. Appl. Phys. Lett. **63**(25), 3470–3472 (1993)
19. M.A. Khan, V. Adivarahan, J.P. Zhang et al., Stripe geometry ultraviolet light emitting diodes at 305 nanometers using quaternary AlInGaN multiple quantum wells. Jpn. J. Appl. Phys. **40**(12A), 1308–1310 (2001)
20. J.P. Zhang, E. Kuokstis, Q. Fareed et al., Pulsed atomic layer epitaxy of quaternary AlInGaN layers for ultraviolet light emitters. Physica Status Solidi a-Appl. Res. **188**(1), 95–99 (2001)
21. J.P. Zhang, M.A. Khan, W.H. Sun et al., Pulsed atomic-layer epitaxy of ultrahigh-quality $Al_xGa_{1-x}N$ structures for deep ultraviolet emissions below 230 nm. Appl. Phys. Lett. **81**(23), 4392–4394 (2002)
22. H.M. Wang, J.P. Zhang, C.Q. Chen et al., AlN/AlGaN superlattices as dislocation filter for low-threading-dislocation thick AlGaN layers on sapphire. Appl. Phys. Lett. **81**(4), 604–606 (2002)
23. J.P. Zhang, H.M. Wang, M.E. Gaevski et al., Crack-free thick AlGaN grown on sapphire using AlN AlGaN superlattices for strain management. Appl. Phys. Lett. **80**(19), 3542–3544 (2002)

24. V. Adivarahan, W.H. Sun, A. Chitnis et al., 250nm AlGaN light-emitting diodes. Appl. Phys. Lett. **85**(12), 2175–2177 (2004)
25. V. Adivarahan, J. Zhang, A. Chitnis et al., Sub-milliwatt power III-N light emitting diodes at 285 nm. Jpn. J. Appl. Phys. **41**(Part 2, No. 4B), L435-L436 (2002)
26. A. Chitnis, J.P. Zhang, V. Adivarahan et al., 324 nm light emitting diodes with milliwatt powers. Jpn. J. Appl. Phys. **41**(Part 2, No. 4B), L450–L451 (2002)
27. W.H. Sun, J.P. Zhang, V. Adivarahan et al., AlGaN-based 280 nm light-emitting diodes with continuous wave powers in excess of 1.5 mW. Appl. Phys. Lett. **85**(4), 531–533 (2004)
28. R.G. Banal, M. Funato, Y. Kawakami, Initial nucleation of AlN grown directly on sapphire substrates by metal-organic vapor phase epitaxy. Appl. Phys. Lett. **92**(24), 241905 (2008)
29. R.G. Banal, M. Funato, Y. Kawakami, Characteristics of high Al-content AlGaN/AlN quantum wells fabricated by modified migration enhanced epitaxy. Physica Status Solidi (c) **7**(7–8), 2111–2114 (2010)
30. M. Imura, K. Nakano, N. Fujimoto et al., High-temperature metal-organic vapor phase epitaxial growth of AlN on sapphire by multi transition growth mode method varying V/III ratio. Jpn. J. Appl. Phys. **45**(11), 8639–8643 (2006)
31. M. Imura, N. Fujimoto, N. Okada et al., Annihilation mechanism of threading dislocations in AlN grown by growth form modification method using V/III ratio. J. Cryst. Growth **300**(1), 136–140 (2007)
32. H. Hirayama, T. Yatabe, N. Noguchi et al., 231-261 nm AlGaN deep-ultraviolet light-emitting diodes fabricated on AlN multilayer buffers grown by ammonia pulse-flow method on sapphire. Appl. Phys. Lett. **91**(7), 071901 (2007)
33. H. Hirayama, T. Yatabe, N. Noguchi et al., 226-273 nm AlGaN deep-ultraviolet light-emitting diodes fabricated on multilayer AlN buffers on sapphire. Physica Status Solidi (c) **5**(9), 2969–2971 (2008)
34. H. Hirayama, S. Fujikawa, N. Noguchi et al., 222-282 nm AlGaN and InAlGaN-based deep-UV LEDs fabricated on high-quality AlN on sapphire. Physica Status Solidi (a) **206**(6), 1176–1182 (2009)
35. Z. Chen, R.S. Qhalid Fareed, M. Gaevski et al., Pulsed lateral epitaxial overgrowth of aluminum nitride on sapphire substrates. Appl. Phys. Lett. **89**(8), 081905 (2006)
36. M. Shatalov, M. Gaevski, V. Adivarahan et al., Room-temperature stimulated emission from AlN at 214 nm. Jpn. J. Appl. Phys. **45**(49), L1286–L1288 (2006)
37. V. Adivarahan, Q. Fareed, M. Islam et al., Robust 290 nm emission light emitting diodes over pulsed laterally overgrown AlN. Jpn. J. Appl. Phys. **46**(36), L877–L879 (2007)
38. S. Hwang, D. Morgan, A. Kesler et al., 276 nm substrate-free flip-chip AlGaN light-emitting diodes. Appl. Phys. Express **4**(3), 032102 (2011)
39. H. Hirayama, J. Norimatsu, N. Noguchi et al., Milliwatt power 270 nm-band AlGaN deep-UV LEDs fabricated on ELO-AlN templates. Physica Status Solidi (c) **6**(S2), S474–S477 (2009)
40. R. Jain, W. Sun, J. Yang et al., Migration enhanced lateral epitaxial overgrowth of AlN and AlGaN for high reliability deep ultraviolet light emitting diodes. Appl. Phys. Lett. **93**(5), 051113 (2008)
41. M.S. Shur, R. Gaska, Deep-ultraviolet light-emitting diodes. IEEE Trans. Electron Devices **57**(1), 12–25 (2010)
42. M. Imura, K. Nakano, T. Kitano et al., Microstructure of epitaxial lateral overgrown AlN on trench-patterned AlN template by high-temperature metal-organic vapor phase epitaxy. Appl. Phys. Lett. **89**(22), 221901 (2006)
43. M. Imura, K. Nakano, G. Narita et al., Epitaxial lateral overgrowth of AlN on trench-patterned AlN layers. J. Cryst. Growth **298**, 257–260 (2007)
44. M. Kim, T. Fujita, S. Fukahori et al., AlGaN-based deep ultraviolet light-emitting diodes fabricated on patterned sapphire substrates. Appl. Phys. Express **4**(9), 092102 (2011)
45. K. Iida, H. Watanabe, K. Takeda et al., High-efficiency AlGaN based UV emitters grown on high-crystalline-quality AlGaN using grooved AlN layer on sapphire substrate. Physica Status Solidi (a) **204**(6), 2000–2004 (2007)

46. H. Tsuzuki, F. Mori, K. Takeda et al., Novel UV devices on high-quality AlGaN using grooved underlying layer. J. Cryst. Growth **311**(10), 2860–2863 (2009)
47. R.T. Bondokov, S.G. Mueller, K.E. Morgan et al., Large-area AlN substrates for electronic applications: An industrial perspective. J. Cryst. Growth **310**(17), 4020–4026 (2008)
48. J.R. Grandusky, J. Chen, M.C. Mendrick et al., Improved efficiency high power 260 nm pseudomorphic ultraviolet light emitting diodes, in *Lester Eastman Conference on High Performance Devices (LEC)*, pp. 1–2 (2012)
49. J.R. Grandusky, J. Chen, S.R. Gibb et al., 270 nm pseudomorphic ultraviolet light-emitting diodes with over 60 mW continuous wave output power. Appl. Phys. Express **6**(3), 032101 (2013)
50. T. Kinoshita, K. Hironaka, T. Obata et al., Deep-ultraviolet light-emitting diodes fabricated on AlN substrates prepared by hydride vapor phase epitaxy. Appl. Phys. Express **5**(12), 122101 (2012)
51. T. Kinoshita, T. Obata, T. Nagashima et al., Performance and reliability of deep-ultraviolet light-emitting diodes fabricated on AlN substrates prepared by hydride vapor phase epitaxy. Appl. Phys. Express **6**(9), 092103 (2013)
52. B. Pantha, *GaN and ZnO-based Materials and Devices* (Springer, Berlin,Heidelberg, 2012)
53. J. Li, K.B. Nam, J.Y. Lin et al., Optical and electrical properties of Al-rich AlGaN alloys. Appl. Phys. Lett. **79**(20), 3245–3247 (2001)
54. J.P. Zhang, H.M. Wang, W.H. Sun et al., High-quality AlGaN layers over pulsed atomic-layer epitaxially grown AlN templates for deep ultraviolet light-emitting diodes. J. Electron. Mater. **32**(5), 364–370 (2003)
55. P. Cantu, S. Keller, U.K. Mishra et al., Metalorganic chemical vapor deposition of highly conductive $Al_{0.65}Ga_{0.35}N$ films. Appl. Phys. Lett. **82**(21), 3683–3685 (2003)
56. M.L. Nakarmi, K.H. Kim, K. Zhu et al., Transport properties of highly conductive n-type Al-rich $Al_xGa_{1-x}N(x \geq 0.7)$. Appl. Phys. Lett. **85**(17), 3769–3771 (2004)
57. K. Zhu, M.L. Nakarmi, K.H. Kim et al., Silicon doping dependence of highly conductive n-type $Al_{0.7}Ga_{0.3}N$. Appl. Phys. Lett. **85**(20), 4669–4671 (2004)
58. Y. Taniyasu, M. Kasu, T. Makimoto, Electrical conduction properties of n-type Si-doped AlN with high electron mobility (>100 $cm^2\ V^{-1}\ s^{-1}$). Appl. Phys. Lett. **85**(20), 4672–4674 (2004)
59. M. Kneissl, T. Kolbe, C. Chua et al., Advances in group III-nitride-based deep UV light-emitting diode technology. Semicond. Sci. Technol. **26**(1), 014036 (2011)
60. L. Hsu, W. Walukiewicz, Theoretical transport studies of p-type GaN/AlGaN modulation-doped heterostructures. Appl. Phys. Lett. **74**(17), 2405–2407 (1999)
61. P. Kozodoy, M. Hansen, S.P. DenBaars et al., Enhanced Mg doping efficiency in $Al_{0.2}Ga_{0.8}N$/GaN superlattices. Appl. Phys. Lett. **74**(24), 3681–3683
62. A. Saxler, W.C. Mitchel, P. Kung et al., Aluminum gallium nitride short-period superlattices doped with magnesium. Appl. Phys. Lett. **74**(14), 2023–2025 (1999)
63. P. Kozodoy, Y.P. Smorchkova, M. Hansen et al., Polarization-enhanced Mg doping of AlGaN/GaN superlattices. Appl. Phys. Lett. **75**(16), 2444–2446 (1999)
64. M.L. Nakarmi, K.H. Kim, J. Li et al., Enhanced p-type conduction in GaN and AlGaN by Mg-δ-doping. Appl. Phys. Lett. **82**(18), 3041–3043 (2003)
65. J. Simon, V. Protasenko, C. Lian et al., Polarization-induced hole doping in wide-band-gap uniaxial semiconductor heterostructures. Science **327**(5961), 60–64 (2010)
66. L. Zhang, K. Ding, J.C. Yan et al., Three-dimensional hole gas induced by polarization in (0001)-oriented metal-face III-nitride structure. Appl. Phys. Lett. **97**(6), 062103 (2010)
67. A. Fujioka, T. Misaki, T. Murayama et al., Improvement in output power of 280-nm deep ultraviolet light-emitting diode by using AlGaN multi quantum wells. Appl. Phys. Express **3**(4), 041001 (2010)
68. S. Sumiya, Y. Zhu, J. Zhang et al., AlGaN-based deep ultraviolet light-emitting diodes grown on epitaxial AlN/sapphire templates. Jpn. J. Appl. Phys. **47**(1), 43–46 (2008)
69. H. Hirayama, Y. Tsukada, T. Maeda et al., Marked enhancement in the efficiency of deep-ultraviolet AlGaN light-emitting diodes by using a multiquantum-barrier electron blocking layer. Appl. Phys. Express **3**(3), 031002 (2010)

70. L. Zhou, J.E. Epler, M.R. Krames et al., Vertical injection thin-film AlGaN/AlGaN multiple-quantum-well deep ultraviolet light-emitting diodes. Appl. Phys. Lett. **89**(24), 241113 (2006)
71. V. Adivarahan, A. Heidari, B. Zhang et al., Vertical injection thin film deep ultraviolet light emitting diodes with AlGaN multiple-quantum wells active region. Appl. Phys. Express **2**(9), 092102 (2009)
72. H.Y. Ryu, I.G. Choi, H.S. Choi et al., Investigation of light extraction efficiency in AlGaN deep-ultraviolet light-emitting diodes. Appl. Phys. Express **6**(6), 062101 (2013)
73. P. Dong, Study on Efficiency Improvement of AlGaN-based UV LED, University of Chinese Academy of Sciences, Ph.D. thesis (in Chinese) (2014)

Chapter 7
III-Nitride LED Quantum Efficiency Improvement Technology

LED quantum efficiency is one of the important indicators to characterize the performance of LED devices, including internal quantum efficiency and external quantum efficiency. The internal quantum efficiency refers to the radiation recombination efficiency of electron-hole pairs injected into the active region, which can be improved by optimizing the epitaxial structure and improving the epitaxial technology. The external quantum efficiency is the ratio of the number of photons incident into the space and number of electron-hole pairs in the active region, which is the product of the internal quantum efficiency and the light extraction efficiency. The improvement of light extraction efficiency mainly depends on the optimization of chip technology and packaging technology. In addition, the current injection efficiency is also an important factor affecting the wall-plug efficiency of the LED, and is related to the bulk resistance of the LED device, leakage of carriers and electrode contact.

7.1 Three Structures of LED

Up to now, Gallium Nitride (GaN)-based LED chip technology has emerged three basic device structures including horizontal, flip-chip, and vertical structure. These three structures have their own advantages and disadvantages in the production and application process. The following describes the structure of these three devices separately.

The first type is a lateral structure LED, which can be fabricated by a conventional planar chip process. Therefore, this is the most easily realized LED chip structure on an epitaxial wafer with sapphire substrate, and is also a commonly used one. Figure 7.1 shows the cross-sectional structure of an LED of this structure. In horizontal structure LED chips with non-conductive sapphire substrate, etching process is needed to expose n-type GaN material. Since p-GaN is very thin, the current spreading capability is poor. As the p-GaN surface acts as a light-emitting surface,

J. Li et al., *III-Nitrides Light Emitting Diodes: Technology and Applications*, Springer Series in Materials Science 306,
https://doi.org/10.1007/978-981-15-7949-3_7

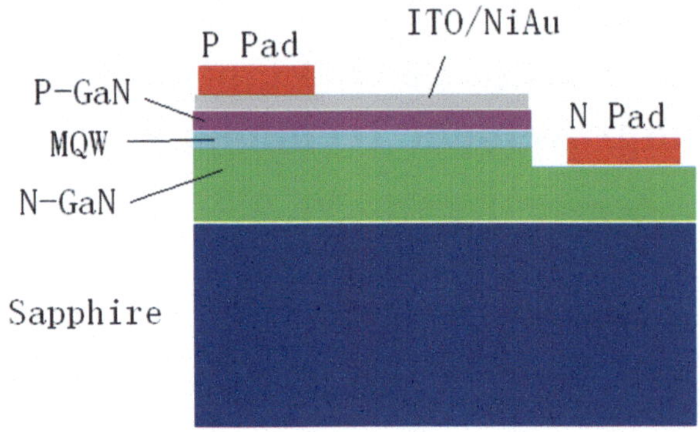

Fig. 7.1 Schematic diagram of GaN-based lateral structure LED

a transparent conductive material needs to be deposited on the surface. The Ni/Au film is an early used transparent film material for GaN-based LED. A Ni/Au film with a thickness of about 30 nm is annealed in an oxygen-containing atmosphere to obtain a transmittance of 80% or more and an Ohmic contact resistance in the order of $10^{-4}\ \Omega\cdot cm^2$ [1]. However, this Ni/Au electrode absorbs light too much and is replaced by an indium tin oxide (ITO) transparent film. For an ITO film, the square resistance is around 10Ω and the light absorption is less than 1%. At the beginning it is difficult to form an ohmic contact between the p-GaN and the ITO film, leading to high operating voltage. By improving the processes of ITO thin films and the GaN epitaxial technique, the contact resistance can be in the order of $10^{-3}\ \Omega\ cm^2$. Considering the optical and electrical properties, ITO is the best choice for GaN-based LEDs p type electrode to date [2, 3]. The light output efficiency of the early horizontal structure LED is very low, mainly due to the large absorption coefficient of the transparent electrode and the total reflection loss. The application of ITO solves the former problem. The Patterned Sapphire Substrate (PSS) technology is a good solution to the second problem. Dielectric and metal composite films are deposited on the back surface of LED chips to form highly reflective mirror to further enhance the light extraction efficiency. In recent years laser cutting technology has been developed to improve the side wall light extraction of LED chips. These processes are relatively easy to implement. The lateral structure LED not only has low manufacturing cost and high yield, but also has high optical efficiency within a certain working range. Combining all these factors, the horizontal structure LED chip has become the mainstream chip structure. However, the lateral structure LED also has its shortcomings, mainly because the thermal conductivity of the sapphire substrate is only 40 W/(m·°C), about one tenth of that of copper. In other words, one of the major disadvantages of this structure is poor heat dissipation performance. In addition, if the electrode structure and size design do not match, it is easy to cause current edge effect, resulting in partial overheating of the chip. The lateral structure LED chips have been widely used in backlight, decoration, display and illumination fields. This structure cannot meet the particularly high-power density application requirements such as the stage lights and projection light source.

The difference between a flip-chip LED and a lateral structure LED is that the light-emitting surface becomes a sapphire surface (there is also a sapphire substrate removed, such as Osram's thin film flip-chip LED), and the p-plane becomes a light-reflecting surface [4]. The LED chip is flip-bonded to another carrier substrate. This carrier substrate may be a chip to improve the current drive circuit, cooling channels, or protection circuits. It's structure is shown in Fig. 7.2. Inverted structure LED chip has several advantages. First of all, ITO transparent electrode is not essential any more. Since the supply of indium element can be an issue, people are trying to find ITO alternatives such as aluminum-doped zinc oxide (AZO) that can act as a GaN-based LED transparent electrode. In the flip LED chip metal electrode can be used as current spreading layer and ITO film for ohmic contact can be thinner or even be replaced by other materials [5]. Although the flip-chip structure LED adopts a new substrate with good thermal conductivity, the thermal conductivity of the actual chip is also limited by the soldering quality and soldering area of the solder joint. Process of inverted structure LED is much more complicated than that of the lateral structure LED. With the developing of ITO and PSS technology, the lateral structure LED chip structure is widely used in solid state lighting industry instead of the flip-structure LED chip. However, a new flip-chip LED process that does not require a substrate is beginning to appear. This chip package does not require gold wire bonding, and the chip is directly flipped on the package substrate by reflow or eutectic soldering. Therefore, the direct contact area between the chip and the package substrate is large. In this case, the heat dissipation problem is well solved, and the process is simplified. At present, flip-chip LED chip technology has been pursued in the industry.

Vertical structure LED chip technology is proposed when there are various problems in the lateral structure LED and flip-chip LED. The p-electrode and the n-electrode of the vertical structure LED are respectively on the upper and lower surfaces of the chip, where the support substrate is a thermally conductive substrate which also functions as an electrode. Therefore, the LED of this structure overcomes the problems of current edge collecting effect and poor heat dissipation capability. At the same time the n-GaN surface has a sufficient thickness as a light-emitting

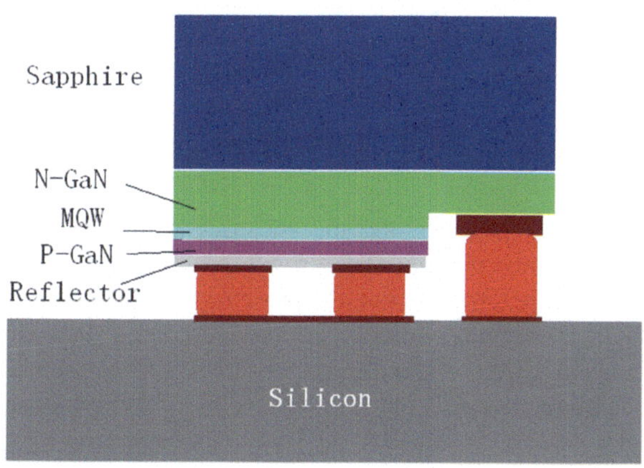

Fig. 7.2 Diagram of GaN-based flip-chip LED

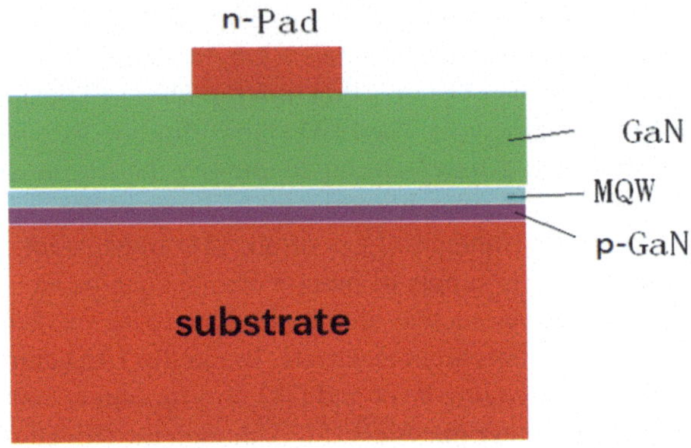

Fig. 7.3 Diagram of GaN-based vertical structure LED

surface for the surface roughening process. Figure 7.3 is a schematic structural view of a typical GaN-based vertical structure LED. Although vertical structure LEDs have many advantages, GaN is difficult to grow directly on this new substrate. The fabrication of vertical structure LEDs requires the removal of old substrates and the attachment to new substrates, which greatly increases the complexity of the chip process and difficulty, resulting in high processing costs of the chip. The chip yield is difficult to reach more than 50%. Despite of this, vertical structure LEDs are favored by domestic and foreign research institutions because of their excellent optical, electrical and thermal properties. In particular, the demonstration of Cree's 303 lm/W high-power vertical structure LEDs has led to more attention and confidence of LEDs in this structure [6].

7.2 Internal Quantum Efficiency Improvement Technology

Internal quantum efficiency (IQE) is one of the key indicators of LED. IQE can be improved by improving the epitaxial quality, optimizing the multi-quantum wells structure and p-GaN.

7.2.1 Homo-Epitaxial Growth of GaN

The epitaxial layer and the substrate material belonging to the same material is called homoepitaxial. The growth of epitaxial layer on different substrate material is called heteroepitaxial. At present, the mainstream substrate for nitride epitaxial layer is sapphire, silicon carbide and silicon. It is well known that heteroepitaxial leads to various dislocations and defects in the epitaxial layer crystal due to problems such as lattice mismatch and thermal mismatch. Problems of heteroepitaxy can be summarized as follows: (1) There is a large lattice mismatch between the epitaxial layer and the substrate. The epitaxial layer growth process will produce dislocation defects. (2) There is thermal mismatch between the epitaxial layer and the substrate. The epitaxial layer is subjected to a temperature change after the epitaxial growth. The growth process can generate thermal stress, resulting in the formation of defects, cracks, problems such as bending of the wafer. (3) Crystal polarity is different, which often causes structural defects such as anti-phase domains. (4) The crystal quality of the substrate directly affects the crystal quality of the epitaxial layer. The dislocations in substrate may extend, multiplicate and re-distribute during the epitaxy growth process.

Due to the above problems, the nitride device based on the heteroepitaxial epitaxy cannot achieve its intended excellent performance. In other words, its performance is severely restricted. In order to improve the crystal quality of the epitaxial layer, researchers have developed different material growth technologies such as low temperature buffer layer and lateral epitaxial overgrown (LEO) technology [7]. These techniques improve the crystal quality of the epitaxial layer in certain degree, but increase the epitaxial growth time and process complexity.

Nitride homoepitaxial uses single crystal GaN as the substrate, also known as freestanding GaN substrate. The freestanding GaN substrate is rapidly grown into a thick GaN film on a sapphire or other material substrate by a hydride vapor phase epitaxy (HVPE) technique. The substrate is then removed by mechanical polishing or laser technology to form a GaN quasi-substrate. Since the dislocation density of the epitaxial GaN layer by HVPE method is reduced as the thickness of the epitaxial layer is increased [8], the crystal quality can be improved as long as the GaN film thickness reaches a certain value. Homogeneous epitaxy can overcome the unfavorable factors caused by heteroepitaxial growth. The epitaxial layer is completely lattice-matched and thermally matched with the substrate, which improves the crystal quality of the epitaxial layer and improves the quantum efficiency of the device. The GaN substrate can be made into a conductive substrate, which is convenient for preparing a vertical structure chip with excellent current spreading performance. GaN epitaxial technique based on freestanding GaN substrate is an important direction for future development.

The freestanding GaN substrate has a Ga polar surface and an N polar surface. The N polar surface is chemically active and easily corroded by acid. However, since the surface energy of the N polar surface is relatively large, the epitaxial GaN tends to form a hexagonal pyramid and a rough surface. The Ga polar surface is easy to form a flat surface due to the relatively small surface energy. Studies have shown that the Ga polar surface is easier to achieve p-type doping. Therefore, the homo-epitaxy generally selects the Ga polar plane as the substrate surface of the epitaxial layer. However, since the Ga polar surface is chemically stable and is not easily corroded by acid, the Ga polar surface is usually obtained by mechanical polishing.

At present, the crystal quality and surface quality of GaN freestanding substrates are not very good, which influents the quality of epitaxial materials. Figure 7.4 is a topography of Ga polar GaN surface. Many scratches can be seen on the substrate surface. The ICP etching and patterning processes are used to treat the substrate surface. In Fig. 7.5 dislocation photo of GaN epitaxial layer grown on freestanding

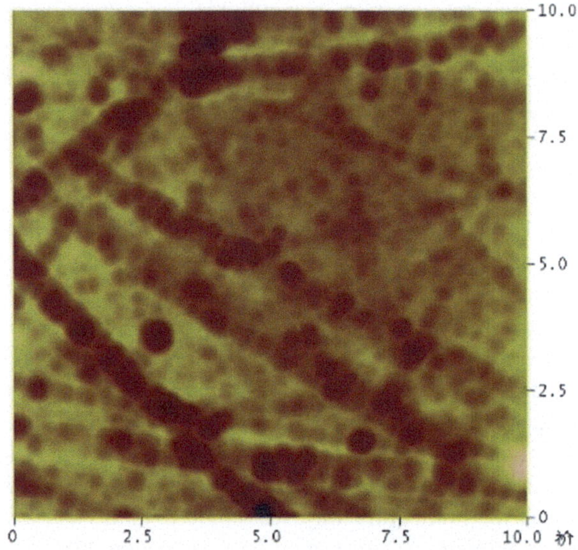

Fig. 7.4 Topography of freestanding Ga polar GaN substrate surface

Fig. 7.5 GaN epitaxial material on freestanding GaN substrate, the dislocation density is 3–5 × 10^7/cm^2

GaN substrate is shown. The dislocation density is about 3–5 × 10^7/ cm^2.

7.2.2 Multiple Quantum Wells

InGaN/GaN multiple quantum wells (MQWs) are generally used as active regions in the InGaN LEDs. Due to the lattice mismatch between InGaN and GaN, the induced positive and negative charge centers do not completely overlap. Such a phenomenon can lead to separation in space. Therefore, a built-in polarization electric field is formed in the active region, which causes spatial separation of the InGaN conduction band and the valence band electron hole wave function, resulting in a decrease in the probability of radiation recombination. This is the Quantum-Confined Stark Effect (QCSE). This is shown in Fig. 7.6. On left side the energy band diagram without electric field is shown as a flat band. The wave functions of electron and hole are overlap completely. When an external electric field is applied, and shown on the right side of Fig. 7.6, the energy band of conduction band and the valence band will tilt. The wave functions of electrons in the conduction band and holes in the valence band will be spatially separated, resulting in greatly reduced probability of wave function overlap. The probability of radiation compounding decreases.

The polarization electric field caused by lattice mismatch in the active region of InGaN is the main cause of the QCSE effect and the reduction of luminous efficiency. For nitride compounds, the spontaneous polarization electric field can be calculated by the following expressions [9]:

$$\begin{aligned} P^{SP}_{Al_xGa_{1-x}N} &= -0.090x - 0.034(1-x) + 0.019x(1-x), \\ P^{SP}_{In_xGa_{1-x}N} &= -0.042x - 0.034(1-x) + 0.038x(1-x), \\ P^{SP}_{Al_xIn_{1-x}N} &= -0.090x - 0.042(1-x) + 0.071x(1-x), \end{aligned} \quad (7.1)$$

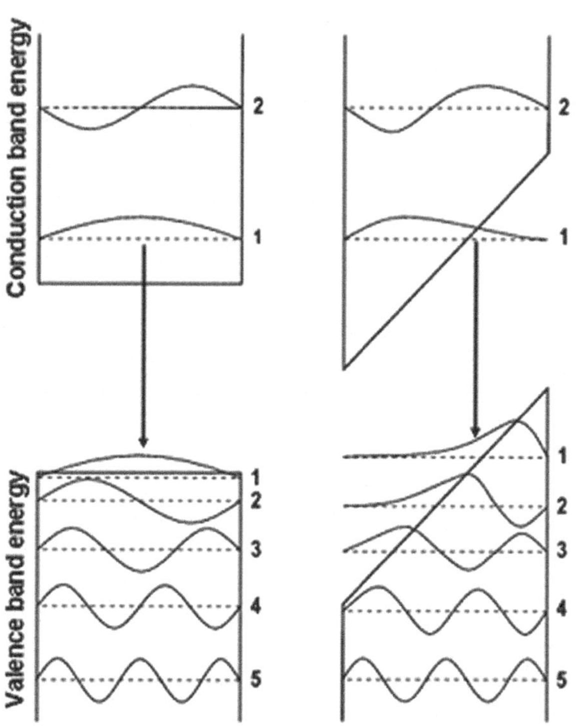

Fig. 7.6 QCSE in InGaN quantum wells

Among them, x is the component of the III element. For the epitaxial nitride layer with a certain degree of lattice mismatch, the piezoelectric polarization electric field can be expressed as [10]:

$$P^{PZ}_{X_xY_{1-x}N} = xP^{PZ}_{XN}[\varepsilon(x)] + (1-x)P^{PZ}_{YN}[\varepsilon(x)] \tag{7.2}$$

Among them, the piezoelectric polarization electric field of binary nitride can be accurately expressed as:

$$\begin{aligned} P^{PZ}_{AlN} &= -1.808\varepsilon + 5.624\varepsilon^2 \text{ for } \varepsilon < 0, \\ P^{PZ}_{AlN} &= -1.808\varepsilon - 7.888\varepsilon^2 \text{ for } \varepsilon > 0, \\ P^{PZ}_{GaN} &= -0.918\varepsilon + 9.541\varepsilon^2, \\ P^{PZ}_{InN} &= -1.373\varepsilon + 7.559\varepsilon^2, \end{aligned} \tag{7.3}$$

Among them,$\varepsilon(x) = \frac{a_s - a_x}{a_x}$. The polarization electric field of the compound epitaxial layer is the result of the combined action of the piezoelectric polarization electric field and the spontaneous polarization electric field. Therefore, the polarization electric field of the epitaxial layer can be obtained by the following relationship [11]:

$$P_c = P_{SP} + P_{PZ,c}(\varepsilon) \tag{7.4}$$

The polarization electric field in the active region of InGaN is calculated as follows [11]:

$$P_c = P_{SP,QW} - P_{SP,QB} + P_{PZ,c}(\varepsilon) \tag{7.5}$$

Based on the above theoretical equation, we can calculate the polarization field in the InGaN/ GaN active region with different In composition.

The electric field polarization is also influenced by crystal defects. Dislocations and other defects may generate stress in quantum well region, and impact piezoelectric polarization field. Exploring different approaches to reduce MQWs polarization electric field and improve the crystal quality has become a major research focus. In addition, factors affecting the quantum efficiency within the nitride device are also associated with very low electron capture efficiencies and low hole injection efficiencies in the multiple quantum well active regions. The activation energy of Mg-doped p-type GaN is very high. As the hole concentration and mobility are low, this will lead to a relatively lower hole injection efficiency. In contrast, the mobility of electron is much higher since the effective mass of the electrons is much smaller than that of the holes. The electrons, therefore, can leak from the active region into p-GaN layer and recombine with the holes non-radiatively. This is especially true in the case of large current injection.

In summary, in order to improve the internal quantum efficiency of the nitride device, it is necessary to reduce the QCSE effect of the multi-quantum structure, improve the crystal quality and increase the carrier collection capability of the multi-quantum wells.

AlInGaN barrier quantum well: AlInGaN offers more freedom to adjust the structure of the LED since it has two independently adjustable compositional parameters of In and Al. By changing the composition and doping level of the quantum well material, the influence of the polarization electric field in the InGaN/GaN quantum well can be reduced. This, therefore, can improve the quantum efficiency of the LED device. By designing InGaN/InGaAlN strain-compensated quantum wells and the doped InGaN/GaN barrier strain-compensated quantum wells, one can manipulate the Fermi level to obtain a higher hole concentration.

According to the simulation results, the LED quantum well using the AlInGaN barrier can have its Fermi level (holes) in the hole potential well. This implies that a higher hole concentration can be obtained. However, the concentration of electrons is reduced because the Fermi level is shifted down compared to the conventional structures. As the electron concentration in the LED (on the order of 10^{18}) is still much larger than the hole concentration (on the order of 10^{17}), the decrease in the electron concentration does not greatly affect the IQE. Thereby, the electron-hole recombination efficiency in the quantum well is improved.

Gradient In component quantum wells is shown in Fig. 7.7. Low In component shallow well is grown in front of the light emitting layer. The Shallow Quantum Well (SQW) model in the gradient well reduces the polarization field caused by lattice mismatch between the light-emitting well and the quantum barrier by introducing a low-In composition InGaN SQW in front of the light-emitting well. This thereby decreases the QCSE effect and improves the luminous efficiency. By growing a transition layer of In gradient InGaN before and after the light-emitting wells, the stress field is reduced and the luminous efficiency is improved [12].

Asymmetric Charge Resonant Tunneling Effect (CART): It is well known that the electron capture rate can be appoximated as the phonon-electron mobility and transmission time divided by the width of the quantum well. Obviously, a wide quantum well favors the electron capture rate, but the carrier confinement capacity decreases as the quantum well width increases. Asymmetric Charge Resonant Tunneling Effect (CART) structure [13, 14], which inserts a very thick electron-emitting layer and a very thin electron blocking layer between the MQW layer and the n-GaN layer,

Fig. 7.7 Shallow quantum well model

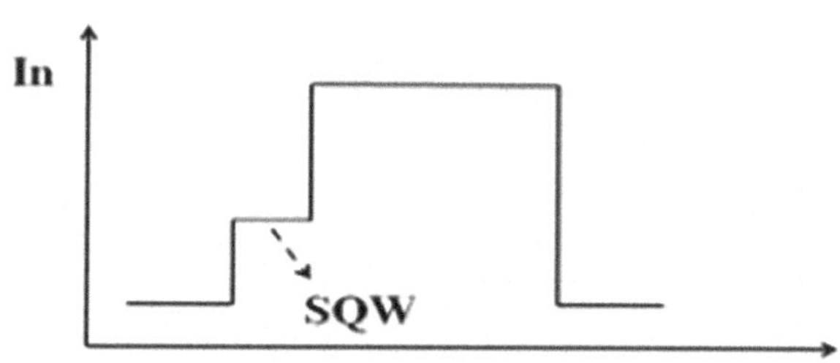

weakens the electron energy and reduces its mobility in the vertical direction. However, a very thick InGaN electron-emitting layer with a CART structure will produce mismatch dislocations in the InGaN electron-emitting layer and lead to the mismatch of InGaN and GaN in the MQW active region. These dislocations increase non-radiative recombination and leakage current [15]. Thus, researchers have proposed double step multi-quantum well LED, i.e., the use of a low In composition of GaN/InGaN MQW active layer as an electron-emitting region. The In composition of the InGaN is increased linearly along the direction of the active region. Such a design can further improve luminous efficiency of the LEDs [16, 17].

7.2.3 Active Region Doping

In nitride LEDs, the hole injection rate and migration ability are an important issue that limits the luminous efficiency of LEDs. By changing the structure of the quantum barrier in the LED and directly doping in the GaN barrier can improve holes distribution and transport. For example, localization effect can be enhanced and the carrier concentration will be increased by lightly Si doping in a multi-quantum well. It can also enhance the lateral spread of current, while heavily doped Si can shield the polarized electric field to some extent [18].

In addition, Mg doping is performed at the quantum barrier position of the multiple quantum well. More holes can be generated after activation by annealing. Holes are injected into the quantum well region to shield the polarization charge generated by the lattice constant mismatch. Such a scheme can effectively suppress the influence of the polarization electric field and enhance the light output power of the LED. By comparing the Mg-doped and undoped samples, it is apparent from the photocurrent test that the polarization field of the sample after Mg doping is significantly weakened [19]. However, doping can efficiently change wavelength of LEDs [20].

7.2.4 Electronic Barrier Layer

As mentioned above, the electron trapping efficiency and hole injection efficiency of the multiple quantum well active region are very low. This makes it easy for electrons to overcome the limitation of the quantum well to achieve non-radiative recombination in the p-region with the hole, especially at high current density. Therefore, reducing the electron leakage rate is an effective way to improve the luminous efficiency of LEDs. Figure 7.8 shows several carrier recombination processes in LED, i.e. indirect (Shockley-Read-Hall, SRH) recombination, the spontaneous radiative recombination and Auger recombination (Auger), and the leakage to the p-region non-radiative recombination.

Component gradient electronic barrier: In order to reduce the overshoot of the electron, electron blocking layer (EBL) is inserted between the LED multi-quantum

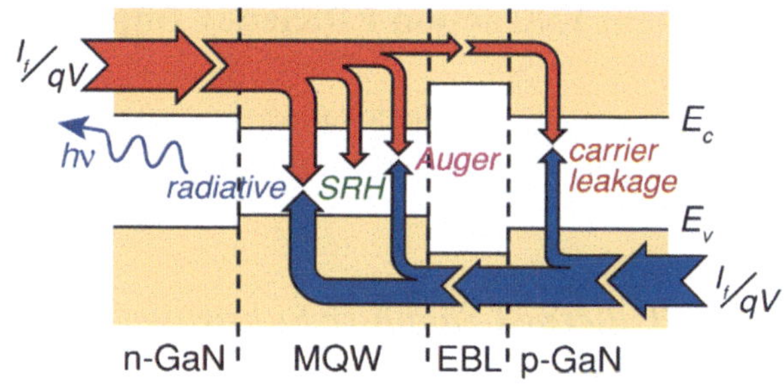

Fig. 7.8 Schematic diagram of the carrier recombination process in InGaN MQWs

well and p-GaN layer. However, this EBL blocks hole injection into the multiple quantum well region due to its higher valence band barrier. The bending of interface band due to the piezoelectric polarization field induced by lattice mismatch at the AlGaN/GaN interface further blocks the injection of holes. In order to eliminate the stress mismatch between the electron blocking layer and the active region, we replaced the traditional electron blocking layer (CEBL) with an electron blocking layer (GEBL) with a gradual increase in composition along the growth direction.

The improvement of the luminous efficiency of GEBL LEDs is attributed to the increase of carrier space coincidence rate, which is due to the reduction of stress-dependent piezoelectric electric field. Figure 7.9 compares the quantum efficiency of two LED samples with GEBL and CEBL. It can be seen that the quantum efficiency is significantly improved.

In addition, the p-InGaN/AlGaN electron blocking layer structure [21] and the graded superlattice AlGaN/GaN EBL [22] have also been studied to further enhance the confinement of electrons and holes.

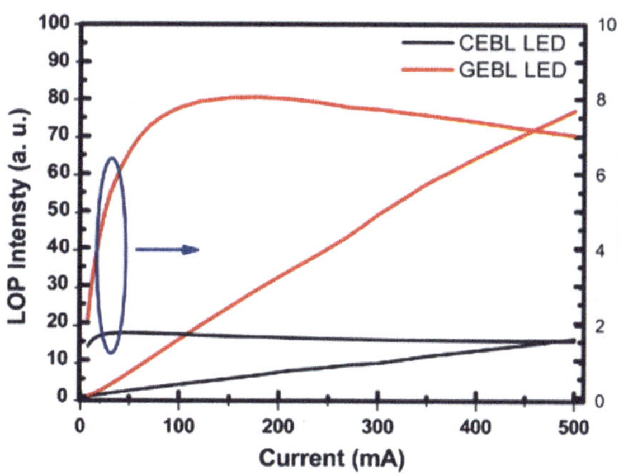

Fig. 7.9 LED quantum efficiency with gradient EBL

7.3 Light Extraction Efficiency Improvement Technology

Approaches to enhance the light extraction efficiency include patterning substrate, surface roughening, utilizing reflector and photonic crystal.

7.3.1 Patterned Sapphire Substrate

Lattice mismatch between the sapphire substrate and the GaN epitaxial layer is about 14% and the thermal mismatch is 25.5%. Such mismatches can lead to high density defects in the GaN epitaxial layer. The defect density can reach 10^8–10^{10} cm^{-2}. Due to the poor crystal quality, the transfer rate of carriers and the injection efficiency of the device are decreased. At the same time, non-radiative recombination is likely to occur due to the impurity level at the defect. The photons generated by the radiate recombination are also easily absorbed by the defects to generate heat. Further, there are multiple interfaces where a large change of refractive index happens in GaN-based LED chip. These interfaces can be from between the GaN and sapphire or GaN layer and packaging materials. The refractive index of different materials is shown in Table 7.1. The photons emitted by the active region are totally reflected at the interface of each film layer. After multiple reflections, the light absorbed by the quantum well and the semiconductor material can be converted into heat energy. Such a process not only affects the light extraction efficiency of the chip, but also shortens the reliability and stability.

Patterned sapphire substrate (PSS) technology can solve the above problems to some extent. Patterning the sapphire substrate is obtained by a photolithography process to the mask pattern and then an ICP etching or wet etching process to form patterns on the sapphire substrate as shown in Fig. 7.10. Using PSS technology, the epitaxial growth mode is changed from ordinary longitudinal epitaxy to lateral epitaxy. This growth mode has been shown to be effective in reducing the defect density of GaN epitaxial materials. Such an approach is therefore effective in suppressing non-radiative recombination and improving the internal quantum efficiency of the device. Since the periodic structures of the patterned sapphire substrate can effectively change the incident angle of light at the interface, more photons are extracted from the chip and the light extraction efficiency is increased. The shape, size, period, pitch, height, arrangement, and curvature of the patterns on the substrate

Table 7.1 The LED layers of structural material refractive index

Epitaxial layer	Refractive index	Absorption rate
Sapphire	1.75	10–6
n-GaN	2.42	0.01
Active layer (InGaN)	2.52	1.25
p-GaN	2.42	0.01

Fig. 7.10 Design of the patterned sapphire substrate

have a great influence on the light extraction efficiency and epitaxial growth. The extraction efficiency of LED and the influence of the growth mechanism of epitaxial materials may have very different outcomes with different patterned substrate topography. Therefore, optimizing the design of the topographic structure is the basis and foundation of patterned sapphire substrate technology.

3D-2D mode transition GaN epitaxial technology based on PSS: Secondary nucleation growth method is proposed for epitaxial GaN on patterned sapphire substrate. Figure 7.11 shows that there are various crystal orientations in the low temperature buffer layer after the annealing. Under such conditions, the merger of the nucleation sites with different orientations can cause crystal twisted and form a large number of dislocations. In order to further improve the quality of the GaN crystal, high temperature secondary nucleation is introduced where the low temperature buffer layer is subjected to high temperature recrystallization. In other words, the second high-temperature nucleation grows GaN at a high temperature on the basis of the first nucleation. After re-annealing, the crystal oricntation of the nucleus becomes more uniform and the lattice quality becomes better.

In order to reduce the penetrating dislocation of the GaN epitaxial layer, a 3D (three-dimensional growth) -2D (two-dimensional growth) growth mode is adopted. After the growth of nucleation layer, the growth conditions are adjusted so that the

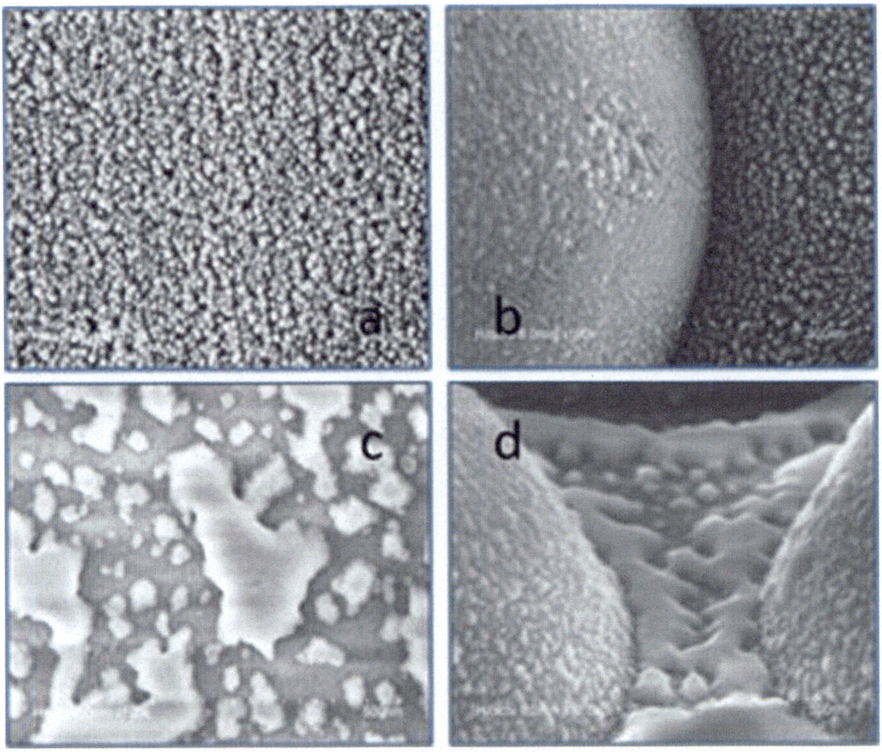

Fig. 7.11 Low temperature buffer layer grown on a planar substrate and a patterned substrate before and after annealing. a and b are the shape of the buffer layer before annealing; c and d are the morphology after annealing

longitudinal growth rate of gallium nitride is much larger than that of the lateral direction. Growing the 3D layer until the thickness is greater than or equal to the height of the substrate pattern. At this stage, the real-time anti-reflectivity approaching 0. The growth conditions are adjusted again such that the vertical growth rate of gallium nitride is less than the lateral growth rate. This will lead to the 2D growth mode until the epitaxial gallium nitride completely covers the substrate pattern and form a flat surface as shown in Fig. 7.12. During the 3D growth process, the longitudinal growth rate is enhanced. When the thickness can be compared with the height of the substrate pattern, the 3D GaN layer encloses the pattern of the patterned substrate. During the 2D growth process, GaN epitaxy rapidly grow along the inclined side surface, which results in the bending of penetrating dislocation, thereby reducing the dislocation defects in GaN epitaxial film.

Figure 7.13 is TEM image of epitaxy LED structure adopted 3D-2D growth mode. Penetrating dislocations (region 1) turn toward the direction of the pattern, where a small amount of penetration dislocations (region 2) still extend upward.

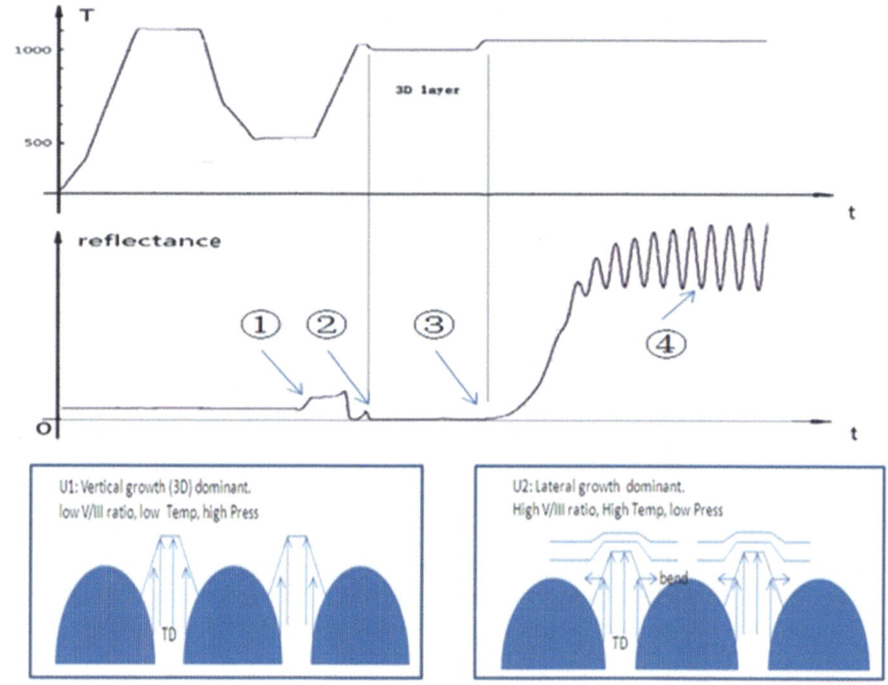

Fig. 7.12 3D-2D mode epitaxial growth of GaN on PSS

Fig. 7.13 TEM image of GaN epitaxial layer in 3D-2D growth mode

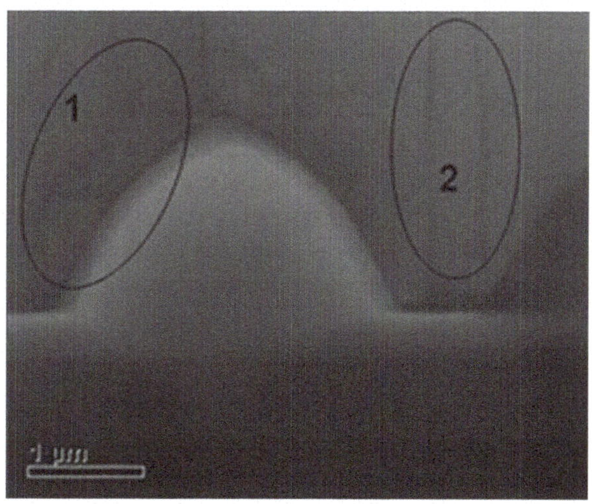

Compared to typical growth method, overall penetration density of dislocations has been substantially reduced [23].

For micro-nano patterned sapphire substrate (NPSS) technology, a nickel film is first deposited on the prepared micro-pattern substrate. Rapid thermal annealing under high temperature conditions is then carried out to form a nickel ball of nanometer size as a mask. A micro-nano mixed patterned sapphire substrate is obtained by dry etching or acid etching.

XRD and SEM analysis shows that the micro-nano pattern substrate has a close influence on the quality of GaN materials. From FDTD simulation, it has found that LEDs with the micro and nano patterned substrate, in comparison with the micro PSS, have higher light extraction efficiency.

7.3.2 Surface Roughening

Extraction efficiency is one important optical characteristic for light emitting diode. When light is emitted from the semiconductor into air, the difference of refractive index acrosst the interface causes the total reflection. Only a small amount of light can be extracted. At the interface of the two media, Snell's law is satisfied.

$$\sin\varphi = n_{air}\Phi \tag{7.6}$$

where n_s, n_{air} represent the refractive index of the semiconductor and air, respectively. Φ is the incident angle of light and φ is the angle of refraction. When $\varphi = 90$, critical angle of the total reflection can be calculated:

$$\Phi_c = \arcsin(n_{air}/n_s) \tag{7.7}$$

Generally, the refractive index of semiconductor materials is relatively high. For instance, the index of GaAs and GaN is 3.4 and 2.5, respectively. Because the critical angle of total reflection is small, only very small part of photons generated by the active region is escaped outside the chip:

$$\frac{P_{escape}}{P_{source}} \approx \frac{1}{4}\frac{n_{air}^2}{n_s^2} \tag{7.8}$$

For GaAs, only 2% of the light can be extracted. GaN has an extraction rate of about 4%. Thus, the LED chip surface must be roughened so that the light extraction efficiency can be effectively improved.

Stealth dicing technology: The epitaxial GaN layer and the sapphire sidewall surface are modified by laser stealth cutting technology to make it more conducive for photon escaping. The method of multiple laser stealth dicing is first proposed for

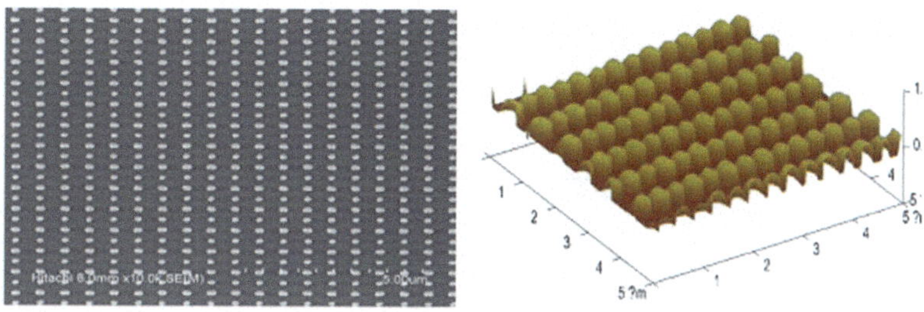

Fig. 7.14 250 nm diameter nanoarray by synchrotron radiation X-rays

LED on sapphire substrate. The sidewalls are modified to make it easier for photons to escape. This will lead to much enhanced the light output power of LED.

GaN microporous etching techniques: To achieve an increase the light extraction efficiency, LED chip surface can be also modified by nano-roughened. For instance, the surface of the LED die is roughened by self-assembled CsCl nanospheres and synchrotron radiation X-rays as shown in Fig. 7.14, which effectively improves the light-emitting efficiency of the LED.

ITO surface treatment technology: ITO surface is roughened in nanoscale to reduce the total reflection loss, thereby improving the extraction efficiency of the LED device. Nano-particles such as CsCl are used as the ICP etching mask layer. The ITO surface with nano-island structures are formed to reduce the total reflection of photons at the interface between ITO and air as shown in Fig. 7.15 [24].

Furthermore, the surface of the ITO can be roughened by using NaCl as a mask. Using such a technique, interconnected ITO nano-network structure is obtained. This structure improves not only the LED light output power but also the LED electrical characteristics. In addition, this method has many practical advantages. The cost is low, and it is simple to make.

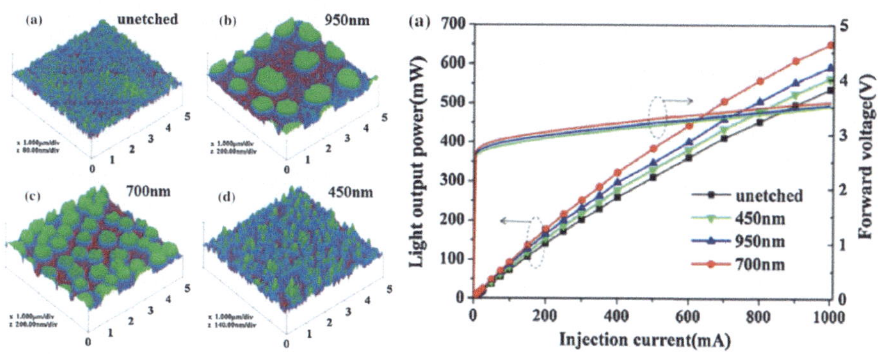

Fig. 7.15 ITO surface roughness and LED optical properties. Copyright (2012) The Japan Society of Applied Physics

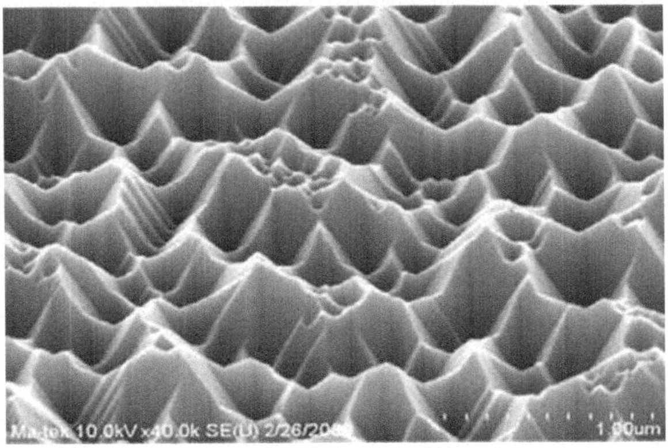

Fig. 7.16 SEM image of in situ growth roughened p-type layer

In situ growth surface roughening technology: Surface treatment technology is an effective way to improve light extraction efficiency. Through the roughening treatment of the light-emitting chip surface, the probability of the light emission across the interface can be significantly improved, and the total reflection loss is reduced. The more obvious of the surface coarsening is, the better of light extraction effect of the LED is. The roughening technology of the surface of the epitaxial material can significantly improve the LED light output power.

The in situ growth surface roughening can be achieved by controlling the epitaxial process parameters of growing the p-type GaN common mirror surface. By lowering the growth temperature and changing the ratio of hydrogen to nitrogen in the reaction chamber, the longitudinal growth rate of p-type GaN is significantly enhanced than the lateral growth rate. Due to the anisotropic crystal growth, V-shape roughened surface is formed during the epitaxial growth process (Fig. 7.16).

When the temperature is too high, the growth rate of the longitudinal and lateral directions is relatively small. This will result in the formation of featureless surface morphology. The roughening effect is poor. When the temperature is too low, the depth of the V-type pit is shallow. The roughening effect is also deteriorated. It is clearly that there is an optimization temperature range that will lead to the best surface roughening morphology. In addition, process parameters such as the hydrogen/nitrogen ratio in the reaction chamber also affect the degree of surface roughening.

7.3.3 Reflector

Having a high reflectivity and high thermal conductivity characteristic properties is important for high brightness, high power LED. From the standpoint of improving the external quantum efficiency of the device and reducing the thermal resistance, a back reflector is required for GaN-based LED device. Therefore, designing a reflector layer with high reflectivity and good stability has become the research focus to improve the LED performance. A distributed Bragg reflection film (DBR) is a good choice. Distributed Bragg reflectors with highly reflective nature and selectivity to the incident spectrum are widely used in semiconductor optoelectronic devices [25]. Internal quantum efficiency of GaN-based LED has reached an optimal configuration of 80% based on MOCVD epitaxial growth techniques and multiple quantum wells structure [26]. However, due to insufficient reflection of the LED chip, the light extraction efficiency of the device is reduced. This will affect the external quantum efficiency of the device. By depositing DBR on the back side of the chip (sapphire surface), the light reaching the back side of the chip is re-reflected. Such a scheme is an important means to effectively improve the light extraction efficiency of the LED.

For the highly reflective DBR film used in the specific band spectral section, there are mainly three film structures: a metal layer, a dielectric layer, or a metal plus dielectric layer. Each of the three reflectors has its own characteristics. (1) Common metal reflective materials for metal-distributed Bragg reflection films include silver (Ag), aluminum (Al), gold (Au), copper (Cu), etc. These materials have relatively high reflectance, broad reflectance spectrum, and good polarization effect. The aluminum film has a high reflectivity from the ultraviolet region to the infrared region. Furthermore, the aluminum film forms a thin layer of aluminum oxide on the surface in the atmosphere to make the film more stable. The silver film has a high reflectivity in the visible and infrared regions. The reflectivity is very high. The polarization effect introduced in the operation is small. However, silver is easy to oxidize and vulcanize. The properties of Ag are not very stable. The gold film has high reflectivity at 600 nm, and the reflectance is high in the infrared region. However, in actual use, the metallic reflective film is rarely used alone due to the disadvantages such as insufficient stability and hardness, and poor adhesion to the substrate. (2) The metal-enhanced distributed Bragg reflection film: Due to the absorption of the metal film, the actual reflectance of metal reflector cannot be maximized. Also, the metal is soft and can be easily damaged. The adhesion to the substrate is poor as well. Therefore, plating a pair of or pairs of dielectric layers with alternating high and low refractive index in a pure metal film can reduce the absorption of metal and improve the reflectivity. At the same time, it can serve as a protective layer. However, due to the interference effect of the dielectric film, the high reflection band of the film layer is thus narrowed. (3) All-media type distributed Bragg reflection film: The all-media type distribution Bragg reflection film is a periodic structure with high reflectivity which is formed by alternately forming a plurality of layers of high-low refractive index materials. The beam has the same phase as it is reflected back from the interface of each layer $\lambda_0/4$ thick back to the front surface, thus producing constructive interference.

For the lateral structure LED, the higher the sapphire surface reflectance, the better the light extraction efficiency of the device is.

After depositing a high refractive index (n_1) film layer with the optical thickness of $\lambda_0/4$ on a substrate having a refractive index n_s, the reflectance is increased. For light with the center wavelength λ_0, the admittance of the single layer film and substrate combination is n_1^2/n_s, the reflectance of normal incidence is:

$$R = (n_0 - n_1^2/n_s)^2/(n_0 + n_1^2/n_s)^2 \quad (7.9)$$

The larger the n_1^2/n_s, the higher the reflectivity is. In practice, however, the refractive index is limited. The maximum achievable reflectivity of monolayer film does not exceed 50%.

If a dielectric multilayer film of high and low refractive index is used for each layer with thickness of $\lambda_0/4$, a higher reflectance can be obtained. This is because the beams reflected from all interfaces of the layer have the same phase when they return to the front surface, resulting in constructive interference. For such a group of dielectric film systems, it is theoretically expected to get close to 100% reflectivity [27].

If the n_H and n_L is the refractive index of the high and low refractive index layers, where the outermost layer on both sides of the dielectric film is a high refractive index layer and the thickness of each layer is $\lambda_0/4$, the maximum reflectance at normal incidence for the center wavelength λ_0 is:

$$R = \left[\frac{1 - (n_H/n_L)^{2S}(n_H^2/n_S)}{1 + (n_H/n_L)^{2S}(n_H^2/n_S)}\right] \quad (7.10)$$

Obviously, the larger the $\frac{n_H}{n_L}$ ratio, or the more the number of layers, the higher the reflectance R is. In theory, as long as the number of layers in the film system can be increased, the reflectivity can be infinitely close to 100% [27].

Table 7.2 lists the refractive index and absorption coefficient of commonly used optical film materials.

Table 7.2 Common optical film materials

Material	Refractive index	Absorption coefficient
SiO_2	1.46488	0.00000
CaF_2(ir)	1.40000	0.00100
Al_2O_3	1.65736	0.00000
CeF_3	1.64000	0.0 0000
Mg_2	1.38723	0.00000
PbF_2	1.46000	0.00000
Gd_2O_3	1.95000	0.00000
Bi_2O_3	1.91000	0.00000
Ge	2.96	2.47000
HfO_2	2.01	0.00049
ITO	2.09200	0.02000
Cr_2O_3	2.2400	0.07000
TiO_2	2.40695	0.00076
ZnS	2.42580	0.00008
ZrO_2	2.07864	0.00029
Ta_2O_5	2.14909	0.00000

In fact, due to the absorption and scattering loss in the film, the continuous stacking more layers does not improve the reflectance after the film system reaches a certain number of layers, Sometimes the reflectance decreases due to an increase in absorption and scattering loss. Therefore, the absorption and scattering losses in the film system and the maximum number of layers of the dielectric film system are limited.

Omnidirectional full-band high-reflection structure design: High-reflection mirror is optimized by high-reflection design to improve LED light extraction efficiency.

7.3.4 Flip-Chip Structure

Most of the LEDs used for illumination are high-current-driven power chips, which require good current spreading and heat dissipation. There are several kinds of deferent flip chip structures such as thin-film type (TF), flip chip (FC) and direct attach (DA) flip chip. The TF structure can be divided into the silicon substrate TF structure chip and the metal-based TF structure chip according to the substrate material used in the chip.

Flip-chip is fabricated by flipping the traditional lateral structure LED on high thermal conductivity substrate. The light emitted from the transparent sapphire substrate can avoid absorption of electrodes and leads, and increase the interface total reflection critical angle. The heat conductive path is changed from sapphire with poor thermal conductivity to materials with higher thermal conductivity (such as silicon, aluminum nitride, etc.). Thus, the thermal resistance is greatly reduced. Flip-chip structure largely solves the problems faced by the lateral LED chip, such as high operating voltage, current crowding, current saturation, low active region utilization, and poor current spread. Flip-chip effectively improves the electrical, optical, and thermal performance of power LEDs.

For thin-film (TF) chip structure, the LED epitaxial layer is transferred from the original substrate to a new substrate by laser lift-off or electrical chemical etching, where the new substrate has good electrical and thermal conductivity. After transferred to a new substrate, n-GaN side is up. Such an approach makes it easier to integrate the chip with metal reflectors and the surface roughening substrate and to enhance device efficiency of light. The main methods to achieve film transfer are bonding and electroplating. The new substrate materials mainly include silicon substrates and metal substrates represented by copper.

Direct attach (DA) flip-chip LED combines the advantages of FC—LED and advanced TF—LED, where wire-bonding is not required during package process. The current spreading and heat dissipation problems of the conventional TS chip are effectively solved. Furthermore, the damage caused by the substrate stripping process to the film can be avoided.

7.3.5 Photonic Crystal

A photonic crystal is a novel micro-structured material in which the dielectric constant changes periodically with space. One of the characteristics of a photonic crystal is that it has a photonic band gap (PBG) and does not allow electromagnetic wave propagation in the band gap. Distribution of the photonic crystal will be influenced by dielectric permittivity difference between the lattice structure and the filling factor. The presence of photonic band gaps allows photonic crystals to suppress spontaneous emission well. Spontaneous radiation is suppressed when the frequency of the spontaneously radiated light of the atom falls in the PBG. Another feature of photonic crystals is the localization of the photons. It is closely connected to the defect level in the photonic crystal. If the symmetry of the photonic crystal is changed or impurities and defects are introduced therein, a corresponding defect state is generated in the photonic band gap. Frequency coincides with the defect states of photons will be restricted at the position of the defect. Once deviated from the defect states the intensity of photons will decay rapidly. The characteristics of the localized photon are mainly determined by the type of defects, including point defects, line defects and surface defects. Line defects are similar to optical waveguides. Light can only propagate along line defects. Plane defects function like a full reflection mirror, capable of reflecting incident light from any direction, where the reflection rate can reach 100%.

There are two main mechanisms for improving the luminous efficiency of LEDs by using photonic crystals. (1) The periodic scattering of light on the surface of the photonic crystal structure causes the light that should be totally internally reflected to escape the LED, thereby improving the luminous efficiency of the LED. (2) The photonic crystal structure forms a special energy band. By adjusting the parameters of the photonic crystal, the LED emission wavelength may fall within the photonic band gap range, thereby suppressing the light emitted from the side surface. The LED light can be extracted within the propagation modes. Both effects have different characteristics. The first effect overcomes the total internal reflection by means of light scattering, similar to surface roughening techniques. The second effect is mainly to extract the propagation mode originally confined within the material by means of the photonic band gap. The second effect can also be applied to different illuminating regions and light extraction regions by adjusting the structural parameters of the photonic crystal. Figures 7.17 and 7.18 are p-GaN surface of the photonic crystal and the n-GaN implanted photonic crystal.

7.4 Current Injection Efficiency Improvement Technology

The main methods to improve the efficiency of current injection efficiency are current spreading technology, current blocking layer technology, and optimized electrode contact.

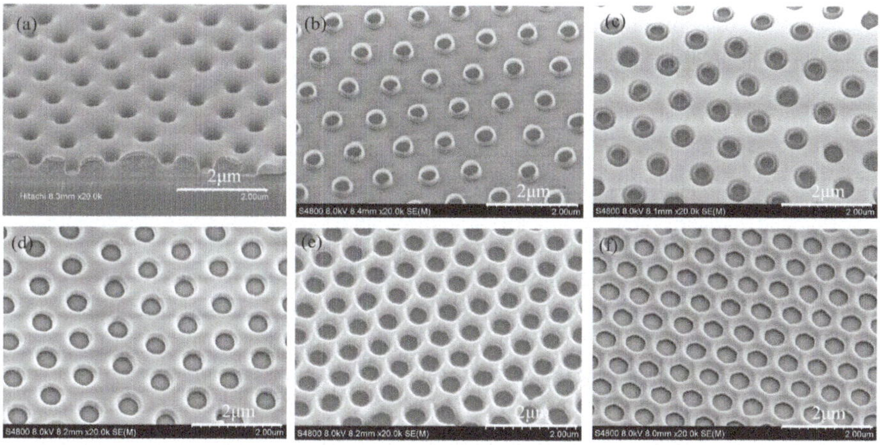

Fig. 7.17 Surface morphology of self-assembled p-GaN photonic crystal LED

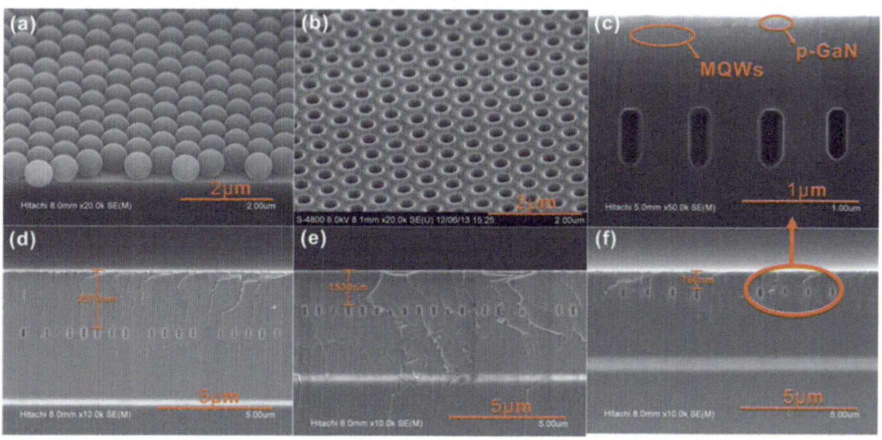

Fig. 7.18 Sectional view of n-GaN implanted photonic crystal

7.4.1 Current Spreading Layer

As a current-driven light-emitting device, the current distribution in the LED during operation can have a large impact on the performance of the LED itself. In one aspect, the injected longitudinal current in the active region of LED is capable of driving LED light emission. The higher the current density within a range not exceeding a certain limit, the larger the number of carriers participating in light emission per unit area is. However, there is no simple linear proportional relationship between them. The proportion of light emitted from different regions of the chip that escapes outside the chip also changes. Therefore, the distribution of current will definitely affect the light efficiency. On the other hand, Joule heat will be generated when current flows inside the chip. Joule heat is generated in places where current is dense. The difference in material and structure will cause different light energy absorbed by different regions of the chip, which will also cause the difference in heat generation. The distribution of current will affect the temperature field of the chip. In many cases, one of the main mechanisms of chip failure is electromigration caused by local overheating, aging of materials, and the like. To this end, it is necessary to conduct in-depth research on various factors that can affect the current distribution of the LED.

Both the p-GaN and n-GaN regions of the LED chip should have a relatively thick layer of electrically conductive material to ensure that the current injected into the chip through the electrode can spread evenly before flowing into the active region. This layer of electrically conductive material is called the "current spreading layer". It is the current spreading layer that determines the difference in current distribution of the LED. For example, in a GaN-based lateral structure LED, the ITO transparent electrode is a current spreading layer on the p-GaN region side. p-GaN is too thin and the conductivity is too poor to function as a current spreading. n-GaN layer is the current spreading layer on n side of the region. In a GaN-based vertical structure LED, n-GaN is a current spreading layer on n type region, and a NiAg metal film on the p side region is almost an ideal conductor. It is, therefore, not necessary to consider the current expansion effect.

The conductivity of the current spreading layer is described by the square resistance in Ω. The smaller the square resistance, the higher the conductivity and stronger the current spreading capability are. E.g. sheet resistance of a commonly used 300 nm thick ITO film on p-GaN is about 10 Ω. The sheet resistance of a 2 μm thick n-GaN is also around this value. The square resistance is inversely proportional to the thickness of the film and proportional to the resistivity of the material. In order to obtain a more uniform current distribution, the square resistance of the current spreading layer should be as low as possible. For a light-emitting device, one should also consider its absorption of light in the design of the current spreading layer. This issue is often related to the square resistance.

7.4.2 Current Distribution Theory

Here are some basic knowledges of the current distribution in LED devices. First, the current-voltage characteristics of the diode need to be introduced since this characteristic is related to the current distribution of the diode.

We know that the current-voltage characteristic curve of an ideal diode can be expressed in the form of Eq. 7.11 [28]. Suppose that GaN-based LED satisfies the current-voltage relationship, the IV curve of LED is shown in Fig. 7.19. Although the actual situation can be slightly different, the trend of voltage changes with current is the same. The dynamic resistance r_{d1} varies with current is also shown in the figure. As the injection current increases, the dynamic resistance r_{d1} gradually decreases.

Fig. 7.19 Diagram of current and voltage of an ideal GaN-based LED

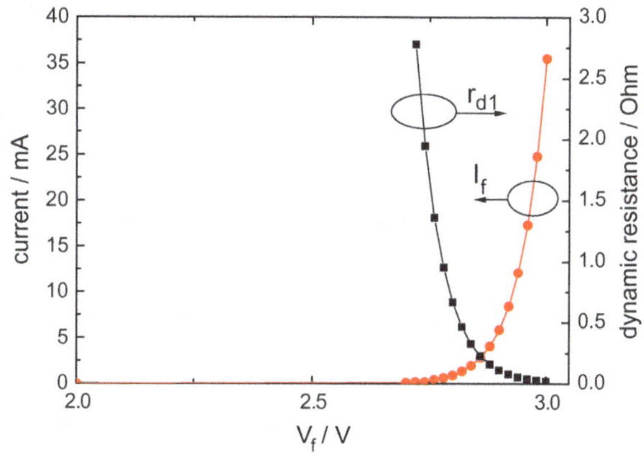

$$J = J_s\left(\exp\left(\frac{qV}{nkT}\right) - 1\right) \tag{7.11}$$

Taking the GaN-based vertical structure LED as an example to calculate the current distribution in the LED chip, the problem can be much simplified because only one current expansion layer needs to be considered while analyzing the current distribution of the vertical structure LED. In the vertical structure LED (Fig. 7.20), since the entire p sides are metal electrodes, only n-type GaN layer acts as a current spreading layer. The current distribution in the current spreading layer is calculated using the current continuity Eq. 7.12 and the drift current Eq. 7.13 [29]. In order to simplify the calculation, it is usually modeled using a two-dimensional structure so that only the x and y directions need to be considered. The following boundary conditions have also to be considered: Longitudinal voltage and current at the boundary of the depletion region of the LED satisfy the PN junction current-voltage characteristic relationship; the current at the nitrogen plane of n-type GaN surface is zero; the voltage at 3 different positions of the n-electrode boundary equals to a specific value.

$$\nabla \cdot J = 0 \tag{7.12}$$

Fig. 7.20 Schematic diagram of GaN-based vertical structure LED chip

$$J_x = q \cdot \mu \cdot n \cdot \frac{dV(x)}{dx} \tag{7.13}$$

The current spreading of the vertical structure LED relies on the expansion of the current in the n-type GaN doped layer. If the longitudinal resistance of n-type GaN is ignored, the longitudinal current distribution of the laterally extended region of the n-type GaN layer electrode can be obtained by mathematical derivation. Since PN junction dynamic resistance is very large at the small current and decreases rapidly with the increase of current, the series resistance of p-type GaN layer and the contact resistance between the electrode and p-GaN can be ignored. With this approximation, the distribution of the injected longitudinal current of the pn junction at a small current density can be expressed by Eq. 7.14, where L_s is the current spreading length representing the current spreading capability which is defined as the lateral distance where the current density drops to a certain percentage of the current density from the edge of the electrode. Its specific situation at small current density is shown in Eq. 7.15 [29].

$$J(x) = \frac{2J_0}{[(x - r_c)/L_s + \sqrt{2}]^2} \tag{7.14}$$

$$L_s = \sqrt{\frac{t_n \cdot n_{ideal} kT}{\rho_n \cdot J_0 e}} = \sqrt{\frac{t_n \cdot r_{d1}}{\rho_n}} \tag{7.15}$$

where J_0 is the current density injected into the PN junction at the same abscissa at the n-electrode side, t_n is the thickness of the current spreading layer, and ρ_n is the resistivity of the current spreading layer. According to Eq. 7.15, in order to obtain a larger extended length, one can reduce the resistivity ρ_n of the current spreading layer, and increase the dynamic resistance r_{d1} and the thickness t_n of the current spreading layer. However, increasing r_{d1} can increases the amount of heat generated by the LED which is not a desirable way to increase the current extension length.

When the vertical structure LED operates at a relatively large current density, the dynamic resistance of PN junction will be very small. When $r_{d1}$is much less than the sum of p-type GaN layer series resistance r_{PS} and the p electrode contact resistance r_{PC}, the series resistance r_{ps} and its contact resistance r_{pc} can be equivalently considered as a constant voltage source of about 2.6 V in series with a resistance of r_{d2}. After this approximation, the longitudinal current distribution of the PN junction injected at a high current density can be derived as shown in Eq. 7.16. In this case, the expression of the current spreading length can be expressed by Eq. 7.17. The dynamic resistance r_{d1} of the ideal PN junction is replaced by the series resistance r_{ps}, contact resistance r_{pc} and r_{d2} [29].

$$J(x) = J_0 \exp(-L_s \cdot x) \tag{7.16}$$

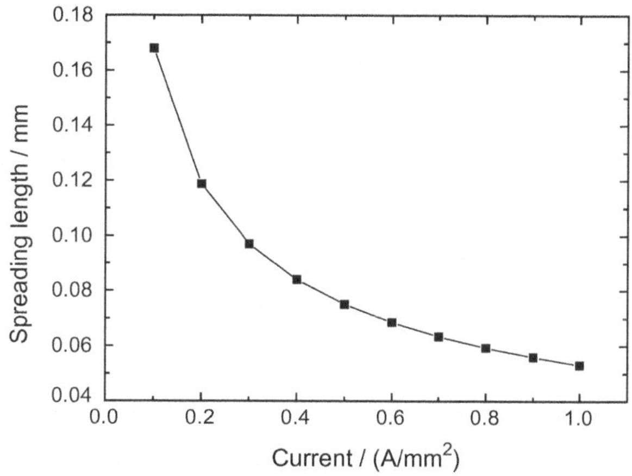

Fig. 7.21 Relationship between current spreading length and current density at low current density

$$L_s = \sqrt{\frac{(\rho_c + \rho_p t_p)t_n}{\rho_n}} = \sqrt{\frac{t_n \cdot r_{d2}}{\rho_n}} \tag{7.17}$$

In the following, the material parameters of the more common GaN-based LEDs and the device process parameters will be substituted into the formula. For a 1 square millimeter chip area with a uniform current injection of 1A, $r_{d1} = 0.168\ \Omega$ is greater than $r_{d2} = 0.103\ \Omega$. The chip still has a working range that is not completely out of the small current operating range. Therefore, for a 1 mm^2 power chip with an injection current of less than 1A, Eqs. 7.14 and 7.15 should be used to calculate the current distribution and current extension length. Figure 7.21 shows the relationship between the current spreading length at a small current density and the longitudinal current density of the pn junction region injected into the boundary of the electrode.

In n-GaN, the current flows laterally away from the n-type metal electrode, and is injected into the quantum well to form a longitudinal current until reaching the boundary of the chip. The remaining current is completely injected into the quantum well. The lateral total current in n-GaN can be obtained by integrating the longitudinal current density from the boundary by distance. Although the distribution of the lateral current is not directly related to the luminous efficiency of the LED, it can generate Joule heat in the chip. The current density is the largest near the n-type metal electrode. If the resistivity of n-GaN is large, it causes severe heat generation and temperature distribution unevenness. Higher current density in active region below n electrode may lead to a considerable proportion of heat generated by non-radiative recombination, Auger recombination, etc. As the LED chip temperature nearby the n electrode is relatively high, the problem of temperature unevenness of the chip is further aggravated.

7.4.3 Current Blocking Technique

There are many discussions about reasons of efficiency droop resulted from Auger recombination [30] and leakage of carriers [31, 32]. In localization in the quantum well is also considered [33]. But there is no mention of the efficiency droop caused by current distribution. From the previous calculation and analysis of the current distribution, it can be seen that the current distribution under the larger current is more uneven. The light absorption from the metal electrode can be large. All of these can lead to LED efficiency droop. The reason for the current fluctuation of the current injected into the multiple quantum wells in the chip changing with the increase of the injection current is that the PN junction dynamic resistance of the LED is not a fixed value which decreases with the increase of the current density. Therefore, the current spreading length in the current spreading layer of the LED is different under different injection currents which have been analyzed in the prior art.

On the other hand, the current density injected into different regions of the multiple quantum wells is different in practical LED devices. Different injection current densities will form different carrier concentrations in the quantum wells. In other words, the internal quantum efficiency is not the same at deferent carrier concentrations. Therefore, there are different internal quantum efficiencies in different regions of the chip. This also indicates that the overall efficiency of the device is related to the distribution of current density [34].

The function of the current blocking layer is to limit the injection current under the n electrode, where the light generated by this partially injected current is difficult to extract. Mathematically, the value of the second term of the denominator in the above formula is reduced to zero, thereby reducing the denominator and increasing the value of the fraction.

A vertical structure LED with a current blocking layer is also modeled to calculate its current distribution. The material structure and parameter of the model are the same as the vertical structure LED model above. Only the structure of the p electrode is changed as shown in Fig. 7.22

As the injection current increases, the dynamic resistance per unit area also changes. As a result, the distribution of current density throughout the chip will change. If the dynamics per unit area decreases as the injection current density

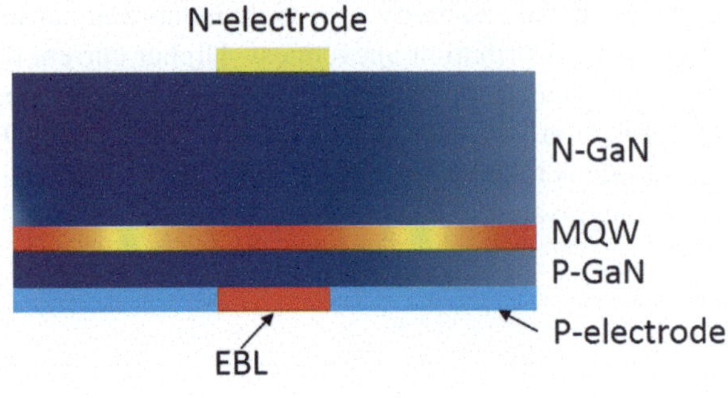

Fig. 7.22 GaN-based vertical LED model with current blocking layer

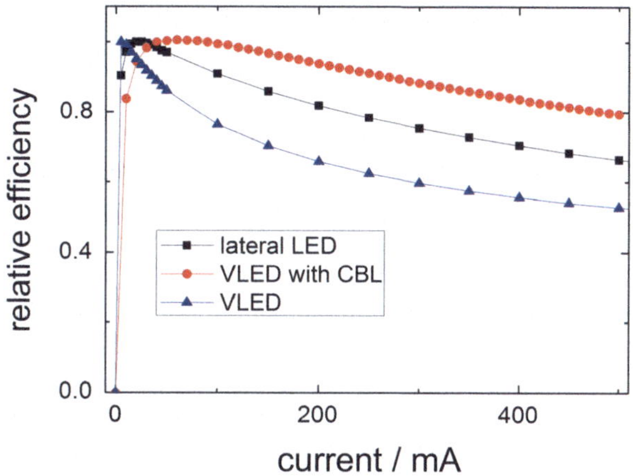

Fig. 7.23 Comparison of LED efficiency with deferent structures

increases, the unevenness of the current distribution will be amplified. In a vertical structure chip without a current blocking layer, a larger proportion of current is distributed under the n electrode as the injection current increases, which is a factor that reduces the internal quantum efficiency. This current distribution varies with the current itself and can cause efficiency droop. In other words, the current blocking layer can limit the effect of this portion of current injection below the n-electrode, which reduces the degree of efficiency droop.

Experiments can prove this phenomenon. Samples based on the same epitaxy structure are fabricated: a sapphire-based lateral structure LED vertical structure LED with p-region current blocking layer, and a vertical structure LED without current blocking layer. The normalization efficiency varies with the injection current as shown in Fig. 7.23. It can be seen that the vertical structure LED with current blocking layer has the least efficiency droop effect. Its efficiency droop is reduced by 43.1% compared with vertical structure LED without current blocking layer. This is because that it has the largest current spreading length, and the current distribution outside the n electrode is more uniform. On the other hand, its current under the n electrode is greatly reduced, and the efficiency droop caused by this part of the current is reduced.

7.5 Droop Effect

The luminous efficiency of InGaN-based LED decreases rapidly as the operating current density increases, i.e. efficiency droop effect [27, 35, 36, 38–44] severely limits the InGaN-based LED applications. In recent years, the efficiency droop effect has received extensive attention from scholars in China and abroad, and has made great progress.

In InGaN MQWs, the radiative recombination of carriers and non-radiative recombination process are schematically shown in in Fig. 7.8, where four processes are highlighted.

(1) SRH recombination: It mainly refers to the recombination process in which the defect level captures carriers. The SRH recombination coefficient A_{SRH} is closely related to the defect density. The SRH recombination term is proportional to the power of the carrier concentration n [35, 36].
(2) Radiation recombination: It refers to the effective recombination process in which electron and hole pairs recombine and produce photons. The radiation recombination B mainly depends on the carrier concentration in the active region and the probability of overlapping of electron and hole wave functions. The recombination term is proportional to the second power of n [37, 38].
(3) Auger recombination: It refers to the recombination of electrons and hole pairs, transferring the generated energy to the third carrier and causing it to transition to a higher energy level, or generating an acoustic wave oscillation to consume the generated energy in the form of phonons. The former is a direct Auger recombination, and the latter is called an indirect Auger recombination. A schematic diagram of these two processes is shown in Fig. 7.24. Since the Auger recombination is a three-part process, the recombination term is proportional to the third power of n. In the InGaN system material, the Auger recombination coefficient theoretically calculated by the first principle is in the order of 10^{-31}–10^{-29} cm^{-6} s^{-1} [39, 40]. Experimentally, Shen et al. reported the coefficient in the order of 10^{-29} cm^{-6} s^{-1} [41] tested by PL. Therefore, in InGaN material devices, Auger recombination is a non-negligible one that plays an important role in the efficiency droop effect [39, 41].
(4) Carrier leakage: This is mainly referred to the process in which part of the electrons injected into the active region crossing the p-type AlGaN electron blocking layer escape into the p-type GaN, and recombine with the holes. Since Mg activation energy in the p-type GaN is very high (>200 meV) [42], the hole concentration of p type GaN is in generally in the range of 1×10^{17}–1×10^{18} cm^{-3}. The p-type layer hole concentration is only 1/100 to 1/10 of the electron

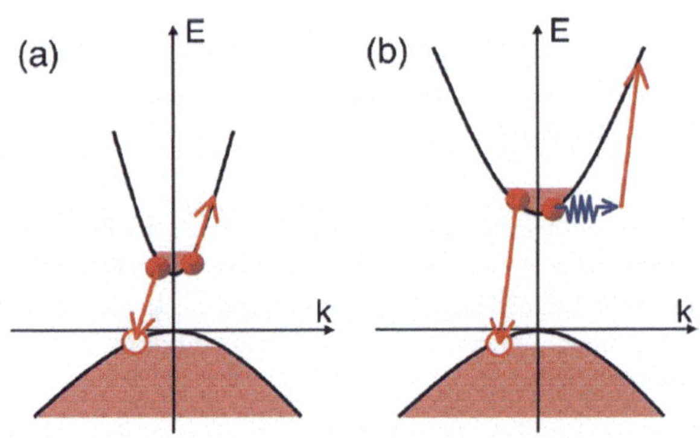

Fig. 7.24 **a** Direct Auger recombination and **b** Indirect Auger recombination Process

concentration (1×10^{19} cm^{-3}) in the n-type GaN. In conventional InGaN-based LEDs, the introduction of a p-type AlGaN electron blocking layer is to restrict excess electrons in the active region from escaping into the p-type layer. However, due to insufficient holes in the active region and excessive electron excess, some electrons will inevitably escape from the active region. The electron blocking layer is restricted to reach the p-type layer. This electron leakage phenomenon is more pronounced under high current drive. Electron leakage also causes insufficient hole injection, which further restricts the injection of holes in the p-layer, resulting in a decrease in luminous efficiency. Recent studies have confirmed that carrier leakage is a key factor in the effects droop effect [43, 44].

7.5.1 Auger Recombination Effect

Shen et al. [45] from Philips Lumileds reported that Auger recombination dominates at high injection current, which is a main cause for LED light emission efficiency droop. Auger recombination is a non-radiative recombination process. The energy generated by the recombination of electrons and holes does not generate photons, but transfers energy to another carrier (electrons or holes), causing this carrier to transition to the higher energy level. This energy is then available in the form of phonons to release. The Auger recombination probability is proportional to n^2p or np^2 (n and p are electron concentration and hole concentration, respectively). Therefore, as the injection current in the LED increases, the carrier concentration also increases, causing the Auger recombination probability to become bigger. In response to this mechanism, Lumileds proposed to increase the width of the quantum well layer or use a double heterojunction instead of a multiple quantum well as the active region to reduce the carrier concentration therein, thereby weakening the efficiency droop [46].

The traditional ABC theory is a traditional model used to describe efficiency droop. In this model, the carrier leakage term is not taken into account, and the carrier recombination rate equation $R\ (n)$ can be expressed as [37]:

$$R(n) = A_{SRH}n + Bn^2 + C_{Auger}n^3 \tag{7.18}$$

$A_{SRH}n$ is SRH recombination, Bn^2 is radiative recombination, and $C_{Auger}n^3$ is Auger recombination. Based on this, Internal Quantum Efficiency (IQE) in the traditional ABC model can be expressed by the following equation:

$$IQE = \frac{Bn^2}{A_{SRH}n + Bn^2 + C_{Auger}n^3} \tag{7.19}$$

For simple conversion, IQE can be expressed as:

$$IQE = \frac{B}{A_{SRH}/n + B + C_{Auger}n} \tag{7.20}$$

This equation shows that the IQE ~ n curve is center-symmetric about $n_0 = \sqrt{\frac{A_{SRH}}{C_{Auger}}}$. In fact, the IQE ~ n curve of typical InGaN blue-green LEDs exhibit non-central symmetry. At $n > n_0$ (the carrier concentration corresponding to the IQE peak), the IQE decays rapidly. We fit the typical InGaN blue and green LEDs using the traditional ABC model. At the higher carrier concentration, the experimentally measured IQE is lower than that obtained by fitting based on ABC model [37]. This deviation is more pronounced as the carrier concentration increases. The above problems occur with the typical InGaN blue and green LEDs. Therefore, the traditional ABC equation is not sufficient to fully describe the IQE characteristics, which is theoretically insufficient, especially in the case of a large carrier concentration [37, 47]. It can be speculated that in the process of carrier radiation recombination, a higher order term $f(n)$ should also be introduced to explain the asymmetry of the IQE-n curve [47].

7.5.2 Electronic Overflow

Schubert et al. from US Leinster Polytechnic Institute believe that the leakage of electrons from the active region is the main cause of efficiency droop [48].

In the upper 7.18 equation, a higher order compound term $f(n)$ is introduced, and $R(n)$ can be expressed as [49, 50]:

$$R(n) = A_{SRH}n + Bn^2 + C_{Auger}n^3 + f(n) \tag{7.21}$$

among them, $f(n) = Dn^4 + En^5 \ldots$

We can get:

$$IQE = \frac{Bn^2}{A_{SRH}n + Bn^2 + C_{Auger}n^3 + f(n)} \tag{7.22}$$

Using the most primitive pn junction theory [50], a higher-order term Dn^4 is introduced in the carrier rate equation. Such a term may be used to describe the characteristics of the carrier leak with certain physical meaning. Using this equation to simulate, the agreement between the IQE ~ n curve and the experimental results has been significantly improved. Theoretically, the non-central characteristics of IQE ~ n are caused by carrier leakage. This theoretical model is called the ABCD model, Dn^4 is the carrier leakage term [50]. In the ABCD model, IQE can be expressed as:

$$IQE = \frac{Bn^2}{A_{SRH}n + Bn^2 + C_{Auger}n^3 + Dn^4}$$

$$= \frac{B}{A_{SRH}/n + B + C_{Auger}n + Dn^2} \tag{7.23}$$

The physical process of introducing high-order terms into the rate equation as a leakage term is derived from the PN junction theory. The leakage current mainly refers to the leakage current from the n-type leakage to the p-type, which is caused by electron leakage which is defined as [51]:

$$J_{Drift}(n->p) = e\mu_n \Delta n_p(0)E = e\mu_n \Delta n_p(0)\frac{J_{total}}{\sigma_p} = e\frac{\mu_n \Delta n_p(0)}{eP_{P0}\mu_p}J_{total} \tag{7.24}$$

The proportion of leaked carriers is $\delta = \frac{\Delta n_p(0)}{n_{QW}}$. Therefore, the leakage current is equal to:

$$J_{Drift}(n->p) = (\frac{\delta\mu_n}{P_{P0}\mu_p})n_{QW}J_{total} \tag{7.25}$$

The carrier leakage $\Delta n_p(0)$ is determined by the following aspects:

(1) The hot carrier leakage term can be expressed as:

$$\Delta n_p(0) = n_{QW}(-\Delta_{Barrier}/KT) = \delta n_{QW} \tag{7.26}$$

At lattice temperature of 300 K and carrier temperature of 425 K [51], the $\Delta_{Barrier}$ is 300 meV, and δ is estimated to be $3 \times 10^{-4.}$

(2) Tunneling current related to defects in the quantum barrier [52];
(3) Electron leakage caused by polarized electric field;

Based on the above considerations, the δ value is approximately about 0.1%. The total current flowing through the PN junction can be expressed as:

$$J_{Total} = ed_{active}R \tag{7.27}$$

Substitute above equation into 7.26 to get the leakage current term:

$$J_{Drift}(n->p) = (\frac{\delta\mu_n}{P_{P0}\mu_p})n_{QW}ed_{active}R \tag{7.28}$$

In the radiation recombination equation of the PN junction active region, $R\,(n)$ can be mainly divided into a radiation recombination dominant region and a non-radiative recombination dominant region.

(1) Radiation recombination dominant part:

$$J_{Total} = ed_{active}R \approx ed_{active}Bn^2 \tag{7.29}$$

Bring the drift current Eq. 7.28 to calculate the drift current:

$$J_{Drift}(n->p) = (\frac{\delta\mu_n}{P_{PO}\mu_p})n_{QW}ed_{active}Bn_{QW}^2 = ed_{active}(\frac{\delta\mu_n}{P_{PO}\mu_p})Bn_{QW}^3$$
$$= ed_{active}C_{DL}n_{QW}^3 \tag{7.30}$$

Among them, $C_{DL} = (\frac{\delta\mu_n}{P_{PO}\mu_p})B$. Combined with the actual situation of the GaN-based PN junction, the value is as follows [50]:

$$\delta = \tag{0.1 \%,}$$

$\mu_n = 300$ cm$^{2/}$(V s), $\mu_p = 2.5$ cm $^{2/}$(V s), $P_{PO} = 5 \times 10^{17}$ cm^{-3}, $B = 1 \times 10^{10}$ cm^3 s^{-1}.

C_{DL} is estimated to be 2.4×10^{-29} cm^{-6} S^{-1}. This value is basically consistent with the theoretical calculation and the measured Auger coefficient [42–44], indicating the accuracy of this method.

(2) n^3 dominant part:

$$J_{Total} = ed_{active}R \approx ed_{active}Cn^3 \tag{7.31}$$

Bring the drift current Eq. 7.28 to calculate the drift current:

$$J_{Drift}(n->p) = (\frac{\delta\mu_n}{P_{PO}\mu_p})n_{QW}ed_{active}Cn_{QW}^2 = ed_{active}(\frac{\delta\mu_n}{P_{PO}\mu_p})Cn_{QW}^4$$
$$= ed_{active}D_{DL}n_{QW}^4 \tag{7.32}$$

Among them $D_{DL} = (\frac{\delta\mu_n}{P_{PO}\mu_p})C$. Similarly, take the value according to the traditional D_{DL} value:

$$\delta = \tag{0.1 \%,}$$

$\mu_n = 300$ cm^2/(V s), $\mu_p = 2.5$ cm ^{2}/(V s), $P_{PO} = 5 \times 10^{17}$ cm^{-3}, $C = 1 \times 10^{-29}$ cm^{-6} s^{-1}.

The D_{DL} result is 1.38×10^{-48} cm^{-9} s^{-1}.

In summary, electronic leakage term Dn^4 plays an important role in the LED efficiency droop effect under high-current densities. Such an effect can be even more pronounced than Auger recombination.

References

1. J. Ho, C. Jong, C. Chien et al., Low-resistance ohmic contacts to p-type GaN. Appl. Phys. Lett. **74**(9), 1275 (1999)
2. J. Ha, J. Jang, M. Joo et al., Investigation of optimized ohmic contact of direct ITO layer with synchrotron radiation analysis. Phys. Stat. Sol. (c), **3**(6), 1828–1831 (2006)
3. S. Chang, C. Chang, Y. Su et al., Highly reliable nitride-based LEDs with SPS + ITO upper contacts. IEEE J. Quant. Electron. **39**(11) (2003)
4. O. Shchekin, J. Epler, T. Trottier et al., High performance thin-film flip-chip InGaN-GaN light-emitting diodes. Appl. Phys. Lett. **89**, 071109 (2006)
5. K. Tamura, K. Nakahara, M. Sakai et al., InGaN-based light-emitting diodcs fabricated with transparent Ga-doped ZnO as ohmic p-contact. Phys. Stat. Sol. (a) **201**(12), 2704–2707 (2004)
6. http://www.cree.com/News-and-Events/Cree-News/Press-Releases/2014/March/300LPW-LED-barrier
7. P. Vennegues, B. Beaumont, V. Bousquet et al, Reduction mechanisms for defect densities in GaN using one or two step epitaxial lateral overgrowth method. J. Appl. Phys. **87**, 4175 (2000)
8. J. Jasinskil, Z. Liliental-Weber, Extended defects and polarity of hydride vapor phase epitaxy GaN. J. Electron. Mater. **31**(5), 230–235 (2002)
9. V. Fiorentini, F. Bernardini, O. Ambacher, Evidence for nonlinear macroscopic polarization in III–V nitride alloy heterostructures. Appl. Phys. Lett. **80**, 1204 (2002)
10. F. Bechstedt, U. Grossner, J. Furthmuller, Dynamics and polarization of group-III nitride lattices: A first-principles study. Phys. Rev. B **62**, 8003 (2000)
11. F. Bernardin, V. Fiorentini, Spontaneous versus piezoelectric polarization in III-V nitrides: conceptual aspects and practical consequences. Phys. Stat. Sol. (b) **216**, 391 (1999)
12. H. Li, P. Li, J. Kang et al., Quantum efficiency enhancement of 530 nm InGaN green light emitting diodes with shallow quantum well. Appl. Phys. Express **6**(5), 052102 (2013)
13. C. Chen, Y. Su, S. Chang et al., High brightness green light emitting diodes with charge asymmetric resonance tunneling structure. IEEE Electron Dev. Lett. **23**(3), 130–132 (2002)
14. T. Wen, S. Chang, L. Wu et al., InGaN/GaN tunnel-injection blue light-emitting diodes. IEEE Trans. Electron Dev. **49**(6), 1093–1095 (2002)
15. X. Cao, E. Stokes, P. Sandvik et al., Diffusion and tunneling currents in GaN/InGaN multiple quantum well light-emitting diodes. IEEE Electron Dev. Lett. **23**(9), 535–537 (2002)
16. H. Hsu, Y. Su, S. Huang et al., Enhanced performance of nitride-based blue LED with step-stage MQW structure. IEEE Photon. Technol. Lett. **23**(5), 287–289 (2011)
17. S. Huang, Y. Su, C. Tseng et al., Improvement of light intensity for nitride-based multi-quantum well light emitting diodes by stepwise-stage electron emitting layer. Appl. Phys. Exp. **3**(12), 122106 (2010)
18. E. Park, D. Kang, I. Ferguson et al., The effect of silicon doping in the selected barrier on the electroluminescence of InGaN/GaN multiquantum well light emitting diode. Appl. Phys. Lett. **90**, 031102 (2007)
19. N. Zhang, Z. Liu, T. Wei et al., Effect of the graded electron blocking layer on the emission properties of GaN-based green light-emitting diodes. Appl. Phys. Lett. **100**, 053504 (2012)
20. Z. Liu, J. Ma, X. Yi et al., p-InGaN/AlGaN electron blocking layer for InGaN/GaN blue light-emitting diodes. Appl. Phys. Lett. **101**, 261106 (2012)
21. Z. Si, T. Wei, J. Ma et al., Modification of carrier distribution in dual-wavelength light-emitting diodes by specified mg doped barrier. ECS Solid State Lett. **2**(10), R37–R39 (2013)
22. J. Kang, H. Li, Z. Li et al., Enhancing the performance of green GaN-based light-emitting diodes with graded superlattice AlGaN/GaN inserting layer. Appl. Phys. Lett. **103**, 102104 (2013)
23. L. Meng, W. Guohong, L. Hongjian et al., Low threading dislocation density in GaN films grown on patterned sapphire substrates. J. Semicond. **33**(11), 113002 (2012)
24. Y. Zhang et al., Enhancement in the light output power of GaN-based light-emitting diodes with nanotextured indium tin oxide layer using self-assembled cesium chloride nanospheres. Jpn. J. Appl. Phys. **51**(2), 1 (2012)

25. Y. Houng, M. Tan, B. Liang et al., InGaAs/GaAs vertical cavitv surface emitting lasers grown by gas source molecular beam epitaxy. J. Cryst. Growth **136**(1–4), 216–220 (1994)
26. H. Shen Tuwei, H.Y. Fei et al., Study on light extraction efficiency of GaN-based LED chips (in Chinese). Optoelectron. Laser **16**(4), 385–389 (2005)
27. L. Yongchang et al., *Principles of Optical Thin Films* (in Chinese), National Defense Industry Press (1990)
28. W. Shockley, The theory of pn junctions in semiconductors and pn junctio n transistors. Bell Syst. Tech. J. **28**, 435 (1949)
29. X. Guo, E.F. Schubert et al., Current crowding and optical saturation effects in GaInN/GaN light emitting diodes grown on insulating substrates. Appl. Phys. Lett. **78**(21), 3337 (2001)
30. Y. Shen, Mueller, S. Watanabe et al., Auger recombination in InGaN measured by photoluminescence. Appl. Phys. Lett. **91** (14), 141101 (2007)
31. X. Ni, Q. Fan, R. Shimada et al., Reduction of efficiency droop in InGaN light emitting diodes by coupled quantum wells. Appl. Phys. Lett. **93**(17), 171113 (2008)
32. Y. Li, Y. Huang, Y. Lai et al., Investigation of efficiency droop behaviors of InGaN/GaN multiple-quantum-well LEDs with various well thicknesses. IEEE J. Sel. Top. Quantum Electron. **15**(4), 1128–1131 (2009)
33. B. Monemara, B. Sernelius, Defect related issues in the "current roll-off" in InGaN based light emitting diodes. Appl. Phys. Lett. **91**(18), 181103 (2007)
34. H. Ryu, J. Shim, Effect of current spreading on the efficiency droop of InGaN light-emitting diodes. Opt. Exp. **19**(4), 2886–2894 (2011)
35. A. David, M. Grundmann, Droop in InGaN light emitting diodes: a differential carrier lifetime analysis. Appl. Phys. Lett. **96**(10), 103504 (2010)
36. M.F. Schubert, S. Chhajed, J. Kim et al., Effect of dislocation density on efficiency droop in GaInN/GaN light emitting diodes. Appl. Phys. Lett. **91**, 231114 (2007)
37. J. Campbell, P. Lenahan, A. Krishnan et al., Identification of atomic scale defect structure involved in the negative bias temperature instability in plasma-nitride devices. Appl. Phys. Lett. **97**, 133507 (2010)
38. M.F. Schubert, J. Xu, J. Kim et al., Polarization matched GaInN/AlGaInN multi-quantum-well light emitting diodes with reduced efficiency droop. Appl. Phys. Lett. **93**, 041102 (2008)
39. K. Delaney, P. Rinke, C. Van de Walle, Auger recombination rates in nitrides from first principles. Appl. Phys. Lett. **94**, 91109 (2009)
40. M. Meneghini, N. Trivellin, M. Pavesi et al., Leakage current and reverse-bias luminescence in InGaN-based light emitting diodes. Appl. Phys. Lett. **95**(17), 173507 (2009)
41. N. Gardner, G. Muller, Y. Shen et al., Blue emitting InGaN-GaN double heterostructure light emitting diodes reaching maximum quantum efficiency above 200A/cm^2. Appl. Phys. Lett. **91**(24), 243506 (2007)
42. D. Iida, K. Tamura, M. Iwaya et al., Compensation effect of Mg-doped a- and c-plane GaN films grown by metalorganic vapor phase epitaxy. J. Cryst. Growth **312**(21), 3131–3135 (2010)
43. J. Xie, X. Ni, Q. Fan et al., On the efficiency droop in InGaN multiple quantum well blue light emitting diodes and its reduction with p-doped quantum well barriers. Appl. Phys. Lett. **93**, 121107 (2008)
44. H. Li, J. Kang, P. Li et al., Enhanced performance of GaN based light emitting diodes with a low temperature p-GaN hole injection layer. Appl. Phys. Lett. **102**, 011105 (2013)
45. http://spectrum.IEEE.org/semiconductors/optoelectronics/the-leds-dark-secret
46. Q. Dai, Q. Shan, J. Cho, et al., On the symmetry of efficiency-versus-carrier-concentration curves in GaInN/GaN light emitting diodes and relation to droop-causing mechanism. Appl. Phys. Lett. **98**(3), 033506 (2011)
47. M. Kim, M.F. Schubert, Q. Dai et al., Origin of efficiency droop in GaN-based light emitting diodes. Appl. Phys. Lett. **91**, 183507 (2007)
48. J. Cho, E.F. Schubert, J. Kim et al., Efficiency droop in light emitting diodes: Challenges and countermeasures. Laser Photon. Rev. **7**(3), 408–421 (2013)
49. G. Lin, D. Meyaard, J. Cho et al., Analytic model for the efficiency droop in semiconductors with asymmetric carrier-transport properties based on drift-induced reduction of injection efficiency. Appl. Phys. Lett. **100**(16), 161106 (2012)

50. E.F. Schubert, *Light-Emitting Diodes*. 2nd ed. Cambridge University Press (2006)
51. X. Cao, S. LeBoeuf, K. Kim et al., Investigation of radiative tunneling in GaN/InGaN single quantum well light-emitting diodes. Solid-State Electron. **46**(12), 2291–2294 (2002)
52. H. Chung, R. Choi, M. Kim et al., Improved performance of GaN-based blue light emitting diodes with InGaN/GaN multilayer barriers. Appl. Phys. Lett. **95**(24), 241109 (2007)

Chapter 8
III-Nitride LED Chip Fabrication Techniques

With the development of epitaxial growth technology and multi-quantum well structure, the internal quantum efficiency of ultra-high brightness LED has been greatly improved. For example, the internal quantum efficiency of the 625 nm AlGaInP-based LED can reach almost 100%. However, defects caused by the lattice and thermal mismatch, stress and electric field which present in GaN-based materials make the internal quantum efficiency of Group III nitride (AlGaInN-based) LEDs relatively low. For instance, a quantum efficiency typically in the range between 35 and 50% is achieved. Therefore, increasing the external quantum efficiency of the LED chips is the key to improve the luminous efficiency. This largely requires the proper designs of a new chip structure to improve the light-emitting efficiency, thereby achieving much enhanced luminous efficiency (or external quantum efficiency). In the development of III-nitride LED technology for semiconductor illumination, many technical and theoretical issues including material epitaxy, p-doping and activation, material etching, ohmic contact, photoelectric characteristics, current spreading, light-emitting structure, and device protection are being solved or continuously advanced. Mature theories and processes such as p-type GaN annealing activation, ICP etching methods, heavily doped top layer preparation for ohmic contacts, electrode structures, etc. have become the standard in current laboratories and factories, driving the science and technology of semiconductor lighting technology. The key fabrication process of the III-nitride LED described in this chapter is a combination of the basic fabrication process and the process technologies.

8.1 Group III Nitride LED Fabrication Process

Since sapphire is an insulator with very stable chemical and physical properties, the p- and n-type electrodes of the GaN-based LED chip epitaxially grown on sapphire substrates are located on the same side of the epitaxial structure. This structure is

J. Li et al., *III-Nitrides Light Emitting Diodes: Technology and Applications*,
Springer Series in Materials Science 306,
https://doi.org/10.1007/978-981-15-7949-3_8

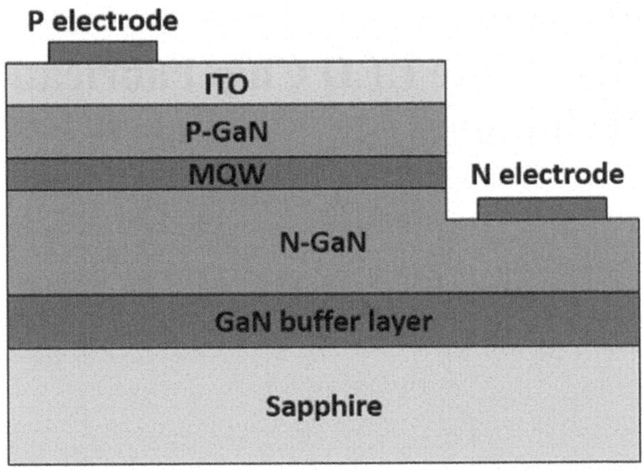

Fig. 8.1 Schematic diagram of a lateral structure GaN-based LED chip

called a lateral structure LED. The III-nitride LED structure includes a sapphire substrate layer, an n-type doped layer (n-GaN), a multiple quantum well (MQW) light-emitting layer, a p-type doped layer (p-GaN), an indium tin oxide (ITO) transparent conductive layer, and metal contacts. In the manufacturing process of the lateral structure LED chip, an ohmic contact metal is deposited on the surface of the p-GaN epitaxial layer to form a p-type electrode. The part of the p-GaN and the multiple quantum well MQW are etched by the ICP process until the n-GaN material is exposed. Then metal contacts are deposited on n-GaN. Figure 8.1 is a schematic diagram of a typical lateral structure LED chip. The chip fabrication process of the lateral structure LED design mainly includes a photolithography process, an etching process, metal evaporation, and sputtering.

The fabrication processes for LED chips mainly include photolithography, etching process, thin film deposition, rapid thermal annealing, grinding and polishing processes. In this chapter, the main basic single-step process mentioned above is introduced, and the problems that need to be addressed in the processing of LED chip are given. The main ohmic contact problem among III-nitride LEDs is also discussed.

8.2 Photolithography

Lithography (Photolithography) is a technique used to remove the area of materials particular defined on the wafer. The essence is to copy the pattern made on the photomask onto the wafer to be etched later. The principle is similar to that of photography, except that the semiconductor wafer and the photoresist replace the photographic film and the photosensitive coating. Lithography is a temporary graphics transfer process. The lithography or etching is used as a photomask to complete the permanent transfer of the pattern. Lithography is one of the most important process steps in LED fabrication. The photomasks and photoresists are the core of the lithography process. All operations in the lithography process are based on the specific photoresist properties and fine-tuning. The design of the photomasks, the selection of the photoresists and the development of the lithography process are a very long process. A complete and mature lithography process needs to be continuously optimized and adjusted. The lithography process is important in the LED fabrication.

8.2.1 Mask and Photoresist

First, the photolithography process requires the wafer to be as close as possible to the design patterns. The graphics of this design are realized by the photomasks containing all the elements from the pattern layout, so the lithographic quality of the photomasks plays a very important role in the lithography process. The pattern with photomasks is the one will be eventually implemented on the wafer surface. Usually LED wafer fabrication process includes multiple photolithography processes. The device pattern on the LED wafer is typically established by a number of specific masks.

Photoresist is a light-sensitive organic compound composed of a photosensitive resin, a sensitizer and a solvent. When such a photosensitive material is subject to the UV exposure, the solubility of the photoresist in the developing solution will change accordingly. The photoresist used in device fabrication is usually applied to the surface of a silicon wafer in a liquid state and then dried to be a gel-like film.

According to the chemical reaction mechanism and development principle of the photoresist, it can be divided into a positive photoresist and a negative photoresist. Negative photoresist is used in the early development of lithography process. Its process cost is low and the output is high. However, since it will expand after absorbing the developer, its resolution is not as good as that of positive photoresist. Therefore, for sub-micron or even smaller size, positive photoresist is most widely sued. The comparison between the negative and positive photoresists is shown in Table 8.1.

A negative photoresist becomes insoluble after exposure. Most negative lithographic photoresist is polyisoprene type which becomes cross-linked polymers after UV exposure. Negative photoresists have good chemical resistance properties, the non-exposed portion will dissolve in the developing solution.

Table 8.1 Comparison of positive and negative photoresist

Negative photoresist	Positive photoresist
Insoluble after exposure	Soluble after exposure
Unexposed in development is dissolved	The exposure is dissolved during development
Cheap	High resolution

Positive photoresist, after exposing to UV light, becomes soluble in the developing solution. The positive gel is generally composed of a novella resin polymer. The cross-linked light sensitizer contained in the resin can decompose the sensitizer and break the cross-linking chemical bond, thereby becoming soluble in the developer. The unexposed positive photoresist is in-soluble in the developer. In other words, the pattern after exposing and developing will be the same as the mask pattern. Due to the lack of expansion, positive photoresist generally achieves higher resolution in LED manufacturing as well as in traditional IC manufacturing applications.

8.2.2 Lithography Process

Lithography is usually done in a series of steps. Each step must be precise to achieve a desired lithography process. The specific experimental steps are shown in Fig. 8.2.

- Surface treatment

Surface treatment is the first step in the lithography process. The main purpose of such a pricess is to treat the surface to enhance its adhesion to the photoresist. The wafer readily adsorbs moisture to its surface during processing and transfer. Photoresist adhesion requires a strictly dry surface. Dewatering baking and adhesives coating should be applied before spin-coating of photoresist. The temperature for dehydration baking is usually between 140 and 200°. HMDS (hexamethyldisilazane) is usually used as an adhesive agent. The main function of the surface treatment is to remove contaminants and particles, and to improve the adhesion between the photoresist and

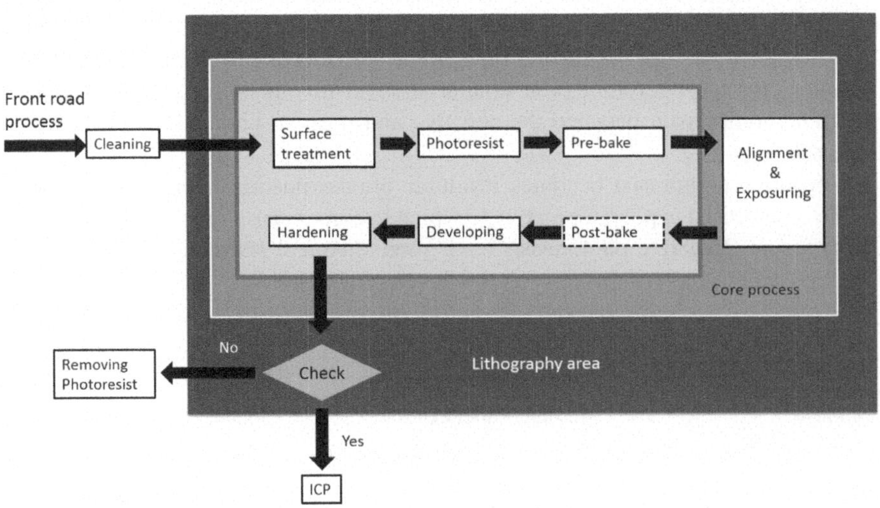

Fig. 8.2 LED lithography process

the substrate so that the photoresist is not infiltrated by the liquid developer during development process.

- Spin coating

In the lithography process, spin-coating is generally used for coating photoresist. The wafer is placed on the vacuum chuck, and the photoresist is sprayed at the center of the wafer. The photoresist spreads outwardly at a high spin speed and is uniformly coated on the surface of the wafer. A photoresist film is finally formed as a result of the surface tension of the photoresist and the centrifugal force of rotation. The higher the rate of rotation, the more uniform of the coating. During the coating process, the photoresist spreads outward on the surface of the rotating wafer. Usually, there is a different rate to determine the thickness of the photoresist. First, there will be a low speed rotation (500 rpm), and then slowly rise to (~3000 to 7000 rpm). The rate is adjusted according to the needs of the photoresist thickness. The final thickness of the photoresist film is determined by the viscosity of the photoresist, the rotation speed, the surface tension, and the dryness of the photoresist.

- Pre-bake

Since the as coated photoresist is fresh and soft, it cannot be directly exposed after the coating. It must be baked to evaporate the solvent in the photoresist. The baked photoresist remains "soft". However, it will adhere to LED wafers more firmly. The purpose of pre-baking is to evaporate the organic solvent component to cure the photoresist on the surface of the LED wafer. The role of the solvent in the photoresist is to make the applied photoresist thinner, but it also absorbs heat and affects the adhesion of the photoresist. After pre-baking, the solvents are substantially removed, thereby the as-coated photoresist becomes thin thickness (a reduction of approximately 25%). This baking process must avoid excessive degree of baking since over-baking will cause polymerization of photoresists and decrease photosensitive sensitivity. This will thus influence the adhesion and exposure.

- Alignment and exposure

Alignment is the positioning or alignment of the desired pattern from the photomask on the surface of the LED wafer. The exposure of a substrate coated with a photoresist to the UV light or other radiation is to initiate a photochemical reaction so that the optical properties of photoresist are changed by the received light. Such a process allows the transfer of the pattern from the photo-mask to the photoresist coating layer. If the photoresist is the "material" core of the lithography process, then alignment and exposure are the "device" core of the process. Accurate alignment is one of the decisive factors in ensuring proper operation of the LED device.

The initial exposure equipment was a contact lithography machine and a proximity lithography machine. The lithography machines have currently evolved into two major types: optical lithography machines and non-optical lithography machines. The optical lithography machine uses ultraviolet light as the light source, while the non-optical lithography machine uses light sources that comes from other components of the electromagnetic spectrum (X-rays, electron beams, etc.). Exposure equipment typically consists of two systems for alignment and exposure: one is to accurately position the pattern on the surface of the wafer (different alignment systems of different alignment types); the other is the exposure system (including an exposure source) and a radiation beam directed onto the wafer. The performance of the alignment machine is reflected in the resolution and registration capabilities. Resolution is the ability of a machine to produce a specific size. The higher the resolution, the better the performance of the machine is. Registration capability is the ability to accurately locate graphics. Light for exposure must pass mirrors and lenses, so it is converted into parallel light beam so as to ensure the required feature size. The most widely used exposure source is a high-pressure mercury lamp which produces ultraviolet light (UV). To achieve higher resolution, the photoresist is designed to react with only a narrow wavelength of light in the spectrum of the mercury lamp (Called the deep ultraviolet region or DUV).

- Development

The development of the photoresist is to dissolve the soluble region of the photoresist by a chemical agent so called developer. The main purpose of the photoresist is to accurately transfer the mask pattern into the photoresist. There are generally three steps in this process: development, rinsing, and drying. Common developing solutions are NaOH (Shipley 351), KOH (Shipley 606), TMAH (Shipley CD-26, MF-321, OCG 945) and so on.

- Post-bake

When the light is irradiated onto the interface between the photoresist and the wafer, a part of the reflected light and the incident light are superimposed to form a standing wave. Post-baking partially eliminates this effect. At the same time, the photoresist side wall can be smoothened to improve the resolution.

- Hard-bake

The hard bake process is to make the film adhere to the surface of the wafer more firmly by heating and baking. Such a process can increase the etching resistance of the photoresist. Hard bake is not a necessary process. This step is to improve the stability in the etching process. It can also improve the adhesion of the photoresist to the wafer surface, which is beneficial to the subsequent wet etching process. It can also improve the pinholes present in the photoresist. Of course, there are certain drawbacks resulted from this step: it may cause the photoresist to reflow, which will reduce the accuracy of the pattern transferred. Also, it will increase the difficulty of removing the photoresist. The post-baking time is therefore critical. For a given photoresist, the hard-bake temperature is usually higher than the pre-bake temperature.

- Pattern check

There will be development failure, poor exposure, uneven coating problems in the lithography process. There are also problems such as misalignment overlap and misalignment during exposure. Furthermore, scratches, pinholes, defects and contaminants on the surface of the wafer itself are fatal to lithography. Lithography is an accurate art that requires precise conditions and operations to achieve near-perfect graphics transfer. After each step, inspections are carried out. The wafers pass the inspections can move into the next process. The unqualified wafers can be re-processed for lithography steps. Through monitoring the pattern, the lithography pass rate can be effectively improved.

- Cleaning

After the pattern transferred, the photoresist has completed its mission and needs to be removed. There are two kinds of photoresist removal methods: wet method and dry method. Traditional LED fabrication processes typically employ a wet chemical process. The wet method uses a variety of acid-based solutions or organic solvents to etch away the photoresist. The most common solvent is acetone, which dissolves most of the photoresist. Dry method is to etch photoresist with an oxygen plasma ashing (strong oxidizing ashing).

8.3 Etching Process

According to the pattern defined by the lithography process, the photoresist is used as a protection layer to achieve the designed pattern on the photoresist to the underlying materials. Etching process is generally divided into wet etching and dry etching.

In the LED chip manufacturing process, etching process is very important. Since GaN materials are very stable at room temperature and have high corrosion resistance, it is difficult to use wet etching at room temperature. Inductively Coupled Plasma (ICP) etching is used for GaN materials. The following section describe some of the basic parameters used in the etching process. The wet etching and the dry etching are then introduced respectively. Finally, the etching of the GaN epitaxial layer, the ITO transparent conductive film and the silicon dioxide (SiO_2) mask material involved in the manufacturing process of the group III nitride LED chip is discussed. The etching technique of the above materials is introduced below.

8.3.1 Etching Parameters

- Etch rate

Etching speed refers to the materials removed per unit time from the etched surface of the silicon wafer.

- Uniformity

Uniformity in etching includes both the uniformity of the entire surface and the lateral erosion of the sidewalls.

- Etch selectivity ratio

The etch selectivity ratio is defined as the ratio of the etch rate of the etched material to the etch rate of the mask material. The etching option can be calculated by the following formula 8.1:

$$S = D/\mathrm{d} \tag{8.1}$$

where S is the etching selectivity ratio, D is the etching depth of the material to be etched per unit time, sand d is the etching depth of the mask material per unit time. In order to fully transfer the pattern onto the material to be etched, it is necessary to select an etching process and a mask material with a high etching selectivity.

8.3.2 Wet Etching and Dry Etching

Wet etching is a technique in which an etching material is immersed in an etchant for corrosion. A specific chemical solution is used to dissolve a material. Wet etching is a purely chemical process that leads to an isotropic sidewall morphology and the high selectivity. Wet etching is divided into three basic steps: corrosion, cleaning, and drying.

Wet etching is a traditional process that has been widely used in IC manufacturing at feature sizes greater than 3 microns. It has been replaced by dry (plasma) etching in the etching of passivation layer (SiO_2) in LED fabrication as well as in some un-patterned film removal, such as the removal of silicon nitride and titanium. Wet etching is also used to remove metal layers during rework experiments.

Dry etching is a technique of performing plasma etching using plasma. The high-frequency RF is used to activate the reaction gas into active ions, where they react with the materials to be etched to form a volatile product and are removed. It has the advantage of a fast etch rate while achieving good physical topography. The characteristics of dry etching are high resolution and strong anisotropy. Dry etching is further divided into physical etching (sputter etching IBE), chemical etching (or

Table 8.2 Comparison of three etching methods for dry etching

Chemical etching	Physical etching	Physicochemical etching
• Purification reaction • The reaction product is a gas • High selection ratio • Isotropic appearance • Cases such as: – Dry method to remove glue – Silicon nitride removal for LOCOS and STI	• Was physical reaction: removing from the surface of sheet material • An inert ions such as Ar + bombardment into the surface line sputtering • Plasma process • Anisotropic morphology • Low selection ratio • Cases such as: – Argon sputter etching	• Combination of physical etching and chemical • Plasma: ion bombardment plus free radical reaction • is an ion assisted etch (IAE) • High-speed controllable etching speed rate • Anisotropic controllable morphology • Good controllable choice ratio • In the LED all graphics are etched using RIE process

plasma etching PE), and physicochemical etching (reactive ion etching—RIE). At present, most of the pattern etching uses RIE. See Table 8.2 is a comparison of the three different etching methods.

Dry etching provides a controlled anisotropic morphology by applying ion bombardment. Anisotropy is caused by damaging management and protection mechanism. Damaging management is the use of a strong intense ion bombardment to break chemical bonds that are exposed on the surface of the atoms. Such a process is more and radical reaction since the ion bombardment is anisotropic. The etching rate in the vertical direction is much greater than the horizontal direction. The protection mechanism is the use of sputtered photoresist and/or chemical reactions to form by-products that are deposited on the surface. Since ion bombardment is along the vertical direction, the deposition at the bottom does not occur. The etching is mainly in the vertical direction. Sidewall deposition can be used to protect the sidewalls.

8.3.3 Etching of GaN Materials

The GaN material is a wide band gap semiconductor material with band gap of 3.4 eV. The bond energy between its atoms reaches 8.9 eV. Its chemical properties are also stable. It is difficult to use an acidic solution or an alkaline solution to etch GaN at room temperature. At 250 °C, etching a GaN material with a melted acidic solution (such as H_3PO_4 solution) or an alkaline solution (such as KOH and NaOH solution) can obtain a certain etching rate. However, this method has the following inconvenience. As the etching is carried out at a high temperature of 250 °C, it is not easy to handle the mixed solution at this temperature in the actual production process. Furthermore, it is difficult to find an effective mask material in the experiment so that it can resist the alkaline solution at high temperature. Although the etching rate is greatly improved compared to the etching rate at room temperature, it cannot meet the production requirements. Due to the above reasons, the wet etching technique has not been applied to the etching of GaN materials.

At present, among the group III nitride LEDs, dry etching is mainly used. The GaN material is etched by the inductively coupled plasma (ICP) technology by using a Cl_2/BCl_3 mixed gas. There are complex chemical etching processes and physical sputtering processes in the ICP etching process. The chemical etching process is a chemical reaction between the active particles and the surface of the material to be etched. The chemical process mainly consists of two parts.

First, when the Cl_2/BCl_3 mixed gas enters the reaction chamber of the ICP etcher, plasma is generated under the action of the RF electric field. The etching gas is decomposed into various neutral particles, electrons (e), and active free radicals (Cl, BCl), positively cssharged ions (Cl_2^+, Cl^+, BCl_2^+) and negatively charged ions (sCl^-). The second major reaction of the chemical process is the interaction of these active particles with the solid surface of the substrate, that is, the etching process is a chemical reaction process at the interface between the gas phase and the solid phase. The positively charged ions under the acceleration electric field bombard the GaN surface of the material, forming the physical sputtering bombardment. While the active particles are adsorbed on the GaN surface of the material, the chemical reaction with GaN can form volatile substance (generating material $GaCl_x$) that is pumped out from the reaction chamber through the pumping system. In the formation of the Cl_2/BCl_3 ions, the neutral group plays a major chemical role. It is generally believed that the reaction product of Cl in GaN etching process is as follows.

$$\mathrm{GaN} + \mathrm{Cl} \rightarrow \mathrm{Ga}, \mathrm{GaCl}_x, \mathrm{Ga} + \mathrm{GaCl}_x + \mathrm{N}_2 (x = 1, 2, 3) \tag{8.2}$$

The main physical sputtering process during the etching process is that high energy ions bombard the surface of the etched material to cause the surface material to be sputtered. In the ICP etching process, the physical bombardment effect is not equivalent to the pure physical process in sputter etching. It has the function of breaking the chemical bond between atoms, increasing the adhesion, accelerating the desorption of the reactants, and promoting the engraving. It can also enhance the chemical reaction on the surface of the etched material and the desorption of non-volatile products attached to the surface of the etched material. Since ion bombardment has a certain amount of energy, it will cause damage to the material to be etched. Therefore, how to choose the appropriate ICP etching process parameters, which has a certain etching rate and reduce the damage caused by etching on the quality of GaN crystal, is the key to manufacture high performance LED chips.

8.3.4 Etching of ITO and SiO_2 Materials

Indium tin oxide (ITO) transparent conductive film has good electrical conductivity and transparency. Compared with other transparent conductive films, ITO films have good chemical stability, thermal stability, and graphic processing characteristics. The two most important performance indicators for ITO films are resistivity and light transmission. The etch rate of ITO is usually low with ICP dry etching. Usually, ICP etching rate for ITO is 20 nm/min or so. Therefore, the common method to etch ITO is to use wet etching. The ITO etching solution (mainly composed of hydrochloric acid and ferric chloride solution) has a higher etching rate of the ITO transparent conductive film. The cost of wet etching is also lower than that of the ICP dry etching.

The silicon dioxide (SiO_2) is a mask material and also a passivation layer commonly used in the LED chip manufacturing process. The SiO_2 mask is etched during the pattern transfer process. When the quality of the edge of the pattern is not high, the SiO_2 may be etched off by a wet etching technique. A commonly used wet etchant for SiO_2 is an HF solution. In order to reduce the penetration of hydrofluoric acid on the photoresist during wet etching and to avoid the loss of hydride, NH_4 F can be used as a buffer. The HF solution added with NH_4 F is called buffered oxidation etchant (BOE) solution.

Compared with the wet etching of SiO_2 with BOE solution, RIE dry etching can provide higher etching precision and ensure the flatness of the etched edge. The commonly used etching gases are SF_6 and CHF_3, etc., which are chemically reacted by active particles (F) and SiO_2 in the plasma to produce volatile etching products for etching. When the SiO_2 mask is etched by CHF_3/Ar mixed gas, CHF_3 can not only produce active radicals (F) to chemically etch SiO_2, but also can be used as a passivating agent. The exposed sidewalls are protected after etching to ensure that the sidewalls are steep, and the Ar ions bombard the SiO_2 to cause physical sputtering. After the dry etching process is completed, the SiO_2 mask material is removed using a BOE solution.

8.4 Evaporation and Sputtering

The evaporation process refers to depositing one or more layers of ITO transparent electrodes and metals such as Cr, Ni, Pt, Au, etc. on the surface of the wafer. In general, the wafer is loaded into a vacuum chamber under a high temperature and metal is deposited on the wafer. In the production of LED chips, an electrode is deposited on the wafer by an evaporation process. The current conducting area of the wafer is increased by vapor deposition of metal layer.

8.4.1 Metal Evaporation

The evaporator is mainly divided into a thermal evaporation and an electron beam evaporation according to the energy source. In view of the high melting point of the metal as the electrode and the requirement of metal adhesion, the electron beam evaporation machine (E-beam evaporator) is commonly used in the LED industry. At present, the evaporator for LED production in China is still mainly imported.

Although the principle of the evaporator has been mastered in the country, and the laboratory evaporator can be manufactured in China, it has not entered the large production line due to challenges such as automation, process repeatability and uniformity. In view of the fact that the electrodeposition for LED electrode deposition is less difficult than the semiconductor PVD device, the domestic equipment manufacturer with experience in semiconductor PVD equipment will be able to realize the localization of the LED vapor deposition station.

Since the multilayer metal material used for the LED chip electrode is not easy to form an electrode pattern thereon by etching, it is necessary to form a metal electrode pattern by a lift-off process. First, a layer of photoresist is coated on the LED chip, and then photolithography is performed to form a desired pattern. The metal electrode material is then evaporated, and finally the residual photoresist and the metal deposited on the photoresist are deposited. The photoresist is removed to obtain an electrode pattern. There are two methods to remove positive and negative photoresists. When using positive photoresist stripping, a layer of polymer film that is insensitive to ultraviolet light and can be corroded by an alkaline solution is applied to the substrate before coating, followed by spin coating a thicker lithography on the polymer film. When the photoresist is developed by using an alkaline solution, the bottom of the photoresist is undercut due to the corrosive action of the alkaline solution on the polymer film, thereby facilitating the stripping process. After development, a metal layer is evaporated on the photoresist. Since the evaporation process does not have a strong covering ability to the step, the vapor-deposited metal film naturally breaks at the edge of the undercut photoresist. Finally, the solution is used to dissolve the photoresist, resulting in the natural stripping of the metal film deposited on the photoresist layer. The metal film deposited on the substrate is the pattern of the electrode. In the manufacturing process of the metal pads of the LED chip where the metal electrode adopts the negative photoresist stripping process, the undercut is naturally formed during the exposure due to the characteristics of the negative photoresist itself. The shape thus facilitates the formation of the peeling.

8.4.2 SiO_2 Passivation Layer

Since dust, moisture, and various impurity ions may adhere to the surface of the LED chip, the number of external electrons at the surface of the chip is not saturated. Dangling bonds, which form surface state energy level in the bandgap, can thereby cause leakage. Moreover, these dangling bonds are highly active and easily adsorb other molecules, atoms and ions. When the adsorbed impurities are ionized, a current path is formed. The leakage problem caused by such surface contamination can be solved by making an insulating protective layer on the surface and sidewall of the LED chip. The insulating layer materials commonly used in the LED chip fabrication process are mainly silicon dioxide (SiO_2) and silicon nitride (Si_xN_y) materials. Usually PECVD deposition of SiO_2 insulating layer on the LED and sidewalls of the chip is used to improve LED chip reliability.

The PECVD (plasma enhanced chemical vapor deposition) process is also used to deposit a layer of SiO_2 or SiN_x on the surface of the wafer after the epitaxial process is completed, where SiO_2 or Si_3N_4 serves as a hard mask layer for electrode etching to increase the etch selectivity ratio of the mask and the GaN epitaxial layer that gives a better etch profile. At present, the mainstream for LED fabrication is generally a flat-plate PECVD (13.56 MHz). The film is formed at a temperature of about 250–300 °C. The capacity can reach more than 40 wafers (2-in. substrate) at one time. The PECVD system for LED production in China is still mainly imported. The key technologies are still temperature control, plasma technology, vacuum system, software system and so on. It is noted that the flat-panel PECVD equipment for passivation layer has been developed in China. This equipment has great compatibility with PECVD equipment for LED. After further hardware improvement, it is easier to realize high-efficiency PECVD equipment that can be used for LED production.

8.5 Ohmic Contacts

The driving current of the III-nitride LED chip needs to be injected into the chip through the metal electrodes, thus requiring a good low-resistance ohmic contact between the metal electrode and the GaN material. Ohmic contact refers to the fact that the contact of the metal with the semiconductor material does not produce significant additional impedance and does not significantly alter the equilibrium carrier concentration within the semiconductor material.

8.5.1 n-type GaN Ohmic Contact

Forming a low-resistance ohmic contact electrode on the III-nitride LED chip can effectively improve the optical and electrical performance of the LED chip. The basic principles of n-GaN ohmic contact electrodes are as follows: (1) selecting a metal material with a certain work function to reduce the barrier height to enhance the thermal electron emission mechanism; (2) selecting the metal material of the electrode reacting with GaN occurring at the interface of the semiconductor material that will lead to the increase of the carrier concentration at the interface and reduction of the thickness of the space charge region to achieve the purpose of enhancing the tunneling mechanism; (3) providing desired bonding property between the metal and the GaN semiconductor to prevent electrode pad falling off during the wire bond process; (4) providing excellent thermal stability and reliability of the LED chip at high temperature [1].

Since the concentration of carriers of the n-GaN material in the chip is high, the ohmic contact is easy to realize. For horizontal structure LED chips, it is necessary to fabricate ohmic contact electrodes on Ga-plane of n-GaN. For the ohmic contact of the Ga-plane n-type GaN, it is relatively easy to form an ohmic contact thereon since the carrier concentration of n-GaN is high (on the order of 10^{18}). When fabricating an n-type ohmic contact electrode, the work function (Φ_{mss}) of the selected metal should be less than the work function of n-GaN [2]. At present, a lower specific contact resistance of 10^{-5} to 10^{-8} Ωcm^2 has been achieved on n-GaN materials, which can meet the requirements of LED chips. There are many n-GaN ohmic contact electrode materials. Commonly used for n-type ohmic contact materials include Al ($\Phi_m = 4.28$ eV), Ti ($\Phi_m = 4.33$ eV), Ta ($\Phi_m = 4.25$ eV), V ($\Phi_m = 4.33$ eV) and other metal elements. At present, many metal alloys such as Ti, Ti/Al, Ti/Al/Ni/Au, Ta/Ti/Ni/Au, V/Al/V/Au, V/Ti/Au, etc. can be also used [3–10].

8.5.2 p-type GaN Ohmic Contact

Since the ohmic contact electrode of p-GaN has a great influence on the performance of the LED, it is necessary to form an ohmic contact electrode having low contact resistance, high light transmittance or high reflectivity, and thermal stability on the surface of p-GaN. However, for the ohmic contact of p-GaN materials, it is difficult to form a low-resistance ohmic contact thereon. On one hand, heavy doping of p-GaN has not made effective breakthrough. This is due to the high ionization energy of the Mg-dopant in p-GaN (up to 170 meV). High ionization energy can result in low ionization rate and low carrier concentration. On the other hand, metal with large work function that is greater than of p-GaN (having a work function of approximately 6.12 eV) is not available, which makes it difficult to fabricate an ohmic contact of low-contact resistance for p-GaN [11–15].

Usually the operating voltage of an LED is related to the series resistance (especially the contact resistance on p-type GaN) and the current distribution. The p-type GaN contact resistance directly affects LED operating voltage. To obtain a low contact resistance, some high work function metal such as of Ni, Pt, Au, etc. are typically used. Some surface treatment process is needed as well. The uneven current distribution causes overload on part of the light-emitting area. Insufficient injection on other active areas makes the total injection efficiency not high. Efforts to homogenize the current usually cause more electrode absorption.

The contact between metal and semiconductor is one of the major topics that began very early in semiconductor science. For an ideal metal and semiconductor contact (regardless of the surface state effect), a barrier is formed if the work function of the metal is greater than that of the n-type semiconductor. Such a contact exhibits rectification characteristic. On the other hand, an ohmic contact will be formed if the work of the metal is smaller than that of the semiconductor. The opposite is true for p-type semiconductor materials. High-quality ohmic contacts are required to have low resistance and thermal stability. High transmittance is required for light-emitting devices so these problems must be considered when selecting a metallization solution. Generally, to obtain a low-resistance ohmic contact, the Schottky barrier height between the metal and semiconductor interface should be small. The method of lowering the barrier height without considering the influence of the surface state is to select a metal having a small work function for the n-type semiconductor and a metal having a large work function for the p-type semiconductor.

Ohmic contact means that the resistance of the metal semiconductor contact is very low, and the current–voltage characteristics of the device are mainly determined by the bulk resistance of the semiconductor. In principle, it can be prepared from a metal whose work function is lower than that of the n-type semiconductor or higher than that of the p-type semiconductor. At present, most practical ohmic contacts are prepared by preparing a highly doped semiconductor thin layer on the surface of the semiconductor material to form an extremely thin depletion layer so that carriers can easily tunnel through the barrier. However, the work function for p-GaN is about 7.5 eV which is much higher than that of the metal material. The carrier concentration of p-GaN is low due to the activation energy of the acceptor with the formation of the Mg and Mg-H complex. It is difficult to obtain a carrier concentration of more than 10^{18} cm^{-3}.

For the design and fabrication of LED chips, a good low resistance ohmic contact can reduce the threshold voltage of the chip. The composition of the ohmic contact alloy is as follows: (1) Adhesive layer metal: Since some metal alloys and GaN semiconductor materials have insufficient adhesion properties, it is necessary to deposit a metal material such as Cr, Ti, Ni as an adhesion layer to reinforce the metal adhesion strength with GaN semiconductor materials; (2) Functional layer metal: important for the formation of ohmic contacts, the selection of a suitable metal as a functional layer is the key to forming a low-resistance ohmic contact; (3) Overlay metal: the metal layer is to facilitate the wire bonding of the subsequent packaging process, and generally requires the metal layer and the lower contact layer to have high adhesion strength, stable performance, and easy soldering. The metals used to form ohmic contact with the p-GaN material are mainly Ni, Pt, Au, and so on. At present, it is common practice in the industry to use Cr as an adhesion layer metal, Pt as a functional layer metal, and Au as a cladding layer metal, that is, a Cr/Pt/Au alloy as an ohmic contact metal material of p-GaN.

8.5.3 Specific Contact Resistivity

GaN-based material has excellent chemical and physical characteristics, exhibiting immeasurable application prospects in the field of microelectronics and optoelectronics [16]. However, GaN-based devices present many difficulties during fabrication, wherein making an ohmic contact to the p-GaN is one of the difficulties. In recent years, many research groups worldwide report various methods to improve p-type GaN ohmic contact [17]. Ishikawa et al. found that the use of a metal with a large work function as an electrode can reduce the Schottky barrier height of the contact, thereby reducing the specific contact resistivity of the ohmic contact [18].

Specific contact resistivity (SCR) is an important parameter to quantitatively reflect the quality of the ohmic contact between the electrode metal and the semiconductor material. Low resistance ohmic contacts are the basis for high quality devices.

According to the metal—semiconductor contact theory, current transport is determined by thermionic emission for low-doping concentration metal–semiconductor contacts. The specific contact resistance is

$$\rho_c = \frac{K}{qA * T} \cdot \exp\left[\frac{q\Theta_{Bn}}{KT}\right] \tag{8.3}$$

where K is the Boltzmann constant, q is the electron charge, A^* is the effective Richardson constant, Φ_{Bn} is the barrier height, and T is the temperature. For higher doped semiconductors, the depletion layer is thin, and the current transport is determined by the tunneling of the carriers. The specific contact resistance is.

$$\rho_c \propto \exp\left[\frac{q\Theta_{Bn}}{E\infty}\right], \ E\infty = \frac{qh}{4\pi}\sqrt{\frac{N_d}{\varepsilon_s m}} \tag{8.4}$$

where ε_s is the dielectric constant of the semiconductor, m is the effective mass of the electron, N_d is the doping concentration, and h is the Planck constant. It can be seen that it is necessary to have a low contact barrier height, a high doping concentration, or both in order to obtain a low resistance ohmic contact.

At present, there are many methods for measuring specific contact resistivity based on different models. The measurement results are quite different. When the specific contact resistivity is small, the results from different authors may be quite different even the same method is used. Accordingly, selecting a relatively accurate model and having the contact resistivity measurement done reproducibly are crucial for the correct valuation of an ohmic contact. The measurement of specific contact resistivity is mainly divided into two categories: one is the measurement of the specific contact resistivity on the bulk material, including the "four-probe method", the "fitting method", etc.; the other one is on the thin-film material. The methods to measure the contact resistivity include "transmission line model method (the TLM, transmission line model) "and" direct interfacial contact resistance assay" [19]. The most common method is the transmission line model TLM, which includes rectangular, circular and dot transmission line models. Experimentally, measurements based on rectangular and circular ring transmission line can introduce larger error, and the measured values are not accurate. On the other hand, dots-based transmission line can eliminate the error introduced by the terminal resistance. Such a method offers good operability, accuracy and reproducibility.

8.5.4 Transparent Electrode Technology

Transparent conductors (TCs), discovered by the Badeker in 1907 and reported for CdO [20] in 1951, has developed into a large family of materials. In particular, indium tin oxide (ITO) transparent conductive film has been widely studied in the past 60 years due to its excellent properties. Currently, typical transparent conductive oxide (TCO) materials are fluorine-doped tin oxide (FTO), aluminum-doped zinc oxide (AZO) and ITO. Although the performance of CdO: In is very good, it is not widely used because of its toxicity. These materials are all made by increasing the carrier concentration by heavy doping in the semiconductor of the wide band gap to reduce the resistivity. Research on further improvement of TCO performance is still an interesting research topic in the field.

With the development of nanomaterials since 1990s, it has been found that nano-mixtures based on conducting carbon nanotubes can also form transparent conductive materials [21]. Although nanocomposite materials have many issues at present, they have new characteristics at the nanometer scale, and have attracted a lot of attention due to their interesting characteristics. For example, such transparent conductive materials like graphene have extremely high mobility. So far, the carbon nanotubes are all disorderly arranged. The nanowire structure of various transparent conductive materials has made considerable progress. The cost is low, the raw materials are rich, and it is safe and harmless. There is still a large room for development. The properties of this broad class of transparent conductive materials are quite different from commonly used transparent conductive oxides and have some special applications.

Since 2000, people began to fabricate GaN-based LEDs using the ITO [22, 23] as the transparent conductive material. Since the ITO material and the p-GaN are difficult to form an ohmic contact, the development is generally in the early stage, A thin metal layer is inserted between ITO and the p-GaN which is used as a transparent electrode. CS Chang et al. found that when epitaxial LEDs are placed, the top layer is inserted into the heavily doped cap layer to effectively improve the contact characteristics [24]. Indium tin oxide materials (ITO) have been widely used in GaN-based LEDs in industrial production to replace conventional nickel-gold transparent electrodes. At the same time, zinc oxide (ZnO)-based transparent conductive materials with excellent performance and low cost have also received extensive attention as substitutes for ITO [25, 26].

The improvement in the conductivity and low absorption of the transparent electrode can simultaneously improve the current injection efficiency and the extraction efficiency. The transparent electrode material also plays an important role as an optical extraction layer for improving the light extraction efficiency. Therefore, the requirements of the performance of the transparent electrode for the III-nitride LEDs are mainly in the following aspects:

(1) Good electrical conductivity, which can make the current of the device expand well;
(2) Having the ability to form good ohmic contact with a GaN material;
(3) The transmittance of the transparent electrode is closely related to the extraction efficiency, and it has a good transmittance;
(4) The refractive index value can be modulated to form an anti-reflection film or a highly reflective film;
(5) The ability to form a structured light-emitting surface.

The characteristics of the transparent conductive oxide material are in good agreement with the above requirements, and thus can be used as a transparent electrode of the group III nitride LED. This section briefly introduces the main theory of transparent conductive oxide formation and its application in GaN-based LEDs [27–30].

From the basic solid-electron gas model, the conductivity of solids is provided by electrons in the conduction band. The conductivity is determined by the product of carrier density, electron charge, and carrier mobility. Carrier mobility is controlled by free carrier relaxation time, and effective mass. That is determined by $\sigma = \mathrm{n} \bullet \mathrm{e} \bullet \mu$, $\mu = \frac{e \cdot \tau}{m*}$, which shows the conductivity of the solid is proportional to the carrier density and mobility. The mobility is proportional to the carrier lifetime and inversely proportional to the carrier effective mass.

The transparent conductive material generally has a high electron density and electron mobility, and thus has a high conductivity. It has a high transmittance in the visible light band due to its wide bandgap. It is very important to study the microscopic electronic structure of transparent conductive oxide (TCO) and its formation mechanism from the perspective of material physics and to understand its chemical and structural origin from the perspective of material chemistry. For example, to clarify the electronic structure, the principle of component-induced electron phase transformation into simple and electronic gas, the mechanism of parasitic electron scattering in degenerate electron gas, all forms of chemical engineering materials-oxides and super-oxides increase the potential performance. Higher conductivity and lower absorption rate are the development direction of transparent conductive oxides. Other functions based on TCO materials such as gas sensing characteristics and piezoelectric characteristics have also been studied.

Taking ITO as an example, the main mechanism of the formation of transparent conductivity of TCO materials is introduced. The ITO material has typical TCO optical and electrical transport properties, and its conductivity is greater than 10^4 S/cm. The visible light transmittance is higher than 80%. This makes it an important candidate in many applications, especially in optoelectronic components with high efficiency.

It is generally believed that the conductivity of ITO is derived from the shallow donors of the Sn_4^+ occupied ions and the oxygen vacancies in the indium oxide structure. These shallow donor and impurity states are very close to the intrinsic conduction band so that the donor electrons are thermally activated at room temperature. After entering the conduction band, by further mixing, a degenerate carrier electron gas is formed, which is characterized by far-infrared absorption and high conductivity, but at the same time the basic band gap is still intact, so the high visibility of the visible region is still maintained. The main advantage of TCO materials applied to Group III nitride LEDs is their transparency and conductivity, especially their transmittance is much higher than that of metal electrodes.

8.6 Flip-Chip LEDs

Flip-chip structure is proposed, where the structure diagram is shown in Fig. 8.3. Flip-chip LEDs are fabricated by flipping the chip and mounting the LED chip on a supporting substrate with high thermal conductivity (such as silicon) to improve the heat dissipation performance of the LEDs. The series–parallel connection between the light-emitting units is realized on the flip-chip substrate. The wiring on the flip-chip substrate breaks through the technical difficulty of the interconnection and climbing between the isolation units, and the reliability is high. Changing the substrate wiring can change the series–parallel relationship between cells and realize different parameters, which is simpler and more flexible than the formal structure. Compared with the formal structure, the flip-chip structure is relatively complicated in manufacturing process, including chip fabrication, support fabrication, and flip-chip bonding. Below we will mainly introduce the most important flip-chip welding process.

Flip-chip bonding is formed with a bump-electrodes facing down, directly bonded to the a wiring layer on the substrate [31]. A schematic structural diagram is shown in Fig. 8.4b. The bonding material may be a metal lead or an alloy solder or a conductive polymer. In comparison with a conventional wire bonding technique (Fig. 8.4a, reference), flip-chip technology offers high mechanical strength. For instance, the

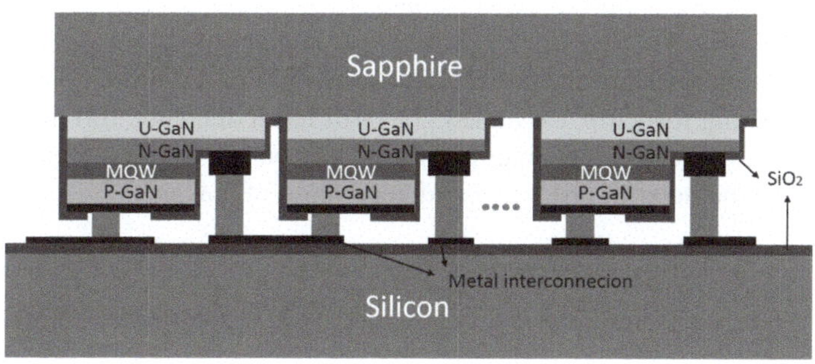

Fig. 8.3. flip-chip LED diagram

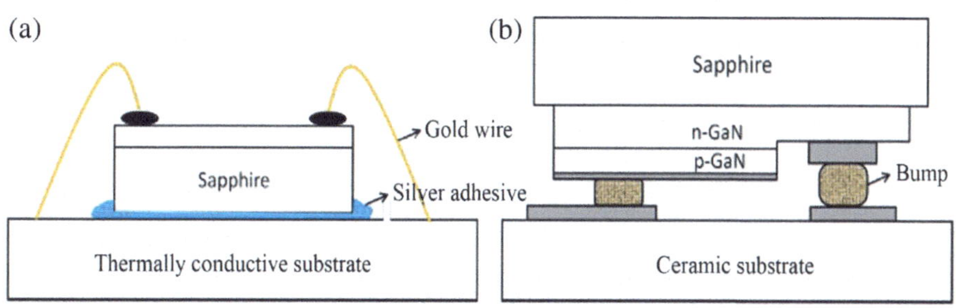

Fig. 8.4. **a** Wire bonding and **b** Flip-chip structure

Table 8.3 Thermal conductivity of each material of 3 conductive channels

Materials	Sapphire	Silver glue	GaN	Metal alloy
Thermal conductivity [W/(m K)]	35–36	2.5–30	130	>200

conventional wire bonding gold wire shows a tension of about 10 g, while the flip chips can have a pressure force up to 2000 g. Conventional wire bonding is a point contact. An instantaneous large current impact can easily melt the wire. On the other hand, flip chip bonding is a surface contact and can withstand large current surges. In addition, the conductive traces of the wire bond chip are sapphire, silver paste and thermally conductive substrate. The thermal conduction channels of the flip chip solder are GaN, metal bumps and flip-chip substrates. The thermal conductivity of each material can be seen in Table 8.3. The sapphire and the silver paste have low thermal conductivity and poor thermal conductivity for the wire bonding structure. For the flip-chip structure, GaN and metal alloys have high thermal conductivity and low heat resistance. Such a scheme will provide better heat dissipation performance.

The flip-chip soldering process generally includes three steps: substrate metallization, metal bump fabrication, and chip flipping. We generally use E-beam vaporator (EB) to achieve metallization of flip-chip substrates. The fabrication of metal bumps generally uses a wafer balling machine, which uses a fully automatic flip-chip bonding device to align the pads on the chip with the corresponding bumps on the substrate. For flip the chip on the substrate, it uses heating effects. The chip and the substrate are welded together by pressurization and ultra-sonication. Figure 8.5 shows the welding model of the gold ball hot-press ultrasonic flip-chip welding [32, 33]. After finishing flip-chip, expanded gold ball has a diameter 100 microns or so and a height of 30 microns. This requires the n electrode on the chip to be larger than 100 microns in diameter in order to avoid generation of leakage.

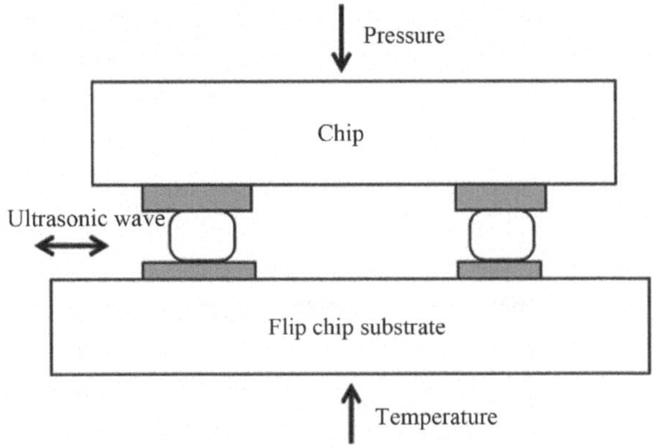

Fig. 8.5 Welding model of flip chip LED

8.7 Vertical Structure LEDs

Since the sapphire substrate of the conventional lateral structure LED is a poor conductor of heat, the LED has a large rise in junction temperature due to difficulty in discharging heat at a large operating current density. This can lead to rapid degradation of its luminous efficiency. The LED lifetime can also be drastically shortened, which makes the conventional lateral structure LEDs more challenging to meet the requirements of the general lighting field.

Compared with the lateral structure LED with poor thermal conductivity, the vertical structure LED adopts a substrate with excellent thermal conductivity. This greatly improves the heat dissipation capability. The material degradation is slower, and the LED lifetime is increased. The vertical structure LED has the upper and lower electrode structures. The injection current can be uniformly distributed in the light-emitting area, avoiding the current crowding phenomenon of the mesa structure LED. In addition, the vertical structure LED upper surface is easy to produce a surface light extraction structure, which is beneficial to improve the light extraction efficiency. On the one hand, the vertical structure LED is particularly suitable for application in the field of general illumination. On the other hand, the vertical structure LED needs to make a new substrate and remove the sapphire substrate, which complicates the manufacturing process and will also produces many process compatibility issues.

The technical solutions for preparing the transfer of substrate mainly include bonding, electroplating, sapphire grinding and so on. In order to carry out the subsequent laser lift-off, ICP etching and other processes smoothly and to maintain high yield, the substrate must be firmly combined with the LED epitaxial layer material. It must be resistant to acid and alkali. The thickness of the substrate should be uniform and the coefficient of thermal expansion should be matched. How to choose the appropriate substrate material and optimize the preparation process conditions to achieve low stress and low damage as well as firmly adhered substrate transfer technology is the most difficult and critical point in vertical structure LED fabrication technology.

Laser lift-off is a one-step process for removing sapphire substrates. The yield of vertical structure chips is generally decided in this step. Adjusting the energy uniformity and energy density of the laser spot is the basic measure to ensure high yield. The extent of the effect of laser lift-off on the GaN epitaxial layer, the degree of damage to the epitaxial layer, and whether it will adversely affect the overall photoelectric performance of the chip, need to be clarified. One advantage of the vertical structure LED technology is that the sapphire substrate can be reused. Usually, the back surface of the sapphire substrate is not polished. To peel off the epitaxial wafer with rough substrate, the process parameters and lining of the laser lift-off are required. The relationship between the roughness of the substrate is well studied. Electroplating technology, bonding technology and laser stripping technology are introduced as below.

8.7.1 Electroplating Technology

In a gallium nitride based vertical structure LED, a copper or copper alloy substrate is generally fabricated by electroplating. This is because the copper plating technology is relatively mature. The cost of copper consumables and plating solution is relatively low. The thermal conductivity and electrical conductivity of copper are ideal. However, the use of pure metallic copper as a substrate also causes a series of problems such as thermal mismatch with gallium nitride materials, low hardness, prone to plastic deformation, and difficult in substrate cutting. It is suggested that a copper alloy plating with a certain composition can obtain an ideal substrate. For instance, copper-tungsten alloy is a hard and brittle alloys. If a suitable composition can be obtained, such an alloy could match well with the thermal expansion coefficient of GaN. However, there is no copper-tungsten plating technology yet.

The electroplating copper process applied to the transfer substrate is discussed below. Common copper plating solutions can be divided into cyanide plating solution and non-cyanide plating solution. The non-cyanide plating solution includes acid copper plating solution, pyrophosphate copper plating solution, citrate copper plating solution, tartrate copper plating solution, HEDP copper plating solution, triacetic acid copper plating solution, diethylamine copper plating solution, and fluorine borate copper plating solution, etc. The cyanide copper plating solution has the characteristics of good uniformity, fine crystallizing layer and good adhesion to the seeding metal. However, the cyanide copper plating solution has strong toxicity, and it is necessary to consider the treatment of waste water and exhaust gas. Copper in the solution is present as a monovalent copper cyanide complex. It is generally believed that there are mainly three forms of copper cyanide coordination compounds: $[Cu(CN)_2]^-$, $[Cu(CN)_3]^{2-}$, $[Cu(CN)_4]^{3-}$. At the cathode, monovalent copper ions are reduced and deposited by discharge. The other side reaction is hydrogen ion reduction to hydrogen that moves out as gas.

$$Cu^+ + e = Cu \tag{8.5}$$

$$2H^+ + 2e = H_2 \uparrow \tag{8.6}$$

The anode copper dissolves into monovalent copper ions. I the anode is passivated, oxygen can move out.

$$Cu - e = Cu^+ \tag{8.7}$$

$$4OH^- - 4e = 2H_2O + O_2 \uparrow \tag{8.8}$$

The cyanide copper plating solution should be placed under ventilated conditions. Masks and rubber gloves must also be worn during operation. The plating solution should not be contacted with acidic substances to avoid the generation of highly toxic cyanide acid.

Acidic sulfuric acid copper plating solution has good leveling and specular brightness, low cost, no volatile highly toxic gas, and relatively easy disposal of waste liquid. Such a solution is widely used. The acid sulfuric copper plating solution has two types of ordinary plating liquid and bright plating liquid. The bright plating solution is obtained by adding some brightener to the ordinary plating solution.

The common plating solution is composed of a salt of copper sulfate, sulfuric acid, a hydroxycarboxylic acid, and a chloride ion. At the cathode, the divalent copper ions are reduced to copper atoms accompanied by a hydrogen reduction reaction. When the anode copper atoms are dissolved into divalent copper ions, the anode is passivated, and oxygen is also precipitated.

$$Cu^{2+} + 2e = Cu \tag{8.9}$$

$$Cu - 2e = Cu^{2+} \tag{8.10}$$

The sulphuric acid copper plating solution is extremely efficient, close to 100%. The chance of hydrogen ion discharge on the cathode is small. On the anode, if the sulfuric acid content in the plating solution is low and the anode current density is large, the anode will be passivated and oxygen will be released.

Copper sulfate is the main salt for providing copper ions, and the content is 150–200 g/L. The low concentration will lower the upper limit of the cathode current density and the brightness of the coating. When the concentration is high, it will be easily crystallized. The role of sulfuric acid is mainly to prevent copper salt hydrolysis, reduce copper powder, improve the conductivity of the plating solution and cathodic polarization. This also improves the dispensability and anodic solubility of the electrolyte. Sulfuric acid is generally controlled at 50–70 g/L. Excessive sulfuric acid concentration will reduce the solubility of copper sulfate. Therefore, when the sulfuric acid content is increased, the content of copper sulfate must be reduced. If the sulfuric acid content is too low, copper sulfate is easily hydrolyzed into insoluble copper oxide, which is deposited on the cathode to cause a loose coating. Sulfuric bright copper plating are mostly combined brighteners which can be divided into main brightener according to their role, leveling agent, and brightener carrier. These three substances can be used together to obtain good results. Chloride ions in the solution not only brighten coating, but also reduce the internal stress generated after adding other brightening agent, generally in an amount of 0.02–0.08 g/L. If the content is low, the plating level is poor and the brightness is low. Furthermore, the internal stress is large and the bright streaks are easily generated. In severe cases, the plating layer is rough. If the content is high, the plating layer is not very bright.

Copper plating conditions require strict control. Increasing the temperature of the plating during electroplating can increase the current density, thereby improving the brightness and leveling of the plating layer. The toughness is also good. The temperature control of acid bright copper plating is related to the brightener used.

When the temperatures of brighteners is higher than 25 °C, the cathodic polarization drops significantly. When the temperature exceeds 35 °C, the coating is completely darkened. The temperature is generally controlled at 5–25 °C. Increased operating temperatures, high copper sulfate concentration and agitation can increase the allowable cathode current density. In the normal current density range, the higher the current density is, the larger the deposition rate is. This can also improve the brightness and flatness of the plating layer. In bright acid copper plating, electrolytic copper is generally not used, and a copper anode containing phosphorus is used. This is because cuprous oxide is easily generated on the surface when an electrolytic copper anode is used. The consumption of the brightener will increase, and the quality of the plating layer is reduced. In order to avoid the generation of copper powder and monovalent copper ions, a copper anode containing 0.04–0.3% of phosphorus should be used.

8.7.2 Bonding Technology

Bonding technology refers to the bonding of two wafers, where the surface is clean under certain conditions, by chemical bonding of the surface directly or through the bonding material of the intermediate layer. Bonding techniques are not limited by the lattice or crystal orientation of the two wafer materials. Therefore, the use of bonding technology to combine new structural materials has great freedom. It can also better utilize the different electrical and optical properties of various materials, and provides a broader design space for device design.

At present, bonding technology has been widely used in the field of SOI material preparation, microelectronic technology, sensor technology, MEMs, optical devices and so on. At present, the bonding technology can be divided into multiple bonding technologies, medium bonding technology, low temperature eutectic bonding technology and so on. From the technical point of view, the bonding technology can be divided into two categories: direct bonding and bonding layer bonding. Different types of bonding technologies have their unique applications in their respective fields. The introduction of bonding technology into the substrate transfer technology of GaN-based LED epitaxial wafers based on sapphire substrates is a very suitable solution for fabricating GaN-based vertical structure LEDs.

It is difficult to bond GaN crystal directly to other materials without passing through the intermediate adhesion layer. The choice of bonded substrate material depends mainly on its electrical conductivity, thermal conductivity, and degree of thermal mismatch with GaN materials. It is currently difficult to find a substrate material that is thermally matched with GaN. Si and Cu are good choices in terms of conductivity, thermal conductivity, and cost. Currently, there are GaN-based vertical structure LED products based on these two materials.

Hot-temperature metal bonding is a bonding technique suitable for fabricating GaN-based vertical structure LEDs: first, a bonding metal layer is deposited on the epitaxial wafer and the surface of the substrate material. Under a certain temperature, vacuum, and pressure conditions, the two metal layers are bonded to complete the bonding of GaN to the substrate material. Heat treatment is a very important step in the wafer bonding process. The heat treatment of the bonded wafer can effectively increase the inter-diffusion between the interface atoms, thereby increasing the bonding strength. However, if the coefficients of thermal expansion between the wafers are relatively large, a relatively large internal strain will be generated between the two layers of materials, causing warpage of the wafer and even cracking and cracking of the wafer during the high-temperature heat treatment. Therefore, the difference in thermal expansion coefficient of the material is a parameter that must be considered during the high temperature bonding process.

The bonding between the wafer and the substrate is achieved by inter-diffusion between the metals after reaching a certain temperature. It is commonly believed that there is no interaction force between the wafer and the substrate parallel to the interface direction at this high temperature. Inter-diffusion between metals after a period of high temperature has caused the wafer to bond with the substrate. During the temperature cooling down to room temperature, interactions between the wafers will be mutually parallel to the interface direction due to the difference in thermal expansion coefficient between the wafers. The force, which is the cause of stress during the bonding process, is also the basis for calculating the stress. Obviously, this view ignores the influence of the applied pressure during the bonding process. The applied longitudinal pressure itself can cause the deformation of the material. The deformation of the material is also related to the mechanical parameters such as the elastic modulus and Poisson's ratio of the material. The difference in material mechanical parameters between the two wafers may cause another part of the stress, which is complicated if the bonding pressure is taken into account when calculating the stress of the wafer bond. In calculating the internal stress after bonding, we can ignore the influence of the bonding pressure on the internal stress if the bonding pressure is not very large.

8.7.3 Laser Lift-Off

The principle of laser lift-off is mainly to utilize the pyrolysis characteristics of GaN materials and the band gap between GaN and sapphire. Using an ultraviolet pulsed laser with a photon energy greater than the GaN bandgap and less than the sapphire bandgap, the GaN material is irradiated through the sapphire substrate. Strong absorption at the interface occurs, causing local temperature to rise as well as GaN gasification and decomposition, thereby realizing separation of GaN material from the sapphire substrate. Equation (8.11) is a gasification decomposition equation of GaN.

$$2\text{GaN} \rightarrow 2\text{Ga} + \text{N}_2 \uparrow \tag{8.11}$$

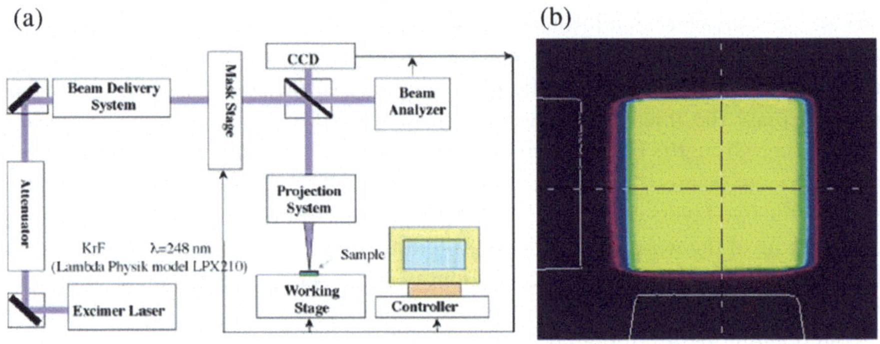

Fig. 8.6. **a** Laser stripper optical path diagram, **b** Adjusted laser spot energy distribution map

The optical path diagram of a common laser lift-off device is shown in Fig. 8.6a. The laser is generated from the laser and passed through the attenuator, mirror, spot shaping component, and optical path transmission system to the processing moving stage. The beam splitter in the optical path transmission system transmits a part of the spot energy to the CCD for use in spot adjustment. Laser lift-off is usually performed using a KrF excimer gas laser with a laser pulse of 10 ns. Figure 8.6b is the energy distribution of the resulting spot, with different colors representing different energy densities. A relatively regular square rectangular spot can be seen from the figure.

In laser lift-off technology, the distribution of temperature field of GaN material is one of the key factors to achieve substrate lift-off. It is of great significance to study the temperature field distribution in laser lift-off technology. Through theoretical calculations, we know that after sapphire irradiation with a single pulse of laser with different energy densities, the temperature variation with depth in the GaN material is shown in Fig. 8.7. As can be seen from the figure, as the depth increases, the temperature rapidly decreases. Studies have shown that the decomposition of GaN

Fig. 8.7 Theoretical calculation of GaN interface temperature distribution after laser pulse irradiation

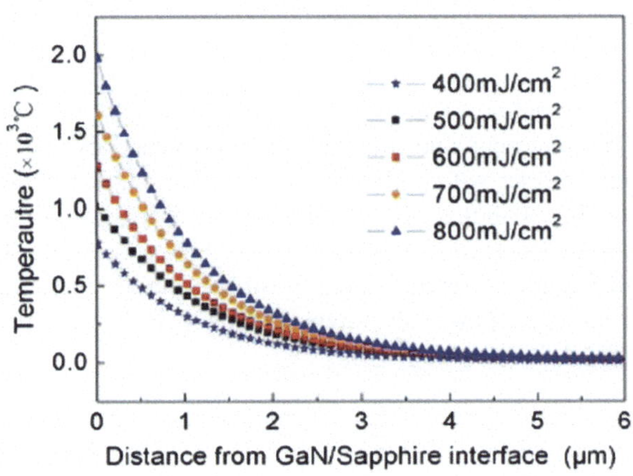

into Ga and N_2 requires temperatures above 800 °C [34]. As can be seen from the figure after a single pulse irradiation, the temperature at the GaN/sapphire interface can reach the GaN decomposition temperature only when the pulse energy density reaches 500 mJ/cm^2. Considering that sapphire reflects and absorbs about 20% to about 30% of the energy, the threshold energy density of the stripped substrate is between about 625 and 715 mJ/cm^2. In addition, it can be seen from the figure that the high temperature region (>1000 °C) gradually increases as the laser energy density increases. When the energy density is 500 mJ/cm^2, the high temperature region is less than 50 nm. The range of the high temperature region rapidly expands to nearly 800 nm when the energy density is increased to 800 mJ/cm^2. Therefore, in order to reduce the damage when the substrate is peeled off, a smaller energy density should be used as much as possible.

When irradiated by a high-energy laser beam, GaN at the interface is decomposed into Ga and N_2 gases. These gases can only be confined to a limited area of the spot size. Therefore, there is usually a high gas pressure generated during the decomposition process. A pressure of 6 GPa is high enough to cause GaN to crack as calculated by Karpinski et al. [35], Tavernier and Clarke [36]. Therefore, strategies to avoid GaN cracking must be considered. To avoid GaN cracking, we can start with the following aspects: firstly, we should choose the appropriate laser energy density to minimize the gas pressure during laser lift-off. Secondly, we can make a channel that so that the generated gas can be released instantaneously. In addition, it is necessary to have a perfect interface without holes or voids between the GaN and the transfer bonding of the substrate to ensure the integrity of GaN.

8.7.3.1 AC/High Voltage LEDs

AC (Alternating Current, AC) is used to deliver power to individual houses, business or public electricity, mainly to avoid long-distance power transmission loss. Since AC power is delivered to the end users, the appliance must be designed accordingly to avoid short circuit due to voltage mismatch.

Conventional light-emitting diodes (Light-Emitting Diode, LED) are required to use DC (Direct Current, the DC) as the drive. The use of an alternating current as a power supply at the same time must be accompanied by the rectifier transformer AC/DC converter to ensure LED normal operation. For applications, it is desirable to save power of LED operation. However, in the AC/DC conversion process, power loss can be as high as 15–30% of the electricity. It can revolutionize the lighting applications if the LED can be driven by an AC source. Such a change will completely get rid of the bottleneck of the driver power lifetime and the chip heat dissipation problems. Furthermore, the product reliability will be improved, and the application cost will be significantly reduced.

AC-LEDs are a class of LED products that integrate various processing technologies. They include a variety of devices or cores that can be directly driven by the AC power of the AC grid without the need of additional transformers, rectifiers or drive circuits. This allows LED products to be directly connected to the home and office

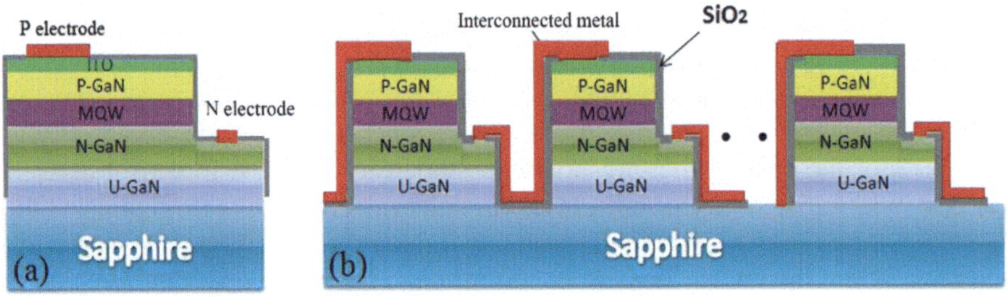

Fig. 8.8. **a** Traditional LED, **b** Array High Voltage AC/DC LED

AC electrical plugs (100–110 V/220–230 V) without the need of converters, which not only significantly reduces the cost of the circuit but also avoids the energy loss during the power conversion process. AC LEDs can be directly driven by AC drive through special circuit design. It can operate normally without the need of rectifier transformer. Therefore, the size and weight of the lighting fixture can have more advantages than the general DC LED lamp.

As shown in Fig. 8.8, the array of high-voltage AC/DC LED is realized by the series and parallel connection of multiple LED micro-pixels in the chip manufacturing process where AC high-voltage power supply is used as energy source. Compared with traditional LEDs driven by small voltage and high current, high voltage LEDs are driven by high voltage and small current. Small currents are more easily controlled by drive circuits and conversion circuits. The high voltage AC/DC conversion circuits are more efficient as well.

Unlike DC LEDs, AC LEDs are driven under AC power, and LEDs are single-phase conduction devices. Therefore, there are always some LEDs that do not emit light during the positive half cycle or negative half cycle of AC power. This introduces the concept of chip utilization which is defined as the proportion of the area occupied by the microcrystals that illuminate in each AC bias direction. Chip area utilization in the AC LED is an important characteristic that determines the AC LED light output efficiency and manufacturing costs.

Special processes for array-based monolithic integrated LEDs include ICP deep etch isolation, sidewall insulation protection, and electrode interconnection. The deep etching isolation process needs to etch the epitaxial GaN material in the isolated runway to the sapphire substrate to obtain a micro-pixel epitaxial sidewall with a certain inclination angle. This is to ensure that the GaN epitaxial layer is not damaged. The sidewall insulation protection process will use an insulating material. The micro-pixel sidewalls of the array LED device are insulated to ensure that micro-pixel short-circuit phenomenon does not occur in the subsequent electrical interconnection. The electrode bridging process uses an electrical interconnection material, spanning the isolation deep trench and the micro pixel. The micro pixels are connected in series or in parallel. The following sections will discuss each item separately.

a. Deep etch isolation and sidewall insulation protection

The methods to isolate materials in microelectronics are laser grooving, plasma etching, ion implantation to form isolated islands, and epitaxial self-isolation. In the production of microelectronic devices, the individual devices must be isolated from each other in order to eliminate punch through and parasitic effects. Such a process is called the isolated pixels. By isolation, the individual pixels are fabricated in isolated areas. After the completion of the pixels, it is necessary to interconnect with low-resistance metal wires in order to achieve the overall function. The interconnection wires must have good contact with the pixels. In IC technology, the integrated circuit is manufactured in three processes: isolation of the device, flow of the device, and wiring. In general, there are three main methods for device isolation: p–n junction reverse isolation, dielectric isolation, and trench isolation [37–40]. Some commonly used MOS components (bipolar transistors, CMOS) usually use a p–n junction isolation or oxidation process to form isolated islands in order to prevent parasitic effects.

GaN-based III–V nitride wide-bandgap semiconductors play an important role in the field of blue light and ultraviolet optoelectronics. They are also important materials for manufacturing high-temperature and high-power semiconductor devices. GaN material is a direct bandgap material with a band gap from 0.7 to 6.2 eV range in accordance with Al or In dopants [41]. The III–V nitride has high thermal stability and chemical stability. On the other hand, it also brings great difficulties to the manufacturing process of the device. The isolation of conventional GaN-based devices can be achieved by wet etching, dry etching, and ion implantation. However, wet etching process has poor controllability, slow rate, and poor heterogeneity. This makes such a process not an ideal method for device isolation. At present, the isolation of GaN devices is mainly achieved by dry etching such as reactive ion etching (RIE), electromagnetic cyclotron resonance (ECR) and inductively coupled plasma (ICP) etching, and so on.

To fabricate a high-voltage chip array, it is necessary to insulate the micro pixels from each other. Therefore, the GaN epitaxial layer between the micro pixels must be etched to the sapphire substrate. Unlike the mesa etching in conventional LED chip, the etching depth is generally 1–1.5 micron. The whole GaN epitaxial material thickness is 6–7 micron, it requires long ICP etching. This makes necessary not only high selective etching ratio of the mask but also the optimum ICP etching conditions. Since the entire epitaxial layer needs to be etched to the sapphire substrate, an etching mask of 4-micron photoresist is far less than the value calculated from the etching ratio of the materials. Because the etching ratio between SiO_2 and GaN is 1:1, it is preferable to design composite layers consisting of SiO_2 and photoresist as a mask for deep etching.

In the microelectronics process, the CVD film can offer both conformal and non-conformal coatings to cover the underlying pattern. However, the specific coverage depends on the type of film, the type of reaction system, and the deposition conditions. The conformal deposition means that the same thickness of film can be deposited

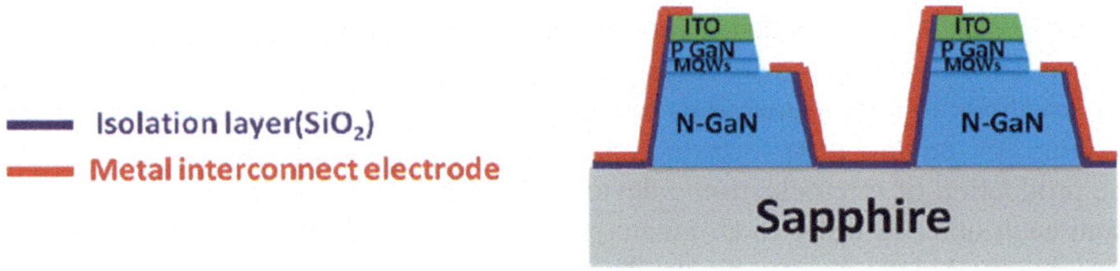

Fig. 8.9 Metal interconnection between microcrystals

on all the patterns regardless of the inclined pattern on the surface of the substrate. Such a deposition is called conformal growth.

If the lithography design for the sidewall silicon dioxide insulation and processing conditions during the process are improperly implemented, the sidewall of the micro-pixel will not be protected by silicon dioxide. The metal electrode can directly contact the GaN epitaxial sidewall, which will cause short-circuit. The micro-pixel is not turned on in this case. The SiO_2 insulating film made by PECVD can well cover the epitaxial sidewall with deep grooves. This will lead to well wrapped edge of the mesa and no broken step coverage at the corners resulted from conformal coating process.

b. Electrode interconnection

In the fabrication of arrayed HV-LEDs, the metal electrode interconnection between the micro-pixels needs to cross the isolation deep trench, from one n-electrode of a micro-pixel to the p-electrode of the next micro-pixel, thereby achieving the tandem micro-pixels. The interconnection diagram is shown in Fig. 8.9. After the insulating layer is grown on the micro-grain epitaxial sidewall, the interconnect electrode can be fabricated. The key to make metal interconnect electrodes is the slope of the epitaxial sidewalls of the micro-grains. The deep trenches between the micro-grains are the main obstacles to metal interconnects. The reliability of the network remains a big challenge since the metal will cover a 6–7-micron level plateau, Fig. 8.9 is the ideal state of the metal interconnect.

Figure 8.10a shows the SEM image of the epitaxial sidewall of the fabricated metal interconnect of gold electrode. Metal electrode system in the experiments employ the Cr/Pt/Au (60 nm/40 nm/1500 nm) structure with a total thickness of about 1.6-micron. It can be seen from Fig. 8.10a that the thickness of the metal layer on the epitaxial sidewall is much thinner than the metal on the surface of the GaN epitaxial wafer [42]. This has a lot to do with the way the metal electrodes are made. We used the electron beam evaporation equipment in this test to make metal electrodes. Figure 8.10c shows the electron beam evaporation model, where the target substrate used in the current equipment is in the rotary pan mode. The angle between the electron beam evaporation source and the rotary pan is matched so that the thickness of the vapor deposition metal of all the wafers on the target substrate is uniform. Therefore, the incident direction of the metal source generated by electron beam

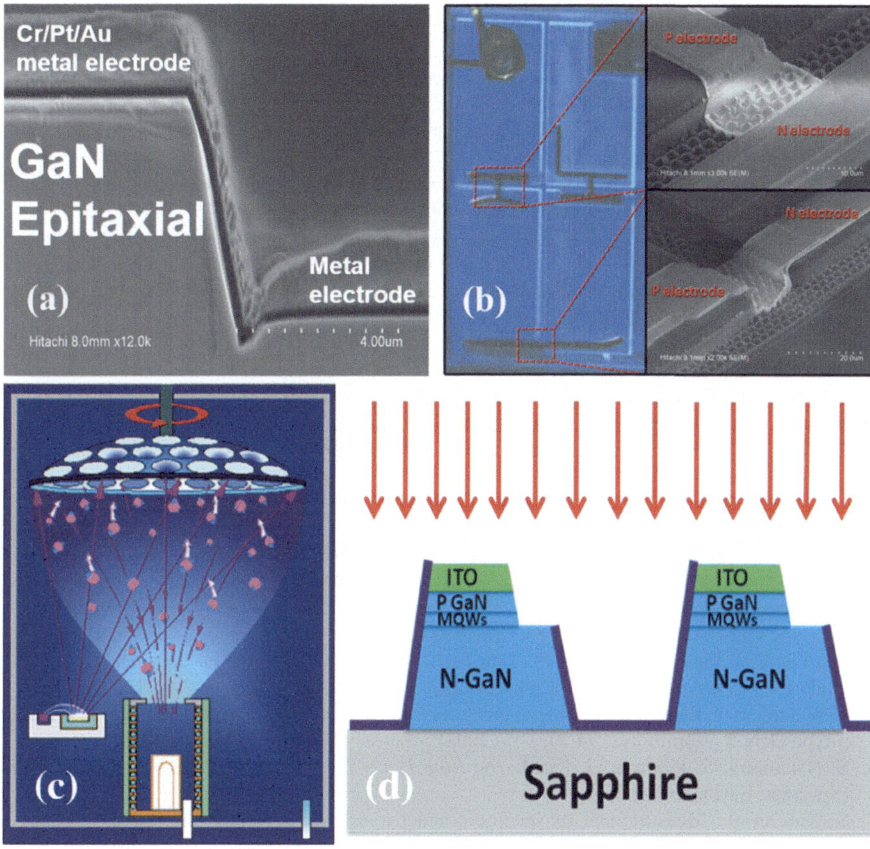

Fig. 8.10. **a**, **b** GaN epitaxial sidewalls after metal interconnection, **c** electron beam evaporation model, **d** metal evaporation angle.

evaporation is perpendicular to the surface of the GaN epitaxial wafer. As shown in Fig. 8.10d, the red arrow represents the metal source, so the thickness of the sidewall metal layer is related to the slope of the micro-pixel sidewall. In theory, the steeper the sidewall, the thinner the metal layer on the sidewall and the lower the reliability of device interconnection are. The overall effect of the metal interconnect electrode is shown in Fig. 8.10b. The reliability of the metal interconnect is determined by the way the metal coating is applied and the slope of the micro-grain epitaxial sidewall.

References

1. T. Margalith, O. Buchinsky, D.A. Cohen et al., Indium tin oxide contacts to gallium nitride optoelectronic devices. Appl. Phys. Lett. **74**(26), 3930–3932 (1999)

2. J.S. Foresi, T.D. Moustakas, Metal contacts to gallium nitride. Appl. Phys. Lett. **62**(22), 2859–2861 (1993)
3. B.P. Luther, S.E. Mohney, T.N. Jackson, Ti and TaN contacts to n-type GaN. Semicond. Sci. Technol. **13**(11), 1322–1327 (1998)
4. M.E. Lin, Z. Ma, F.Y. Huang, Low resistance ohmic contacts on wide band-gap GaN. Appl. Phys. Lett. **64**(8), 1003–1005 (1994)
5. L.F. Lester, J.M. Brown, J.C. Ramer, Nonalloyed Ti/Al ohmic contacts to n-type GaN using high-temperature premetallization anneal. Appl. Phys. Lett. **69**(18), 2737–2739 (1996)
6. B.P. Luther, S.E. Mohney, T.N. Jackson, Investigation of the mechanism for ohmic contact formation in Al and Ti/Al contacts to n-type GaN. Appl. Phys. Lett. **70**(1), 57–59 (1997).
7. S. Ruvimov, Z.L. Weber, J. Washburn, Microstructure of Ti/Al and Ti/Al/Ni/Au ohmic contacts for n-GaN. Appl. Phys. Lett. **69**(11), 1556–1558 (1996)
8. Z. Fan, S. Mohammad, W. Kim, Very low resistance multilayer ohmic contact to n-GaN. Appl. Phys. Lett. **68**(12), 1672–1674 (1996)
9. A. Motayed, K. Jones, M.A. Derenge, Electrical, microstructural, and thermal stability characteristics of Ta/Ti/Ni/Au contacts to n-GaN. J. Appl. Phys. **95**(3), 1516–1524 (2004)
10. J.O. Song, S.H. Kim, J. Kwak, Formation of v-based ohmic contacts to n-GaN. Appl. Phys. Lett. **83**(6), 1154–1156 (2003).
11. J.S. Jang, I.S. Chang, H.K. Kim et al., Low-resistance Pt/Ni/Au ohmic contacts to p-type GaN. Appl. Phys. Lett. **74**(1), 70–72 (1999)
12. H. Cho, T. Hossain, J. Bae et al., Characterization of Pd/Ni/Au ohmic contacts on p-GaN. Solid State Electron. **49**(5), 774–778 (2005)
13. J.O. Song, D.S. Leem, J.S. Kwak et al., High-quality nonalloyed rhodium-based ohmic contacts to p-type GaN. Appl. Phys. Lett. **83**(12), 2372–2374 (2003)
14. J.K. Ho, C.S. Jong, C.C. Chiu et al., Low resistance ohmic contacts to p-GaN achieved by the oxidation of Ni/Au films. J. Appl. Phys. **86**(8), 4491–4493 (1999)
15. H.W. Jang, S.Y. Kim, J.L. Lee, Mechanism for Ohmic contact formation of oxidized Ni/Au on p-type GaN. J. Appl. Phys. **94**(3), 1–5 (2003)
16. S. Nakamura, S. Pearton, G. Fasol, *The Blue Laser Diode: The Complete Story*, 2nd edn. (Springer, Berlin, 2000)
17. S. Xue, et al., Measurement of specific contact resistivity of ohmic contact on p-GaN. J. Semicond. **26**(5) (2005)
18. H. Ishikawa et al., Effects of surface treatments and metal work functions on electrical properties at p-GaN/metal interfaces. J Appl Phys **81**(3), 1315 (1997)
19. D. Wu, B. Yan, *Principle, Measurement and Processing of Ohmic Contacts on Metal-Semiconductor Interface* (Shanghai Jiaotong University Press, Shanghai, 1989). (in Chinese)
20. G. Lewis Brian, C Paine David, Applications and processing of transparent conducting oxides. MRS Bull. (2000)
21. Z. Wu, Z. Chen, X. Du, et al., Transparent, conductive carbon nanotube films. Science **305**, 27 (2004)
22. H.Y. Liu, V. Avrutin, N. Izyumskaya et al., Transparent conducting oxides for electrode applications in light emitting and absorbing devices. Superlattices Microstruct. **48**(5), 458–484 (2010)
23. J.H. Lim, E.J. Yang, D.K. Hwang et al., Highly transparent and lowresistance gallium-doped indium oxide contact to p-type GaN. Appl. Phys. Lett. **87**, 4 (2005)
24. C.S. Chang, S.J. Chang, Y.K. Su, et al., High brightness InGaN/GaN LEDs with indium-tin-oxide as p-electrode. Phys. Stat. Sol. **7**, 2227 (2003)
25. K. Nakahara et al., Jpn. J. Appl. Phys. **43**(2A), 180–182 (2004)
26. Li. Jinli, D. Hong, L. Caikun, Structure and optoelectronic properties of AZO transparent conductive film. Electron. Components Mater. **26**, 1 (2007). (in Chinese)
27. P.P. Edwards, A. Porch, M.O. Jones, et al., Basic materials physics of transparent conducting oxides. Dalton Trans., 2995–3002 (2004)
28. M.R. Graham, C.J. Adkins, H. Behar, R. Rosenbaum. J. Phy. Condens. Matter **10**, 809 (1998)

29. A. Ioffe, A. Rogel, *Progress in Semiconductors*, 4th edn. A.F. Gibson, F.A. Kroger, R.E. Burgess (eds.) (Heywood, London, 1960)
30. K. Nomura, H. Ohta, K. Ueda et al., Thin-film transistor fabricated in single-crystalline transparent oxide semiconductor. Science **300**, 1269 (2003)
31. T.Y. Wang, X.Y. Li, et al., Flip-chip technology. In *Proceedings Radar Technology Forum in China*, pp. 207–210 (2006). (in Chinese)
32. S. Du, Research on hot-press ultrasonic flip-chip welding process. In *Proceedings of the 14th National Conference on Hybrid Integrated Circuits*, pp. 364–368 (2005). (in Chinese)
33. Y.X. Wu, Z.Q. Long et al., Development status of thermal ultrasonic flip-chip welding for new processes of chip package interconnect. Electron. Process Technol. **25**(5), 185–188 (2004). (in Chinese)
34. M.K. Kelly, O. Ambacher, R. Dimitrov et al., Phys. Status Solidi **159**, R3 (1997)
35. J. Karpinski, J. Jun, S. Porowski, Equilibrium pressure of N_2 over GaN and high-pressure solution growth of GaN. J. Cryst. Growth **66**, 1 (1984)
36. P.R. Tavernier, D.R. Clarke, Mechanics of laser-assisted debonding of films. J. Appl. Phys. **89**(3), 1517–1536 (2001)
37. S. Wolf, *Silicon Processing for the VLSI Era* (Lattice Press, California Sunset Beach, 1990)
38. C.Y. Chang, *SM Sze* (ULSI Technology, The McGraw-Hill Companies, 1996)
39. Wu. Dexin, Q. He, *Modern Microelectronics Technology* (Chemical Industry Press, Beijing, 2002) (in Chinese)
40. Z. Zhengyong, *Semiconductor Integrated Circuit* (Tsinghua University Press, Beijing, 2001) (in Chinese)
41. S. Nakamura, G. Fasol, *The Blue Laser Diode* (Springer, Berlin, 1997)
42. Z. Teng et al., The design and fabrication of a GaN-based monolithic light-emitting diode array. J. Semicond. **34**(9), 094010 (2013)

Chapter 9
Packaging of Group-III Nitride LED

Group-III nitride LED encapsulation is one of the key technologies of solid-state lighting. The technology aims to improve light extraction efficiency, beam profile, color conversion, reliability and thermal management of the LEDs. This chapter will mainly introduce the LED related packaging materials, design, process and related technology. In the end, a brief description of the latest LED packaging trends is discussed.

9.1 Group III Nitride LED Packaging Materials

Encapsulation of solid-state lighting has its own characteristics. One of the essential issues is to control the raw materials. Considering the application environment of the devices, especially for those LEDs operated outdoor, the device will be vulnerable unless its reliability is optimized. Therefore, the control of the encapsulating material is important.

9.1.1 LED Chip

The first material used in LED packaging is LED chip itself. The function of LED is to convert electricity into light. LED can be categorized into three main types according to the number of surface electrode on the chip such as single, double and multi-electrodes chip.

LED chip with single electrode usually has vertical structure as illustrated in Fig. 9.1.

LED chip with double electrodes are commonly based on insulating substrate, e.g. sapphire. Hence the device has planar structure as shown in Fig. 9.2.

J. Li et al., *III-Nitrides Light Emitting Diodes: Technology and Applications*,
Springer Series in Materials Science 306,
https://doi.org/10.1007/978-981-15-7949-3_9

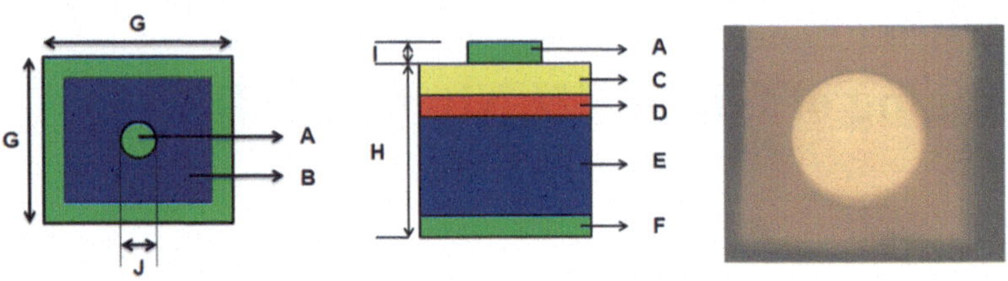

Fig. 9.1 Illustration of LED with single electrode. (A: p-pad; B: luminous region; C: p-type layer; D: n-type layer; E: n-type crystalline substrate; F: n-pad; G: chip dimension; H; chip height; I: p-pad thickness; J: p-pad diameter)

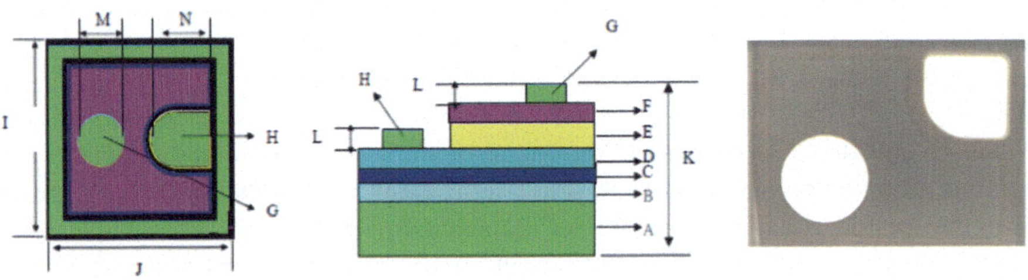

Fig. 9.2 Schematic of LED with double electrodes. (A: insulating substrate; B: buffer layer; C: n-type layer; D:active region; E: p-contact; F: transparent conductive layer; G: p-electrode; H; n-electrode; I: chip length; J: chip width; K: chip height; L: electrode thickness; M: p-electrode diameter; N: n-electrode diameter)

9.1.2 Lead Frame of LED

LED lead frame is the bottom base used in LED packaging. Its function is to obtain conduction and heat dissipation. It is one of the most important raw materials in LED packaging.

The chip is fixed on the lead frame, where the gold wire ball is welded with positive and negative electrodes and then sealed with packaging glue. This is the basic packaging process of LED. Because copper conducts electricity and heat well, LED stands often use copper [1, 2]. The inside of the lead frame includes lead wire, which is used to connect LED chip electrodes. After the packaging is formed, the packaging body is removed from the lead frame. The copper feet at both ends are positive and negative electrodes, which are used for welding to LED products.

Generally speaking, fabrication process of LED lead frame consists of the following steps: stamping, electroplating, injection molding, cutting and packaging. Low power-LED lead frame is normally copper silver plated while high-power LED lead frame is generally made of copper-silver-plated structure with plastic reflector cup. Among them, copper plays the role of connecting the circuit, reflection, welding, etc. Plastic mainly plays the role of reflection, providing the interface combined with glue and so on. The plastic material of Surface Mounted Devices (SMD) supports is mainly white Polyphthalamide (PPA) material. The material has the advantages such as high temperature welding resistance, high reflection, and combination with silica gel and reliability. On the other hand, high power lead frame is generally plastic reflector cup and rivet type heat dissipation structure.

9.1.3 LED Die Bonding Glue

There are two main types of glue for LED die bonding [3–5]: conductive silver glue and insulating glue. The conductive silver paste is mainly composed of a resin matrix, conductive particles, and a dispersion additive. It is an adhesive with certain conductivity after curing or drying. It can be cured at room temperature to 150 °C, which is much lower than that of the tin–lead soldering. This would prevent material deformation caused by high soldering temperature, thermal damage of electronic devices and the formation of internal stress. The requirements of the conductive silver paste are good electrical conductivity, good thermal conductivity, high shear strength, and strong adhesion. As far as insulating paste is concerned, insulating rubber is mainly used for heat dissipation and fixing of chips. The composition is mostly resin.

The application procedure and process of silver glue and insulating glue are very different. For example, silver glue needs to be stirred, the insulation glue does not need to be stirred; silver glue has good heat dissipation, and the insulation rubber has poor heat dissipation; and silver glue has stronger light absorption and weaker reflectivity. Therefore, the brightness of silver glue in LED products is lower than that of insulating glue.

9.1.4 Bonding Wire

In the LED packaging, two electrodes of the chip and the LED lead frame need to be connected by bonding wires so that electricity can be injected into the LED chip. LED bonding alloy wire is widely used in various packaging occasions since it has the advantages of high electrical conductivity, oxidation resistance, corrosion resistance, good toughness, etc. The alloy wire is mainly made of Au with a purity of more than 99.99%, which contains trace elements such as Ag/Cu/Si/Ca/Mg [6]. In addition, the diameter of the alloy wires on the market is in the range of 16 um to 50 um according to the requirement of applications.

9.1.5 LED Packaging Adhesive

There are two main types of LED package adhesives: epoxy resin and silica gel. Epoxy resin generally refers to an organic polymer compound having two or more epoxy groups in its molecule. It has the following advantages: (1) no by-product and the volume shrinkage during solidification process; (2) less restriction on the amount of filling agent, (3) good adhesion for most of the materials; and (4) excellent electrical and mechanical characteristics. However, epoxy resin has disadvantages such as weak ozone resistance and yellowing effect on the light transmission [7]. The basic structural unit of silica gel is composed of silicon-oxygen chain links. The side chains are connected with other organic groups. The special composition and molecular structure make it good performance compared with other polymer materials [8]. For instance, it has good thermal stability, stable chemical bonds of the molecules at high temperatures (or radiation), and operational over a wide temperature range. In addition, electrical properties of the materials are not significantly affected by temperature and frequency. In comparison with the same molecular weight hydrocarbon, it is a stable electrical insulating material with lower viscosity, weak surface tension, small surface energy and strong film forming ability. Currently, methyl silicone-based material is widely used in LED packaging.

9.1.6 Thermal Interface Material

LED heat dissipation could be affected by various factors including chip structure, thermal interface material, heat sink substrate material, and package structure. In order to improve the LED heat dissipation, besides the effort to reduce the thermal resistance of the chip and packaging structure, the thermal interface material between LED chip and heat sink is needed to fill the voids between the contact surfaces of the two materials to reduce the thermal resistance. Since air has a thermal conductivity of only 0.024 W / (m • K), existence of air voids between LED chip and heat sink would lead to relatively high thermal resistance if they are directly mounted together. Therefore, it is necessary to fill in these voids with a thermal interface material with high thermal conductivity to eliminates air and to establish an effective heat conduction path between the electronic component and the heat sink. Such a design would greatly reduce the contact thermal resistance, and fully utilizing the function of the heat sink. Good thermal interface materials should have high thermal conductivity, low thermal resistance, compressibility and softness, surface wettability, proper viscosity, good stability during thermal cycling, etc. Meanwhile, development of low cost and preparation technology should be explored. Currently, LED chip packaging generally includes: conductive silver paste, thermal paste, solder paste and tin-gold alloy eutectic solder.

Conductive silver paste is made by adding silver powder into epoxy resin. Its hardening temperature is low, and its heat conduction coefficient is about 20 W/(m·K). It has good thermal conductivity and good adhesion strength, but its light absorption is relatively large, resulting in a decrease in light efficiency.

The thermal conductive adhesive combines the matrix with some high thermal conductivity fillers such as SiC, AlN, Al_2O_3, SiO_2, carbon tubes, graphene to improve its thermal conductivity [9–13]. The thermal conductive adhesive has the advantages of low price, simple process, insulation performance, etc. Meanwhile, curing temperature of the material is low. It can even be cured at room temperature.

Gold tin alloy eutectic soldering is also used as thermal interface material. Adding a certain component to the metal can lower the overall melting point. The eutectic bonding, also known as low-melting alloy bonding, has many advantages such as high thermal conductivity, low electrical resistance, high reliability, and high reliability. Its basic feature is that two different metals can form alloys at a certain temperature, which is lower than their respective melting points. There are two main cases in LED eutectic soldering: one is to make a eutectic pad on the heat sink substrate, but the substrate production process is complex. The other is to prepare the bottom of the eutectic alloy directly at the bottom of the chip. Gold tin solder has better thermal conductivity and higher thermal fatigue resistance than tin-based or lead-based solder, and is an excellent packaging material.

For high-power LED packages, besides the requirement of high thermal conductivity, ideal thermal interface materials would have thermal expansion coefficient and elastic modulus that match the material of the chip substrate. There are also requirements for good mechanical properties, high operating temperatures, low material and process costs, and sometimes high requirements for the optical properties of the materials.

9.1.7 Substrate Material

Ideal material for making package substrate should have high thermal conductivity to transfer heat from the LED chip, while achieving heat exchange and electrical interconnection. Nowadays, high thermal conductivity material such as metal, ceramic, graphite and composite materials are widely used. In addition, in LED thermal design and thermal management, materials with low expansion properties and high thermal conductivity properties such as W/Cu, Mo/Cu, Cu/Mo/Cu, Cu/Mo70Cu/Cu are used. By adjusting the composition of the material, a good thermo-mechanical match can be formed with ceramic material, semiconductor material and metal material [14].

It is also a new type of heat-conducting composite material to evenly fill the polymer matrix material with heat-conducting filler. Generally, it can be divided into composite heat-conducting plastics and filling heat-conducting plastics, while filling heat-conducting plastics can be divided into heat-conducting insulating plastics and heat-conducting conductive plastics [15].

Due to its high thermal conductivity, silicon carbide matches well with the thermal expansion coefficient of the LED chip. Meanwhile, thanks to its high thermal conductivity, graphene has great potential to be used as heat dissipation substrate material. Recently, high thermal conductivity graphene/carbon fiber flexible composite film has been developed in combination with carbon fiber [16, 17], which has made important progress in the field of flexible heat dissipation of graphene.

At present, the LED package substrate mainly includes printed circuit board (PCB), metal-core printed circuit board (MCPCB), direct-bonding copper substrate, ceramic substrate, copper-plated substrate, aluminum and silicon substrate.

PCB is widely used in LED packaging. However, thermal conductivity of the material is relatively low. PCB with vertical structure for better heat dissipation channel is the development trend. MCPCB consists of metal plate, insulation layer and copper. Thermal conductivity of the structure can be further improved by replacing the organic insulating layer with high thermal conductive material. Direct bonding copper substrate is ceramic plate with copper-cladding layer with good electrical and thermal conductivity. Thermal expansion coefficient of such material is very similar to that of sapphire substrate, which enables high reliability. However, application of the material is limited by circuit fabrication accuracy and production cost.

The low temperature co-fired ceramic (LTCC) substrate has simple structure and little thermal interface. Direct copper plating substrate has the advantages of high thermal conductivity, low process temperature, low cost, fine line and high reliability, which is very suitable for high-power LED packaging requirements with high alignment accuracy. Aluminum substrate surface is insulated by applying anodic oxidation treatment and the substrate has advantages such as simple process, low cost, high thermal conductivity, high temperature resistance, good impact resistance performance. Disadvantage of the substrate is that the anodic oxidation layer strength is insufficient, which limits the practical application. Silicon substrate has high thermal conductivity and good thermal mismatch with LED chip material. Most importantly, processing technology of the material is mature. Especially through-silicon via (TSV) technology can greatly improve the integration and the cooling capacity of the device. In recent years, transparent ceramic material technology is developed. It can achieve high heat dissipation efficiency, better matching of thermal expansion coefficient, excellent electrostatic protection and other excellent thermal and electrical properties. It is expected to make breakthroughs in optical properties.

9.2 Group III Nitride LED Encapsulation Process

Although LED packaging forms may vary according to their application situations, shapes, heat dissipation design and luminescence mode, their packaging processes basically are the same, mainly comprising raw material preparation, die bonding, sealing, testing and other processes.

(1) Raw materials preparation: as illustrated in Fig. 9.3, LED chips are examined by microscope. During the process, any chip surface damage, chip size, electrode size can be confirmed. As far as the lead frame of LED is concerned, it is necessary to check the size and surface of the lead frame for abnormalities.

(2) Die bonding: as shown in Fig. 9.4, adhesive is placed onto the LED lead frame for the following die attach process. The adhesive generally consists of silver paste and insulating paste. For GaAs, SiC and other conductive substrates, silver paste is generally used for die bonding while enables good electrical conductivity. On the

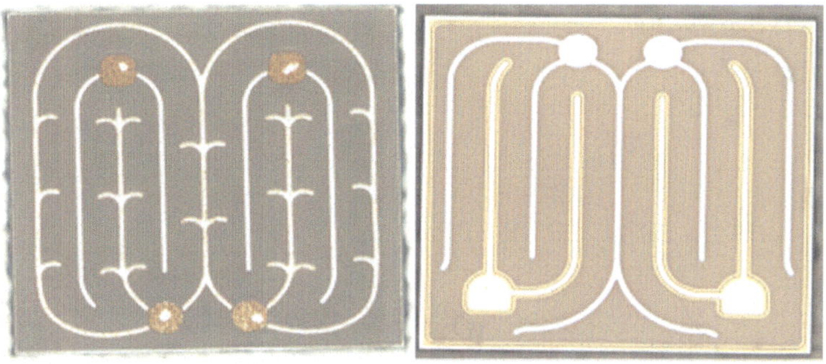

Fig. 9.3 Microscope image of high-power LED chips

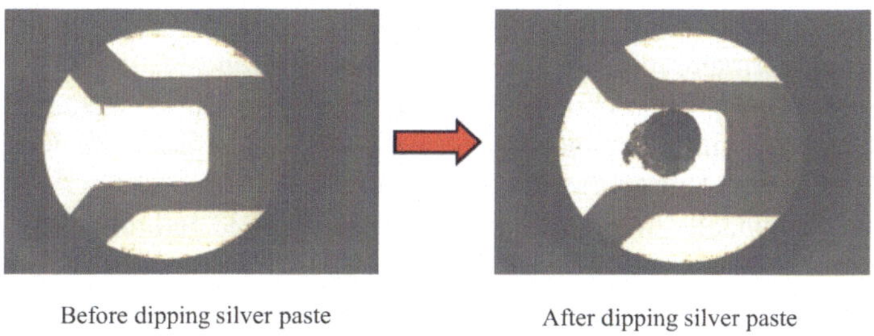

Fig. 9.4 Lead frame before and after dipping silver paste

other hand, for LED chips grown on insulated substrate such as sapphire, high-power chips would use silver paste in order to achieve good thermal performance. For small chips, insulating paste is chosen. This is mainly driven by cost consideration.

In actual production stage, automatic die bonding equipment is commonly used, which is shown in Fig. 9.5. The equipment dispenses adhesive onto the lead frame before grabbing and placing LED chip onto the lead frame.

(3) Curing of die bonding adhesive: die attach adhesive is cured primarily to make sure LED chips being firmly bonded onto the lead frame. The curing condition

Fig. 9.5 Automatic die bonding equipment is grabbing LED chips

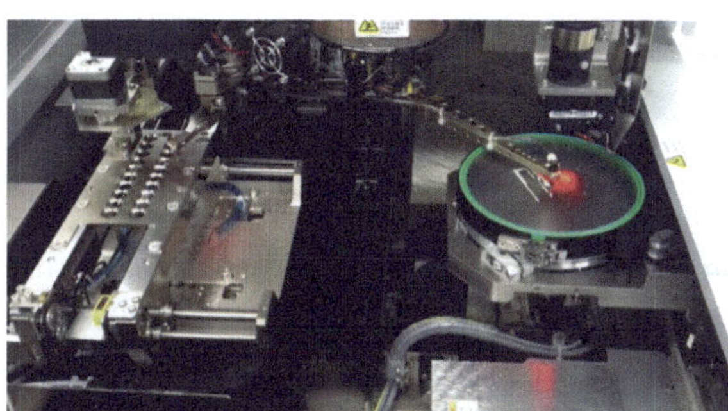

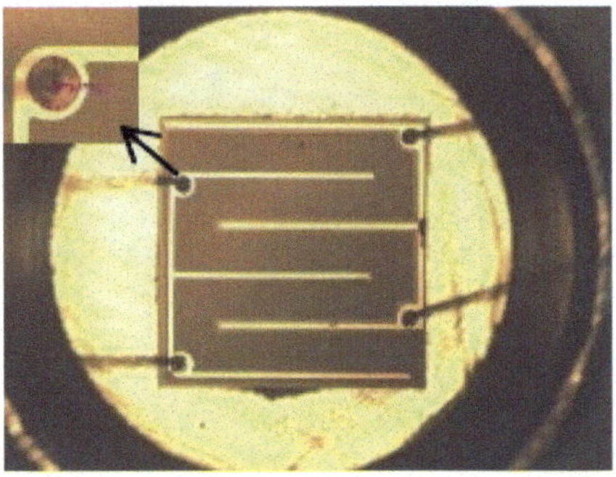

Fig. 9.6 LED chip wire ball bonding

of silver paste would generally require 150 °C baking for 2 h. The curing of the insulating paste would require 1-h bake at 150 °C. Furthermore, in order to prevent contamination, curing requires the use of a dedicated oven.

(4) Wire bonding: this process is a key link in LED packaging technology, which aims to achieve circuit interconnection between the LED chip and the lead frame as illustrated in Fig. 9.6. During the process, it is necessary to monitor wire arch shape, solder joint, tension, etc. in order to improve packing yield and reliability.

(5) Sealing: LED packaging and sealing mainly include dispensing and molding. During dispensing, it is essential to control and prevent the formation of air bubbles and impurities. For white LEDs particularly, problems could arise during the process such as light and chromatic aberration caused by phosphor precipitation. In molding process, LED lead frame is loaded into the mold, clamping the upper and lower molds with a hydraulic press and vacuuming, and putting the solid epoxy resin into the injection molding before entering into each LED forming groove along the glue path and cured.

(6) Cutting, testing and packaging: since a number of LED lead frame units are connected together, dicing machine is used to complete the separation in order to obtain individual devices. Then the LEDs are tested to obtain their optical parameters and dimension. The devices can then be sorted and categorized under specific requirement. Finally, the devices are packaged and the entire LED encapsulation process is completed.

9.3 LED Packaging Technology

In order to better ensure the photoelectric performance and reliability of power LED, LED packaging technology should meet the following requirements: the packaging structure should have a high light efficiency, and the thermal resistance should be reduced as far as possible. At present, according to different LED chips and application needs, the LED packaging technology used is also different, including high power density packaging technology, white light packaging technology, ultraviolet packaging technology and wafer level packaging technology.

9.3.1 White LED Package Technology

White LED packaging technology is the basis for the development of solid-state lighting technology. Different approaches are employed to realize white LED, such as phosphor conversion and color-mixed approach. Taking phosphor conversion as an example, the method was first invented by Nichia, the blue light emitted from the LED chip excites phosphor to obtain yellow-green light, yellow-green light and blue light synthesize white light. Color rendering index of the design can be further improved by adding red and green phosphors and the colorimetric parameters needs to be optimized. Combination of ultraviolet LED and trichromatic phosphor is another promising method, which should provide better light color parameters. However, the efficiency is still not high enough for practical application.

The structure and technology of phosphor coating can be classified as coating and prefabrication. The coating method includes mixing phosphors in the encapsulating gel, which covers the LED chip. When the light from the LED chip passes through the encapsulating mixture, the phosphors can be excited and producing white light. This is the traditional packaging structure because the packaging method is simple and widely used in industry.

With the development of technology, a new type of phosphor coating structure is proposed. In such design, fluorescent material is directly coated on LED chip. The advantage of this structure is fluorescence material directly coated onto the chip, which gives the possibility of controlling the thickness and uniformity of fluorescent materials and improving the luminescence efficiency and white light quality. It needs to be pointed out that due to density differences between phosphor powder and silica gel, phosphor powder would gradually precipitate in the mixture and may lead to the phenomenon of precipitation or concentration stratification in the phosphor layer. In addition, the viscosity of silica gel will change obviously at different temperatures. In the curing process of phosphor glue, the viscosity of silica gel will first drop and then rise as the temperature rises, which will lead to the precipitation of phosphor powder in silica gel to different degrees. Therefore, pre-prepared fluorescent film has been proposed and used during the process to improve the issue.

As far as color-mixed approach is concerned, the method is the trend of future development of solid-state lighting. Such design would enable a wide range of dynamic color adjustment. However, it would require complex control and feedback system.

9.3.2 UV Packaging Technology

With the maturity of the visible light field, short-wavelength ultraviolet (UV) light has gradually become the focus of research. According to the wavelength, ultraviolet light can be divided into near-ultraviolet UVA (320–400 nm), UVB (275–320 nm) and far-ultraviolet UVC (100–275 nm). UV-LED has the advantages of compact size and full solid state. In contrast with UVB and UVC LEDs, UVA-LEDs have higher luminous efficiency. The luminous efficiency of UVB and UVC-LEDs still has great potential due to the use of different materials [18–22]. Ultraviolet light plays an irreplaceable role in many areas including optical sensors and instruments (230–400 nm), ultraviolet identification barcode (230–280 nm), ultraviolet water sterilization and purification (240–280 nm), identification and body liquid detection and analysis (250–405 nm), protein analysis and pharmaceutical invention (270–300 nm), medical light therapy (300–320 nm), polymer and ink printing (300–365 nm), money detection (375–395 nm), sterilization (390–410 nm), and so on. In the production and application of UV LEDs, efficient and reliable UV packaging are the important links. Due to the high photon energy of UV light, the raw materials, processes and testing methods used in conventional white light LED packaging may not be suitable for UV-LEDs. UVA band LED has higher luminous efficiency, and its wavelength is close to that of visible light band. Its packaging form is similar to the general LED packaging, with more consideration in packaging raw materials and reliability. Due to the fact that the efficiency of UVB and UVC-LED is still not high enough, light source with module packaging form is an important way to enter the application market.

TSLC recently published a series of complete UV-LED component products, UV wavelength ranging from 365 to 420 nm. Using aluminum nitride (AIN) substrate packaging technology to improve the UV-LED components and heat dissipation, it can effectively conduct chip heat into the substrate. Furthermore, the light output has adopted Quartz glass lens to reduce optical loss and increase the uniformity [23]. Sensor Electronic Technology, Inc. (SETi) has launched a protective case for mobile devices for reducing cross-infection [24]. High Power Lighting Corp. (HPL) introduces the whole series of UV-LED (365–410 nm) package products currently designed for different curing, printing, exposure and sterilization applications by using the patented technology of copper substrate [25].

Encapsulation or packaging technology has an important impact on the improvement of UV-LED light power, heat dissipation and other performance. At present, there have been a lot of researches on optics and heat [26, 27]. The main functions of general package include mechanical support, heat dissipation, signal transmission, chip protection and so on. The package of LED needs to protect the chip from mechanical, heat, moisture and other external impact under normal current flow, so that the chip working temperature does not exceed the permitted range.

UV-LED encapsulation consists of several types, transistor outline (TO-can), surface mounted devices (SMD), and chip on board (COB) package. In 2002, Nichia introduced ceramic packaged UV-LED products with wavelengths ranging from 365 to 385 nm. Seoul Semiconductor's SVC and firms in Taiwan have launched UV LED products in metal substrate SMD packaging. Stanley Electric developed an ultraviolet LED using a glass package and demonstrated it on "CEATEC JAPAN 2010". SETi UVTOP series has successfully launched 240–355 nm UV-LEDs products, which are packaged in TO-39 and TO-18. In 2014, the firm announced the launch of its latest SMD UVC-LED products. Packaging manufacturers in mainland China have also introduced SMD and COB packaged UV-LEDs products. Despite the diversity of packaging methods, it still has a long way to meet customer needs. It is urgent to find a stable and reliable packaging method.

Packaging raw materials also have a critical impact on the performance and reliability of UV LEDs. The choice of encapsulant and lens material directly affects the luminous efficiency and lifetime of UV-LEDs. Generally, the encapsulant should have high transmittance, high refractive index, good thermal stability, good fluidity and easy for spraying. Meanwhile, in order to improve the reliability of the UV-LED package, it is ideal for the material to have low hygroscopicity, low stress, weather resistance and environmental protection. For UV-LED packaging, quartz glass has the highest transmittance (up to 95%), followed by silicone. The UV transmittance of epoxy resin decreases sharply with wavelength. However, despite the high ultraviolet light transmittance of quartz glass, its high thermal processing temperature is not suitable for sealing. Therefore, in UV-LED packaging process, quartz glass is generally used as the outermost lens material. Epoxy resin has a poor resistance to UV and heat resistance, while silicone resin has an obvious advantage in UV region. Therefore, epoxy resin is generally not used as the encapsulating lens material of UV-LED, but silicon resin and glass are used as the encapsulating material. In order to improve performance and reliability of UV-LEDs, glass encapsulated UV-LED is currently one of the trends to avoid undesired materials aging issues induced by UV exposure.

9.3.3 High Power Density Packaging Technology

A large amount of heat is generated during the operation of high-power LED. The generated heat should be transferred to the heatsink as soon as possible to avoid heat accumulation. Otherwise, excessive heat will reduce luminous efficiency and reliability of the LED. In 1998, Lumileds in the US launched high-power LED (1 W Luxeon device), which enables the possibility of using LEDs for general lighting, rather than just working as indicating light source [28].

In order to meet the requirements of surface mount technology, Cree uses ceramic substrate as the carrier of LED chip instead of the original metal plastic lead frame and heat sink, which has good heat dissipation effect, small size and is suitable for SMD process. From 2007 to 2009, Cree company launched xp-c, xp-e and xp-g series of optical lenses with the size of 3.5 mm × 3.5 mm and the direct modeling of optical lenses on ceramic substrates, eliminating metal reflecting cups and optical glass lenses.

Early power LED suffers from low degree of integration and productivity due to the use of conventional lead frame structure rather than the planar structure. The recently introduced high-power LED packaging process uses a planar, array- arrangement, together with encapsulation of the silicone optical lens structure in a planar area, which can be called a planar array packaging process. During the processing, the

Fig. 9.7 300 W LED light source module and luminescence performance

substrate can be processed into a plurality of matrix arrangements as needed, which is beneficial to the realization of mass production.

COB packaging was introduced into LED encapsulation for further improvement in luminous flux and color quality. Meanwhile, the proposed packaging design would enable high integration, compact size and cost effectiveness. Multiple LED chips are directly bonded to the substrate, packaging and assembly process has been simplified. In addition, the entire surface of the packaged LED array is covered by phosphor lead to improve output uniformity. Figure 9.7 shows the schematic diagram and luminescence characteristics of 300 W high-power density LED light source module.

At present, the input electric power density of most COB LED light sources is below 0.3 W/mm^2. The power density of a few single-die packages can reach 1–2 W/mm^2 [22]. With further development in high thermal conductivity substrates and high-intensity technologies, the injected electric power density and output light intensity of the LEDs would increase, which would expand the application horizon of the light source.

9.3.4 Wafer Level Packaging Technology

There are several interrelated concepts within the wafer-level packaging technology industry, such as wafer-level packaging (WLP), chip-scale packaging (CSP), flip-chip, etc.

As far as WLP is concerned, packaging and even testing are directly processed on the substrate before the wafer is scribed to obtain individual devices. Such technology has a list of advantages such as compact packaging size, small parasitic parameters and low cost. Therefore, it has been widely used in integrated circuit industry. Main technical difficulties of wafer level packaging technology are high precision machining technology, electrical connection establishment, structure of optical conversion layer and the most important sealing reliability of thin protective layer. CSP technique refers to the package size is similar to that of the chip. Generally speaking, the area ratio of the chip and package is less than 1.2. Flip-chip refers to a chip that can be flip-chip soldered, generally refers to a chip that has not been packaged. In some cases, it is also used to refer to a package that can be used

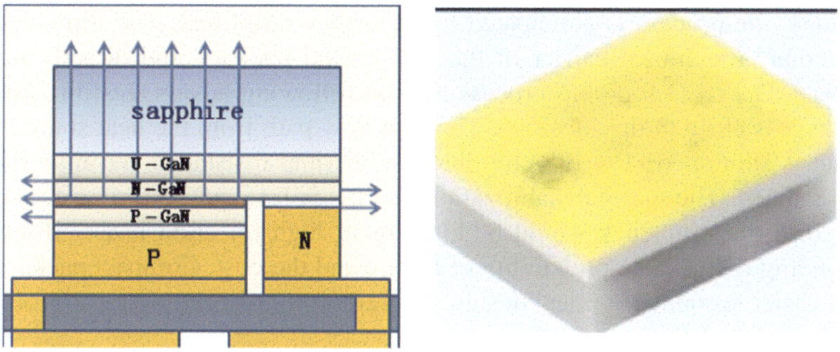

Fig. 9.8 Schematic diagram of the unpackaged chip structure, using flip chip, to achieve the package in the chip process

directly using a flip-chip process. IC packaging technology has certain reference and enlightenment meaning to the LED packaging. In recent years, LED packaging has also taken WLP technology as an important development trend. Figure 9.8 shows the structure and device illustration of WLP LED chip implemented by flip-chip technique.

In 2009, Mitsubishi Electric Co., Ltd. announced that it has successfully developed wafer-level bonding [29]. In the design, wafers with encapsulated LEDs and wafers equipped with photosensors were bonded, together with another two wafers. The package contains three-color LEDs, a heat-dissipating substrate layer, a light-reflecting substrate layer, a monitor wafer equipped with a sensor (for sensing LED light), and a light-absorbing substrate having a dual function of light diffusion. The heat-dissipating substrate extracted heat from the bottom of the package by using a via wiring. The company also prototyped an illumination system by arranging the components in an array. The light monitor can sense the LED light and feedback to control the brightness of the LED. Therefore, it can emit light with uniform color and brightness, and can also set the color and brightness. In 2011, VisEra Technology Co. Ltd. and NCSIST jointly developed wafer-level aluminum nitride LED technology and developed 20 W ultra-high power VISES series LED aluminum nitride package. In 2013, Taiwanese semiconductor lighting company announced packaging products on 8-inch silicon wafers developed using SMD and inkjet phosphor coating technology. In 2014, Toshiba developed a white LED with 0.65 mm illumination using wafer-level chip-scale packaging technology. The package area was reduced by more than 50%. The small light source increased the flexibility of the lighting design and expanded the process space of the package [30].

LED flip chip refers to a chip that can be directly bonded to a substrate to complete electrical connection, sometimes referred to a direct patch chip. Flip-chips have also undergone major changes. Early flip-chips were first flipped onto silicon or other material substrates, and then the entire package was soldered as a light-emitting chip. The recent flip-chip process has been simplified, and the chip can be directly attached to the package substrate. In the future, with further development of the technology, the process is anticipated to be further simplified. The flip-chip itself may become a technical carrier of the wafer level package and directly used as a package. The LED flip-chip process has the following advantages: the flip-chip active layer is close to the substrate. The heat flow path from the heat source to the substrate is shortened. The flip-chip has low thermal resistance and high thermal response speed. The thermal stabilization process is fast and performance changes are small. Light efficiency would be high under high injection level. Absence of wire bonding reduces the probability of failure and the cost. Compact package size enables easier secondary optical design. Therefore, it is a necessary path for major development in LED encapsulation.

9.4 Package and System Cooling Technology

9.4.1 Packaging and System Cooling Technology

Chip cooling technology is a key issue that must be addressed in high power density LED packages. When a high-power LED is in operation, a large amount of heat is generated. Excessive heat accumulation will cause an increase in the temperature of the LED chip, resulting in a series of problems that reduce luminous efficiency and reliability of the device. High-power LED heat dissipation technology mainly includes chip placement, packaging material selection (substrate material, thermal interface material) and process, heat sink design and so on.

LED package thermal resistance mainly includes material internal thermal resistance and interface thermal resistance. The function of the heat dissipation substrate is to absorb the heat generated by the chip and conduct it to the heat sink to achieve heat exchange with the outside world. The commonly used heat-dissipating substrate material is a copper-clad ceramic plate, which is sintered by a ceramic substrate and a conductive layer (Cu) under high temperature and high pressure. It has good thermal conductivity, high strength, strong insulation, and reduces thermal stress of the packaging material.

At the same time, the package interface of the LED has a great influence on the thermal resistance. If the interface cannot be processed correctly, it is difficult to obtain a good heat dissipation effect. For instance, a well-contacted interface at room temperature may have gaps at high temperatures. The warpage of the substrate may also affect soldering robustness and local heat dissipation. The key step of improving LED packaging is to reduce interface and interface contact thermal resistance and enhance heat dissipation. Hence, the thermal interface material must be carefully selected to meet the technical requirements. The use of low temperature or eutectic solder, solder paste or conductive paste with nano-particles as the thermal interface material can greatly reduce the interface thermal resistance.

High-power LED luminous intensity and power dissipation are high, efficient way of cooling is much needed. In addition, stable encapsulating material and advanced packaging technology are needed to solve the long-term reliability problems. From the actual needs of LED lighting applications, the development of power-type packaging technology with low thermal resistance, high reliability, micro, and excellent optical characteristics, as well as high-power LED packaging and system integration technology are the key to realize the industrialization of LED lighting. This is also an important development trend.

The cooling technologies involved in LEDs include a list of methods, which will be discussed below.

Natural cooling: natural convection cooling is often used for low-power devices and single high-power devices to dissipate heat. Fin cooling is the most common method.

Air-cooling: ordinary fans have low noise, high power consumption, and poor reliability. The rotating magnetic field around the fan drives the normal operation of the surrounding electronic components due to magnetic leakage or electric spark, so it is rarely used in LED heat dissipation.

Liquid cooling: it consists of water cooling, microchannel cooling, micro spray cooling, porous micro heat sink cooling, liquid metal cooling.

Heat pipe cooling: it is a phase change heat transfer element with high thermal conductivity. It can be also considered as a heat conduction element that relies on internal working phase transformation for efficient heat transfer. It has a large heat transfer rate with a small temperature difference, simple structure, and a unidirectional heat conduction. Therefore, there are many advantages to apply heat pipe technology in SSL. Ordinary aluminum profiles, brazing and other types of heat sinks could not meet the thermal resistance requirements, and need to adopt a new type of radiator with superior performance.

Micro-slot group composite phase change heat dissipation technology: this is a new generation of high-intensity micro-size fine-scale phase-change heat transfer technology, which is developed on the basis of heat pipe technology. It avoids technical defects of heat pipes, and significantly improves technical capabilities. In micro-scale micro-groove geometry, there is very small amount of working fluid. The fluid flows along the micro-groove under the capillary pressure gradient formed by the micro-groove group's own structure. No additional power consumption is required [31].

Semiconductor thermoelectric refrigeration: It is also called thermoelectric refrigeration or heat pump. It has the advantage of no sliding parts and is used in applications where space is limited. The reliability is high, and there is no refrigerant contamination. Using a semiconductor material Peltier effect, when the galvanic series into a direct current through two different semiconductor materials, at both ends of the couples can respectively absorb heat and release heat, cooling purposes can be achieved. It is a refrigeration technology that produces negative thermal resistance, which is characterized by no moving parts and high reliability.

9.4.2 LED Thermal Testing Technology

The generation and extraction of heat is one of the essential factors that affecting the performance of LED devices. Thermal management study in LED encapsulation includes material selection, process design and interface like plurality of aspects. Poor thermal management of LED devices will lead to a list of undesired results such as rising junction temperature, decreased luminous efficiency, wavelength drift, color temperature variation, electrical performance degradation, thermal stress increase, component lifetime loss, and phosphor quality degradation. One of the main indicators for evaluating the quality of thermal management is the thermal resistance of the device. It is an indicator mainly reflects the heat dissipation capability caused by the medium used by the device and the related device structure. Variety of factors would affect the thermal resistance, which comprises a LED chip structure and materials, thermal interface materials, heat sink material, sectional area of heat path, thermal structure design, temperature change of the thermal conductivity, ambient temperature, etc. In order to maximize LED output power per unit area, higher power density LED package shape is the future development trend, and LED's thermal management becomes increasingly important.

At present, LED thermal testing mainly includes infrared thermal imaging, electrical parameters, transient thermal testing, spectral and optical power methods, and pin temperature methods. These methods measure or derive the junction temperature and thermal resistance of the LED based on different principles.

Infrared thermal imaging refers to the equipment or method that detects the infrared energy emitted by an object, converting the infrared energy into a temperature value, and thereby the real-time temperature distribution of the measured objected can be obtained in the form of image. The measurement can be made in a distance, without making direct contact with the object.

The principle of electrical parameter method is based on a good linear relationship between the forward voltage drop VF of LED and the temperature of pn junction, commonly known as K factor. After obtaining the K factor, the junction temperature of LED can be obtained by measuring the voltage change of LED. During the whole measurement process, LED is not only the measured object, but also the sensor.

Transient thermal test method: the thermal characteristics of LED can be described by equivalent RC circuit. The relationship between the transient response of LED to thermal step input and the physical characteristics of each part of the package is used to obtain the thermal resistance-heat capacity curve of each part of the LED package [32].

9.5 Development Trend of LED Encapsulation Form

According to the application requirements, LED packaging technology has gone through four stages: lead-type LED package, surface mount electronic SMT package, COB chip-on-board LED package and SiP system electronic package-like LED package [33]. Among them, the lead-type electronic package is a commonly used 3–5 mm electronic package structure, which is commonly used for indicator. The LED packages usually have low current (20–30 mA) and low power (less than 0.1 W). The disadvantage of this type of package is that it has a large thermal resistance (generally higher than 100 K/W) and short lifetime. Surface mount technology is an electronic packaging technology that can directly attach and solder electronically packaged devices to designated position of PCB. One of its main development trends is currently applied to wafer level LED packages. COB is the abbreviation of Chip-on-Board. It is an electronic packaging technology that directly bonds LED chips onto PCB through adhesive or solder, and then electrically interconnects the chips and PCB boards through wire bonding. In contrast with SMT, COB greatly increases the power density of the electronic package while reducing the thermal resistance of the electronic package. SiP (System in Package) is a new type of electronic package integration method. It is developed on the basis of system on chip (SOC) in order to meet the requirements of portable development and system miniaturization. SiP-LEDs can be used to assemble multiple light-emitting chips in an electronic package. They can also be integrated with other functional devices (such as power supplies, control circuits, optical microstructures, sensors, etc.) for building a more complex system.

References

1. L. Zhang, Y. Zhu, W. Wang et al., Study on Ag-plated Cu lead frame and its effect to LED performance under thermal aging. IEEE Trans. Device Mater. Reliab. **14**, 1022 (2014)
2. L. Zhang, Y. Zhu, H. Chen et al., Failure analysis on reflector blackening between lead frame electrodes in LEDs under WHTOL test. Microelectron. Reliab. **55**, 799–806 (2015)
3. J.W. Park, Y.B. Yoon, S.H. Shin et al., Joint structure in high brightness light emitting diode (HB LED) packages. Mater. Sci. Eng. A **441**(1), 357 (2006)
4. F.P. McCluskey, M. Dash, Z. Wang et al., Reliability of high temperature solder alternatives. Microelectron. Reliab. **46**, 1910 (2006)
5. C. Chen, C.M. Chen, R.H. Horng et al., Thermal management and interfacial properties in high-power GaN-based light-emitting diodes employing diamond-added SAC305 solder as a die-attach material. J. Electron. Mater. **39**(12), 2618–2626 (2010)
6. Y. Tseng, F. Hung, T. Lui, Microstructure, tensile and electrical properties of gold-coated silver bonding wire. Microelectron. Reliab. **55**, 608 (2015)
7. D. Barton, M. Osinski, P. Perlin et al., Life tests and failure mechanisms of GaN/AlGaN/ InGaN light emitting diodes, in *Proceedings of 35th Annual IEEE International Reliability Physics Symposium*, 8–10 Apr, 1997 (Denver, 1997)
8. E.F. Schubert, *Light Emitting Diodes*, 2nd edn. (Cambridge University Press, 2006)
9. M. Li, J. Zhang, X. Hu et al., Appl. Phys. A Mater. Sci. Process. **119**(2), 415 (2015)
10. M. Jeong, S. Jeon, S. Lee et al., Effective heat dissipation and geometric optimization in an LED module with aluminum nitride (AlN) insulation plate. Appl. Thermal Eng. **76**, 212 (2015)
11. K. Zhang, G.W. Xiao, C. Wong et al., Study on thermal interface material with carbon nanotubes and carbon black in high-brightness led packaging with flip-chip technology, in *Proceedings of 55th Electronic Components and Technology Conference*, May 31–Jun 4, 2005 (Lake Buena Vista, 2005).
12. Z. Yan, D.L. Nika, A. Balandin, Thermal properties of graphene and few-layer graphene: applications in electronics. IET Circuits Devices Syst. **9**(1), 4–12 (2015)

13. N.H. Hashim, P. Anithambigai, D. Mutharasu, Thermal characterization of high-power LED with ceramic particles filled thermal paste for effective heat dissipation. Microelectron. Reliab. **55**(2), 383–388 (2015)
14. D. Wang, D. Chen, Y. Yang et al., Interface of Mo-Cu laminated composites by solid-state bonding. Int. J. Refract Metal Hard Mater. **51**, 239 (2015)
15. Y. Mamunyaa, V. Davydenkoa, P. Pissisb et al., Electrical and thermal conductivity of polymers filled with metal powders. Eur. Polymer J. **38**, 1887 (2002)
16. L. Li, C. Wang, Study and Application of Thermal Conductive Plastic. Chin. Polym. Bull. **7**, 25 (2007). ((in Chinese))
17. Q. Kong, Z. Liu, J. Gao et al., Hierarchical graphene-carbon fiber composite paper as a flexible lateral heat spreader. Adv. Funct. Mater. **24**, 4222–4228 (2014)
18. A. Yang, J. Wang, J. Wang et al., Fluorescence spectrum of organic light transform films. J. Optoelectron. Laser **16**(1), 17–20 (2005)
19. T. Tsunemasa. Japanese semiconductor lighting project based on ultraviolet LED and phosphor systems, in *Proceedings of Conference on Solid State Lighting and Displays*, Jul 31–Aug 1, 2001 (San Diego, 2001)
20. M.A. Khan, Deep-ultraviolet LEDs fabricated in AlInGaN using MEMOCVD, in *Proceedings of 4th International Conference on Solid State Lighting*, 3–6 Aug 2004 (Denver, 2004).
21. D. Purnendu, Q. Li, H. Temkin et al., Applications of deep UV LEDs to chemical and biological sensing, in *Proceedings of 4th International Conference on Solid State Lighting*, 3–6 Aug 2004 (Denver, 2004)
22. M. Hubert, C. Dimas, P. Orava et al., High-power UV LED array for curing photo-adhesives, in *Proceedings of 6th International Conference on Applications of Photonic Technology*, 25–29 May 2003 (Montreal, 2003)
23. https://www.tslc.com.tw/news. Accessed 9 Jun 2020
24. https://www.semiconductor-today.com/news_items/2013/NOV/SETI_121113.shtml. Accessed 9 Jun 2020
25. https://www.hplighting.com.tw/products.php. Accessed 9 Jun 2020
26. L. Kim, M. Shin, W. Moo, Thermal analysis and design of high-power LED packages and systems, in *Proceedings of 6th International Conference on Solid State Lighting*, 14–17 Aug 2006 (San Diego, 2006)
27. I. Moreno, Configurations of LED arrays for uniform illumination, in *Proceedings of 8th Latin American Meeting on Optics, Lasers, and Their Applications*, 3–8 Oct 2004 (Porlamar, Venezuela, 2004)
28. D.A. Steigerwald, J.C. Bhat, D. Collins et al., Illumination with solid state lighting technology. IEEE J. Sel. Top. Quantum Electron. **8**(2), 310–320 (2002)
29. Z. Cong-fu, Wafer-level packaged LED by Panasonic. Semicond. Inf. 5, (2009) (in Chinese)
30. https://compoundsemiconductor.net/article/91810-toshiba-reveals-ultra-small-chip-white-leds.html. Accessed 9 Jun 2020
31. N. Xuelei, T. Wang, X. Hu et al., Laser cooling with phase-change heat transfer in microgrooves and TEC. J. Eng. Therm. **30**, 8 (2009). ((in Chinese))
32. G. Farkas, Q. Vader, A. Poppe et al., Thermal investigation of high power optical devices by transient testing. IEEE Trans. Compon. Packag. Technol. **28**(1), 45–50 (2005)
33. J.R. Lin, T.Y. Ng, Z. Wang et al., Wafer-level LED-SiP based mobile flash module and characterization. Microelectron. Reliab. **52**(5), 916–921 (2012)

Chapter 10
Reliability Analysis of Group III Nitride LEDs Devices

In addition to productivity and energy saving, another outstanding advantage of III-nitride LEDs is its long lifetime predicted to reach a record of over 10,000 h. Under normal circumstances, the LEDs would not be instantly total failure, but decays after a period of operation due to the improper design or production process which may not meet the operating requirements. In the IES LM-80, the LEDs' lifetime was defined as the time required by initial value of light intensity decaying to 70% or 50%. In practical applications, LEDs reliability is closely related to various factors such as material growth, preparation technology, and packaging process. Therefore, the overall device performance will be affected by any weakness involved in these steps. It's of great importance to raise luminous efficiency and cut the cost for the further application of III-nitride LEDs. Moreover, how to ensure high reliability and predict the lifetime of LEDs is becoming increasingly prominent.

Reliability analysis is a multidisciplinary performance test, comprising semiconductor optics, electrics, thermodynamics, chemistry, etc. Reliability analysis generally studies the photoelectric decay properties of LEDs devices through aging tests. Based on such test, people could predict the lifetime of the LEDs, analyze its decay mechanism, and find weak steps. People could apply these results in material growth, chip preparation, packaging technology, etc., to improve the overall reliability of LEDs devices.

10.1 Failure Mode and Failure Analysis

Because of the complex structure and process of power-type white LEDs devices represented by GaN-based heterostructures, researchers have also studied the failure mechanism of GaN-based white LEDs while further improving the luminous efficiency. Studies show the failures that affect GaN-based white LEDs devices can be divided into two categories: one is the decay or failure of light due to factors such

J. Li et al., *III-Nitrides Light Emitting Diodes: Technology and Applications*,
Springer Series in Materials Science 306,
https://doi.org/10.1007/978-981-15-7949-3_10

as defects in the epitaxial material itself, ohmic contact degradation, etc. the other one is the packaging process. The potential problems can come from the aging and discoloration of the lens, the reduction of the excitation efficiency of the phosphor, and the failure caused by improper solid-state bonding wires.

10.1.1 Light Decay

After GaN-based LEDs work for a period of time, the light output power decays as the result of the degradation of the active area, the p-type GaN layer and the ohmic contact.

(1) Degradation of the active region

Because there is no matching substrate material, epitaxially grown films often contain a large number of defects. When the device is working, these defects, under the influence of external temperature or current, can act as non-radiative recombination centers and carrier tunneling channels. This can lead to the formation of low-resistance ohmic channels, which cause LEDs light attenuation. As the injected current becomes larger and the stress time becomes longer, the output optical power degradation will become more obvious as shown in Fig. 10.1 [1]. The electrical characteristics of the LEDs after failure are mainly manifested by the same magnitude increase in the leakage current at low voltages in the reverse and forward directions [2].

Research and analysis show that there are three main sources of non-radiative recombination centers in LEDs devices: ① Existing non-radiative recombination defects (point defects, dislocations, etc.) in the device material have migrated or expanded to the active region [3],② Mg impurities atoms diffused into the active region [4] and acted as non-radiative recombination centers,③ New N vacancies were generated in the active region [5].

(2) Degeneration of ohmic contact

The degradation of ohmic contacts and the current crowding effect can lead to light attenuation. Comparing the I-V characteristics before and after aging, the increased parasitic series resistance of the degraded LEDs makes the forward current decrease at the same voltage bias. It is generally believed that the semi-transparent ohmic contact and the upper surface of the p-type GaN layer become deteriorated due to the high current and high temperature, which leads to an increase in series resistance. Thus, the current crowding effect will reduce the optical power [6].

In the Ni/Ge/Au ohmic contact, Ga in GaN diffuses outward through Ni/Ge to reach the Au layer, which will form a non-stoichiometric region. At the same time, Au diffuses inward to form a high-resistance alloy, which makes contact resistance higher [7]. Failure modes caused by electrical contact metallurgical inter-diffusion of LEDs devices include degradation of light output, increase of parasitic series resistance and short circuits.

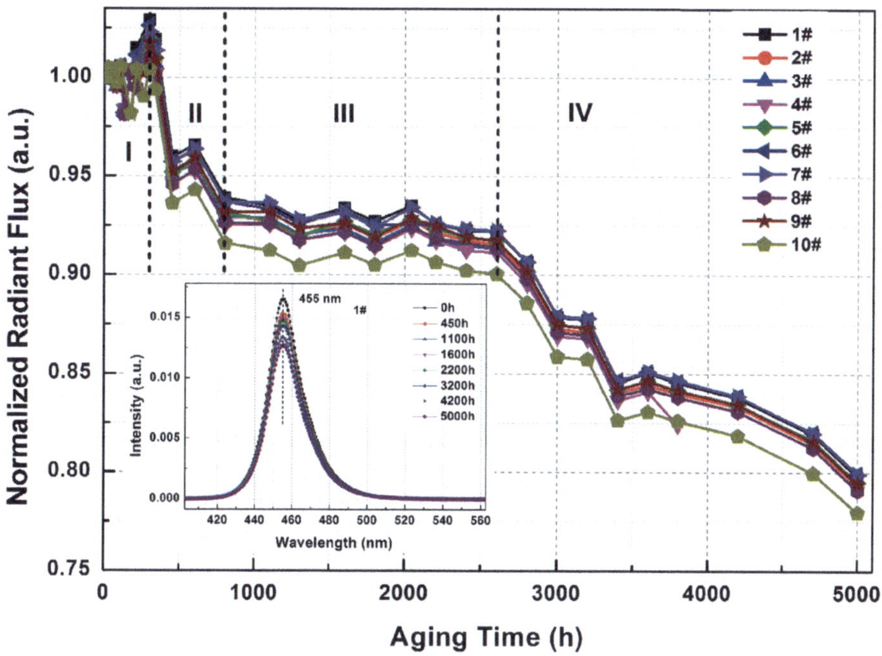

Fig. 10.1 The normalized radiant fluxes of ten commercial vertical GaN-based LEDs. The inset shows the typical electroluminescence at different stress times

(3) Electro-migration

Unlike metal diffusion, electro-migration refers to the phenomenon in which many metal atoms migrate and gather to the location where crystal defects or defective tubes are formed. The higher the current density and the higher the temperature, the more severe the electro-migration phenomenon is. Serious electro-migration may cause a short circuit between the contact electrode and the LEDs die. Due to defects such as substrate defects, electro-migration, and incomplete soldering, areas with high and low thermal resistance may be formed on the LEDs chip. Conductive areas with non-uniform thermal resistance will cause current crowding effect, heat dissipation, and finally sharp increase of temperature of package layer, which degrade device performance [8]. Kim et al. found that electro-overstressed devices from the electro-migration can lead to the formation of spots on the electrode surface and higher leakage currents of the devices at both forward and reverse bias [9].

(4) Degradation of p-type GaN

For the p-GaN layer, Mg is often selected as the p-type doped acceptor impurity to improve the conductivity for large current injection into the LEDs. During growth of the p-GaN layer, the growth temperature is high. Therefore, the effect produced by the diffusion of Mg impurities in the quantum well is called LEDs impurity diffusion degradation. Failures caused by diffusing impurities that are non-radiative recombination centers include an increase in series resistance and forward operating voltage, current crowding effect, reverse leakage current caused by the tunnel effect, and decrease in luminous efficiency. Kohler et al. found that high concentration of Mg near the active region at low growth temperature can enhance luminous efficiency due to the increase in hole concentration. However, it increases the non-radiative recombination rate in the opposite condition. In addition, the use of gradual doping can also effectively reduce the diffusion of Mg to the active layer and improve device performance [10].

10.1.2 Sudden Failure

The sudden failure of LEDs devices mainly manifests as an open circuit or a short circuit. Most flip-chip LEDs packages fail suddenly due to short circuits. Lift-off of leads and ball solder is open circuits. Sudden failures can be divided into intrinsic failures and wear-out failures. If we describe the distribution law of the failure rate of a product throughout its life cycle, we get the bathtub curve. The intrinsic failure corresponds to the flat part of the bathtub curve. For wear failures, the failure rate increases over time.

(1) Die cracking

Due to the different characteristics of the materials such as the difference in thermal expansion coefficient, the LEDs packaging layer will suffer from a greater impact of thermal and electrical stress when the device is suddenly placed in a high-temperature environment or a large current injection. If such an effect is large enough, it will directly damage the chip of the GaN-based LEDs device. Whether the thermal shock or the electric heating effect caused by high current density leads to the rapid heating or cooling of the material, the huge temperature difference causes a large amount of heat exchange in a short period of time. This can produce great mechanical stress on material. Thus, in order to prevent the die from cracking under high stress, the thermal expansion coefficients of the substrate and the epitaxial layer need to be adjusted to make them match as much as possible as shown in Fig. 10.2. An appropriate intermediate layer can be grown between the substrate and the epitaxial layer or using flexible substrates, which can both effectively reduce the chance of die cracking.

(2) Fatigue or fracture of wire ball bonding caused by electric heating overstress

Wire bonding is the most commonly used method to connect the LEDs die and external electrodes, i.e. to complete the current path between the chip and the external pins of the package. Welding bonds are formed between these metal interfaces. The bonding wire is the key to connecting the LEDs chip and the substrate. The failures in wire bonding mainly include wire breakage and fracture, bond point disintegration, and wire ball fatigue. Most of the failures are abrupt. The annealing process is used to solder the gold wire. The area above the solder ball point processed by annealing is called the heat affected zone (HAZ) which is the most vulnerable part in the entire

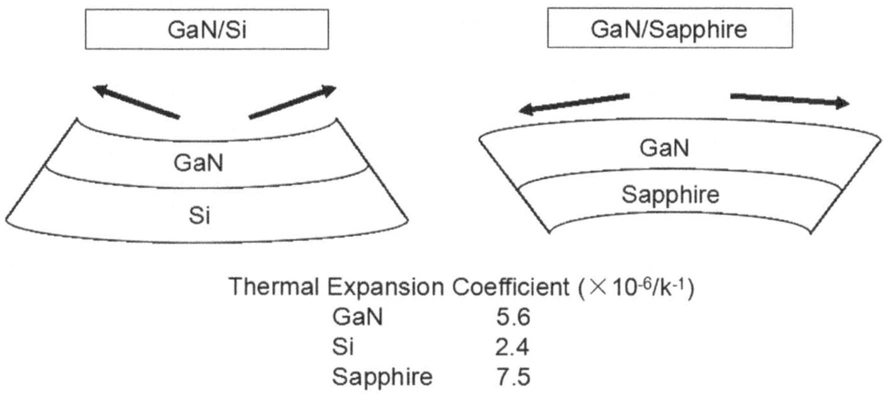

Fig. 10.2 The thermal expansion coefficient of GaN/Si and GaN/sapphire

wire. When it receives enough mechanical stress or repeated stress (stretched and released), the led will break. What's more, the heat affected zone is broken after excessive fatigue after half a year over use of the LEDs device as repeated stretching and release of the mismatch stress between the encapsulation layer and the chip [11].

The cause of wire bonding failure is a variety of metal compounds formed by different chemical potentials between the bonding wire and the pad metal (such as Al/Au). Different metal compounds with different lattice constants and thermal expansion coefficients generated a large stress at the bonding point. Therefore, crack lines are more likely to appear along the intermetallic compound area, even cause contact with the solder joints disconnect in severe cases. For example, Al-Au bonding of ball cracked after three weeks of accelerated high-temperature thermal storage experiments (175 °C). Qbvious micro-cracks appeared in a direction parallel to the intermetallic compound layer [12, 13]. Due to the formation of the intermetallic compound layer, the adhesion at the interface decreases, resulting in a semi-short-line state and an increase in contact resistance. Such an effect will lead to degradation of device performance. Higher bonding reliability would be realized by optimizing the pad metal bonding process, obtaining the best soldering conditions, and reducing the damage to the chip under the same bonding strength.

(3) Electrostatic damage

GaN wide bandgap materials have high resistivity, but low thermal conductivity and conductivity on the substrate Al_2O_3 or Si. The charges on the two different electrodes of the LEDs were accumulated to a certain degree due to irregular operation and other reasons during the production and preparation process. When charges were accumulated high enough, high electrostatic voltage would exceed the material's capacity. What's more, the charges are released in a very short time and generate power Joule heat, resulting in LEDs static breakdown, p–n junction short circuit and leakage. In short, poor antistatic properties often lead to the sudden failure of LEDs.

In a mounted LEDs device, the p-type and n-type electrodes are generally located on the same side of the device. Due to the low thermal resistance and insulation of the sapphire substrate, GaN/InGaN devices are very vulnerable to electrostatic discharge damage. In general, in order to avoid permanent damage to the device by ESD, a suitable reverse voltage regulator diode can be connected in parallel. This structure allows voltage spikes in two directions to pass without damaging the LEDs device, [14] or connecting a GaN Base Schottky diodes to improve the overall ESD characteristics of nitride LEDs devices [15, 16], or anti-parallel shunt GaN-based ESD protection diodes to improve the ESD reliability of GaN LEDs devices [17]. In addition, the use of SiC and GaN substrates with small lattice mismatch and Si substrates with high thermal resistance can also improve the antistatic capability of the device.

10.1.3 Packaging

(1) Lens and potting layer

The materials used for the lenses and potting layers in LED devices are mostly based on polymeric materials such as epoxy, silicone polymers and polymethyl methacrylate (PMMA). When the junction temperature is too high or the injection current density is high, the encapsulation layer and the phosphor layer prepared from these materials will accelerate aging, resulting in color drift.

When a large amount of Joule heat is generated under a large current, or directly in a high temperature environment, the surface encapsulation layer would be locally carbonized to form a conductive path across the LED device. This effect can directly destroy the device, which is called carbonization of the encapsulating material. After the encapsulation layer is carbonized, the insulation resistance of the encapsulation layer is lowered, and the electrical insulation between the bond alloy wires and between the gold wires and the leads is severely damaged. When the temperature of the working environment of the device is high enough, it will cause thermal dissipation of the device. In this process, the bond alloy wire is partially melted due to the high temperature and high current, where part of the current in the gold wire passes through the encapsulation layer in the encapsulation layer. A large amount of Joule heat is generated, which further reduces the insulation resistance of the encapsulation layer and ultimately leads to carbonization of the encapsulation layer. Under sustained, high current and high temperature environment, it is extremely easy to form a conductive layer due to the carbonization at the contact area of package layer and sealing layer region, resulting in shorting out of LED device. If continuous high electrical stress is applied, the abrupt failure caused by the encapsulation layer will occur. Figure 10.3 shows the cross-sectional view of the LED device undergoing a high temperature stress aging test. The trace of carbonized reflective cup can be clearly seen [18].

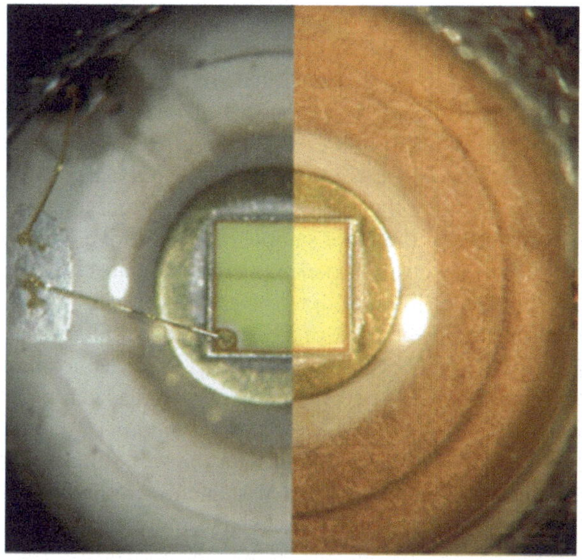

Fig. 10.3 Micrograph of a white LEDs; left: before stress; right: after stress at 100 A cm^{-2}, 100 °C

Studies have shown that the use of high refractive index, high UV resistance and heat aging resistance, low stress silicone and other packaging materials can significantly improve the light output power and service life of lighting devices.

(2) Fluorescent powder

A white LED device using yellow-blue two-color luminescence is composed of YAG:Ce^{3+} phosphor and a blue LED chip, wherein the thermal stability of the YAG fluorescent agent is strongly influenced by its composition. With regard to the fluorescent agent, the more compact the crystal structure is, the smaller the heat quenching temperature is. The heat quenching temperature is defined as the temperature at which the fluorescent agent is heated to and then cooled to room temperature, where the luminous intensity at room temperature becomes the 50% of that at the quenching temperature. The main reasons of thermal quenching of phosphor powder are the occurrence of multi-phonon relaxation and forming long afterglow luminescent material, i.e. namely the enhancement of non-radiative transition. The reason for the increase in the probability of non-radiative transition is the introduction of an intermediate state between the excited state and the ground state (conduction band and valence band), so that the original radiation recombination becomes a non-radiative recombination [19]. This will reduce the luminous efficiency. Jia et al. reported the blue shift and spectral broadening phenomenon as the temperature rise, indicating the existence of the interactions of electrons and phonons. The long afterglow illuminator is mainly caused by the thermal ionization process. The electrons whose energy level in the forbidden band is close to the conduction band absorb the phonon energy and then ionized. The generated electron trap captures the activated state electrons to form a long afterglow illuminator. The thermal ionization will deteriorate the thermal quenching characteristics of fluorescent agents [20].

Thermal quenching properties of rare-earth phosphor powder is influenced by doping component and crystal structure. Take three different components of Lan-based dopant of Ce^{3+} compound (La-SiON: Ce^{3+}) as an example, Fig. 10.4 shows the PL spectra of the three samples at different temperatures [21]. The emission peak wavelengths of the three samples did not move with increasing temperature, but the FWHM of spectrum broadened. This indicates that the thermal stability of the color coordinates of the nitrogen–oxygen fluorescent agent was good. In addition, in the relationship of the peak of PL spectrum with temperature, the emission intensity of $La_5 Si_3 O_{12} N$: 2% Ce, $La_3 Si_8 O_4 N_{11}$: 6% Ce phosphor agent change uniformly with temperature, while $LaSiO_2 N$: 4% Ce decreased rapidly with increasing temperature. Figure 10.5 is a normalized PL as a function of temperature. Among three samples, $LaSiO_2 N$: 4% Ce was found with the worst thermal stability, $La_3 Si_8 O_4 N_{11}$: 6% Ce showed the highest thermal quenching temperature. The corresponding reason is that $La_3 Si_8 O_4 N_{11}$: 6% of Ce has a more compact crystal structure and the highest N/O ratio [21] in comparison with other two materials.

Sublayer

When high-power LED device is working at high temperature thermal cycling environment for a long time, delamination may take place between the device and the encapsulating layer, between an encapsulation layer and the lead frame of the package, as well as between the LED die and the substrate. When delamination occurs in the LED optical path of the device, the output light flux of the device will be reduced, and the LED's illuminating color may also change. The common place can be between the die and the encapsulant layer, between the die and the phosphor layer, as well as between the lens of high-power LED and the encapsulation layer. Zhou et al. used GaN-based high-power LED devices, aged at 350 mA forward current for 256 h (25 °C, 54 is% RH), and subsequently discovered appearance of division layer around the die since a large amount of light is restricted by reflection in the sub-layer inside, and "rainbow belt" phenomenon appeared [22].

10.2 The LED Aging Test and an Aging Mechanism

Unlike other electrical elements, the LED device have a long service life. Except for sudden failure, light attenuation to a certain extent beyond the requirement in most cases is the reason for the undesired service life. In order to quickly understand the reliability and life of LED, accelerated aging tests are applied generally. This section will primarily describe the aging test and the mechanism where the temperature, humidity, and electrical stress are used to conduct the measurement.

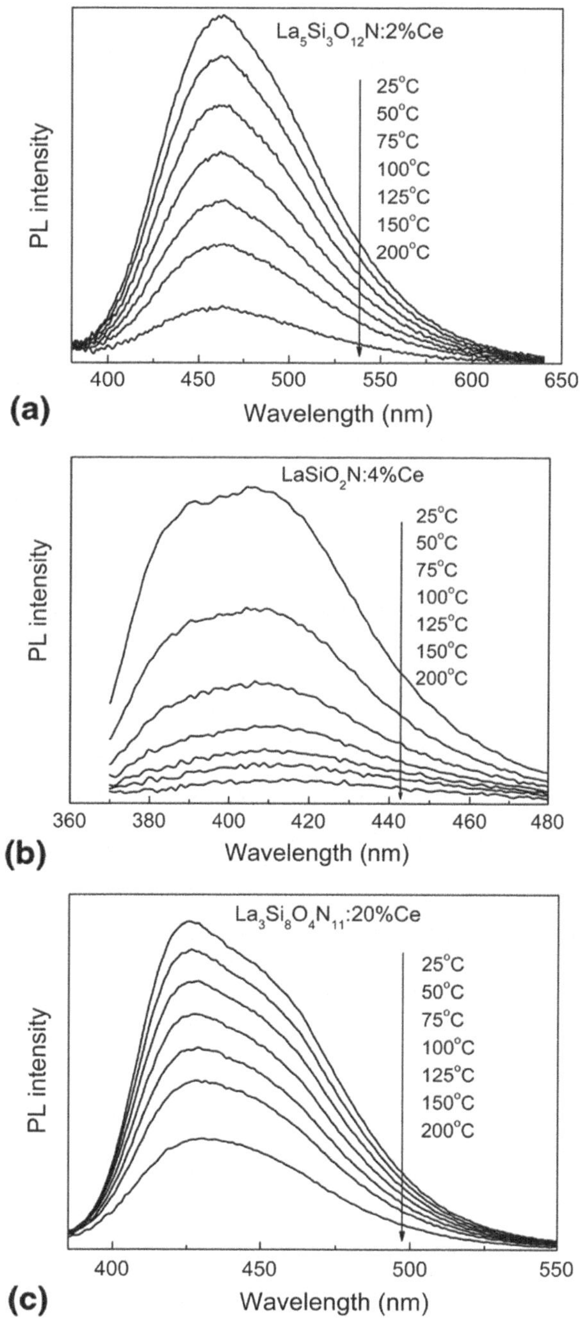

Fig. 10.4 Luminescence spectra of lanthanide siloxanes doped with Ce^{3+} as a function of temperature

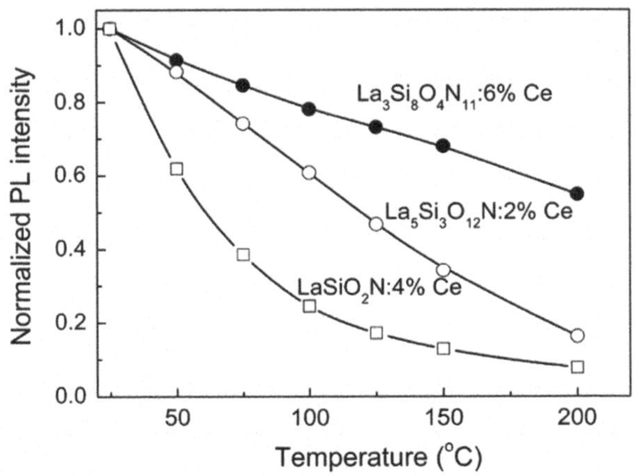

Fig. 10.5 Relationship between emission intensity and temperature of three lanthanide-doped Ce $^{3+}$ samples

10.2.1 Aging Experiment and Acceleration Factor

With the help of accelerated testing, one can quickly identify the cause of failure of the product and quickly assess the reliability of the product. According to different experimental purpose, accelerated test can be divided into two categories: accelerated life test and accelerated stress testing. Accelerated life test accelerates the testing cycle to obtain the necessary life distribution, thereby predicts the life or efficiency loss under normal circumstances. When we use accelerated life test to determine life time, it is critical to determine the acceleration factor. Accelerated stress testing applies one or several accelerated stresses to identify problems in design or fabrication process which can result in the failure.

Accelerating factor (Factor Accelerated, the AF) is the exponential factor that uses a particular failure mechanism and accelerated stress conditions to predict the failure time under normal stress conditions. Under linear accelerator circumstance, the acceleration factor can be expressed as ratio of the average life under the normal stress and average life under accelerated stress:

$$AF = \frac{Time\ to\ fail(stress1)}{Time\ to\ fail(stress2)} \quad where\ stress2\ > stress1 \tag{10.1}$$

where, stress1 represents the stress during normal use, stress2 is the stress under accelerated test.

The acceleration factor reflects the rate of the failure mode of the LED device when it is applied with high and low stress, respectively. Acceleration factors and life indicators can generally be calculated using different predictive models based on different accelerated stress conditions. For example: Arrheniu S model can be used when temperature is taken as the accelerated stress; Inverse Power Law model can be used when the voltage, large current, pulse current and other non-thermal stress are taken as the accelerated stress; Peck model can be used when high temperature and high humidity are taken as the accelerated stress; Eyring accelerated life test model can be used when compound stress is taken as the accelerated stress to estimate the relationship between lifetime and accelerated environmental stress. In the following, the mechanism of accelerated aging test under different temperature, electrical stress, high temperature and humidity, and other composite stress as well as aging mechanisms will be introduced.

10.2.2 Temperature Acceleration Test.

The failure of the LED device is related to the materials, interfaces, electrical connections, and various physical and chemical reactions that exist in the package structure. As the temperature increases, the physicochemical reaction speed inside the LED increases, and the failure process will be accelerated. The Arrhenius model can be used to study the corresponding law between the reaction speed and the temperature as shown by the formula:

$$\frac{dM}{dt} = Ae^{-\frac{E_a}{kT}} \tag{10.2}$$

where:
dM/dt is the reaction rate of the chemical reaction;
A is the pre-factor constant;
E_a is the activation energy (eV) that causes the failure or degradation process;
k is the Boltzmann constant (0.000086174 eV/K);
T is the absolute temperature (K) under certain acceleration conditions;

Degradation rate of the device performance is exponentially proportional to the activation energy inversely and the reciprocal of the temperature. At given constant temperature, the failure state number is M_0 when the device is at the t_0. At time t_1, the device is in a failed state number M_1. The obtained integral

$$M_1 - M_0 = Ae^{-\frac{E_a}{kT}}(t_1 - t_0) \tag{10.3}$$

Let $M_0 = 0$, $t_0 = 0$, $t_1 = t$, yield

$$t = \frac{M_1}{A}e^{\frac{E_a}{kT}} \tag{10.4}$$

Taking the natural logarithm of both sides of the equation

$$\text{lnt} = \ln\frac{M_1}{A} + \frac{E_a}{kT} \tag{10.5}$$

The above equation is the acceleration equation based on the Arrhenius model with the temperature stress as the acceleration variable. Where t represents the time at which a device reaches a cumulative failure probability F(t). Its logarithm and the inverse of the absolute temperature show a linear relationship. Using the ln(t) ~ 1/T coordinate map, one can get a straight line. You can use the graph estimation method or numerical method to calculate the lifetime value of the LEDs at different temperatures, and figure out the activation energy of the device.

With regard to the acceleration factor, it is assumed that the time required to reach the cumulative failure probability F_1 at the reference temperature is t_1 (F_1). The time required to reach the same cumulative failure probability after applying the temperature stress is t_2 (F_2), then the ratio of the two is the acceleration factor AF. With respect to the reference temperature T_1, the acceleration factor at the high temperature T_1 can be expressed as

$$AF = \frac{t_1(F_1)}{t_2(F_2)} = \exp\left[\frac{E_a}{k}\left(\frac{1}{T_1 - T_2}\right)\right] \tag{10.6}$$

Temperature accelerated testing process can be divided into the thermal shock test and the temperature cycling test if it uses the changeable temperature stress. The thermal shock test requires the test sample to have the ability to withstand large rates of temperature change. The acceleration factor is given by the Coffin-Manson model as the following formula:

$$AF = \left(\frac{\Delta T_{stress1}}{\Delta T_{stress2}}\right)^{-n} \tag{10.7}$$

where the ΔT is the range of the whole temperature cycle during the working process of the device, and n is a material-related parameter. The value of the parameter n ranges from 1 to 5, and the typical value is 2.

Coffin-Manson model does not consider the influence of time. To this end, Norris and Landzberg introduced a frequency factor to improve the Coffin-Manson model in 1969. Based on Norris-Landzberg model, the corresponding acceleration factor is:

$$AF = \left(\frac{\Delta T_{stress1}}{\Delta T_{stress2}}\right)^{-n}\left(\frac{f_{stress1}}{f_{stress2}}\right)^{o} AF_{Arrhenius} \tag{10.8}$$

where the ΔT is the range of temperature cycle, and n is the material-related parameter. The o is the frequency-related parameter, its range is from 0 to 1, and the typical value is 1/3. $AF_{Arrhenius}$ is the Arrhenius model acceleration factor.

In addition, corrosion caused by moisture in the environment is one of the failure mechanisms commonly seen in electronic products. Humid environments cause the interface layer failure. This sublayer is caused by the absorbent moisture together with the thermal stress. The stress can accelerate the penetration of water molecules in the protective material, causing the degradation of material and device performance. Peck Model was used in this situation:

$$AF = \left(\frac{RH_{stress2}}{RH_{stress1}}\right)^{q} AF_{Arrhenius} \tag{10.9}$$

where RH is the relative humidity size. Q is the related humidity parameter that is in the range of (0, 3). The typical value is 2.5.

In addition to the consideration of the temperature and humidity, other stress environment can also be applied and a Generalized Eyring model needs to be adopted:

$$AF = \left(\frac{L_{stress2}}{L_{stress1}}\right)^{r} \left(\frac{RH_{stress2}}{RH_{stress1}}\right)^{q} AF_{Arrhenius} \tag{10.10}$$

where T is the temperature (K), L_{stress} is the additional stress, r is the additional stress-related parameter, and q is the humidity-related parameter.

10.2.3 Accelerated Electrical Stress Test

Failures like ion migration, mass migration, electro-static discharge (ESD), electrical over stress (EOS), short circuit and breakdown can be caused by electrical stress as well. The device will have a higher failure rate and a shorter lifetime due to the strong electrical stress which might be provided by current, voltage or electrical power. The relationship between device life and electrical stress can be studied by Ellen model:

$$t = \frac{1}{k \cdot ES_{stress}^{c}}. \tag{10.11}$$

In the equation, t represents the lifetime of the electronic component, k is a constant, c is the empirical constant as a function of Joule heat, and ES_{stress} represents the electrical stress applied to the electronic component. Take the natural logarithm of the equation, and use the values of $\text{ES}_{\text{stress}}$ at different time to obtain the values of k and c:

$$\ln t = -c \ln ES_{stress} - \ln k. \tag{10.12}$$

After determining k and c, the lifetime or failure rate of the device under the different electrical stress can be evaluated by the graph estimation method or numerical calculation method according to the equation.

The acceleration factor AF can be calculated by the following equation:

$$AF = \frac{t_1(F_1)}{t_2(F_2)} = \left(\frac{ES_{stress2}}{ES_{stress1}}\right)^c. \tag{10.13}$$

For LEDs, the $t_2(F_2)$ can be determined by an accelerated life test under the high voltage V_2. If c and AF are known, the failure time $t_1(F_1)$ under the normal electrical stress V_1 can be obtained according to the equation.

The researches about the generation and the protection of ESD have also been the focus of the reliability research of III-nitride LEDs. The ESD can be generated by three common modes: machine mode, human body mode, and charged device mode. The ESD sensitivity test of LEDs is a destructive test. Its main purpose is to obtain the threshold voltage of the ESD, and to improve the ability of LEDs to avoid the ESD. Electrical testing, sonic scanning, TEM, SEM, XRD and other testing methods are needed to screen and detect ESD-failure devices.

In addition to the temperature and electrical stress, humidity can also be applied as an accelerated variable in the LED accelerated aging test to evaluate the moisture-resistant lifetime of the device, such as the temperature, humidity, bias (THB) accelerated test.

$$t = A\exp(\frac{E_a}{kT}) \cdot f(\mathrm{ES}_{stress}) \cdot \mathrm{g}(RH). \tag{10.14}$$

In the equation, t, Aexp (E_a/kT), f($\mathrm{ES}_{\mathrm{stress}}$), and g(RH) represent the average lifetime, the equation factor of Arrhenius model, the function of electric stress influence, and the function of relative humidity, respectively.

$$f(\mathrm{ES}_{stress}) = \frac{1}{k \cdot ES_{stress}^c}. \tag{10.15}$$

$$g(\mathrm{RH}) = B\frac{1}{RH^m}. \tag{10.16}$$

In the equation, m represents the moisture related factor.

10.2.4 Other Factors Affecting the Lifetime

(1) Vibrations variable frequency (VVF) test

The vibration test is performed to detect the robustness of the LED structure and the stability of the electrical characteristic within the specified vibration frequency range. The damages of the structure can be caused by the vibration, such as deformation, bending, cracks, fractures, and collisions between components, which will lead to the fatigue damages to the product. These damages will cause poor contact of components and affect the process performance such as loose adhesive and in chaos. There are five major mechanical vibration tests: sweep vibration, fixed vibration, resonance, vibration fatigue, and random vibration. According to International Electro-technical Commission (IEC) standards, the full amplitude is 1.5 mm under the test frequency of 20 ~ 100 Hz. The peak value of the acceleration is about 50 g under the test frequency of 100 ~ 2 kHz. The experiment can be performed according to JESD22 B-103. Performance test is required before and after the VVF test.

(2) Salt spray aging test

The salt spray test is performed to evaluate the corrosion resistance of exposed parts of the component under salt spray, humidity and hot environment. The salt spray test includes natural environment exposure test and artificial accelerated salt spray environment test. The artificial salt spray test includes neutral salt spray, acetate salt spray, copper-salt-accelerated acetate spray, and alternating salt spray. The condition and method of the salt spray test is described in more detail in the GB/T 10,125–1997 national standard.

(3) Gas environment accelerated test

The atmosphere contains corrosive components and corrosive factors such as oxygen, humidity, temperature changes and pollutants. Therefore, under the influence of the atmosphere, free radicals and peroxides will be generated in the polymerized organics of packaged LEDs, which may destroy the organics. Under a certain temperature and humidity, an accelerated corrosion of packaged LEDs or packaging materials by H_2S, SO_2, NO_2, Cl_2 and other harmful gases can be used to evaluate the resistance to gas corrosion of the device. The condition and method of the gas environment accelerated test and mixed gas test are described in more detail in the IEC 60,068 standard.

(4) Temperature and light exposure (TL) test

Organic decomposition reactions like photo-oxidation and photo-hydrolysis can be caused by sunlight, short wavelength radiation from the device itself, or other random light sources. Thus, the polymer of packaged LEDs will be destroyed. The light conversion efficiency and transparency of the resin-based polymer commonly used in packaged LEDs will be reduced by short wavelength irradiation, which is the result of the aging of high-molecular materials. After aging, the high-molecular materials often suffer from surface chalking, discoloration, blistering, cracking, and shedding. Therefore, a basis for improving device reliability can be provided by the aging test of light conversion materials and light transmitting materials by performing temperature and light tests on LEDs. An alternative wavelength of 445 ± 5 nm is used in the TL test, while the irradiation level is 350 ~ 1000 mW/cm^2. The test period is 1000 h if the test temperature is higher than 20 °C of the operating temperature and 10 °C below the glass transition temperature. If the operating temperature is 20 °C above the temperature of 10 °C below the glass transition temperature, the temperature of 10 °C below the glass transition temperature should be chosen as the test temperature. The material is considered to be failed when the light transmission efficiency or light conversion efficiency is reduced by 10%.

10.3 LED System Reliability

With the rapid development of semiconductor lighting technology and the increasing number of semiconductor lighting products, the reliability of LED products has gradually become one of the most important technical bottlenecks that hinder the development of semiconductor lighting industry and the application of large-scale markets. The semiconductor lighting is expected to achieve its inherent high efficiency and high reliability. However, LED lighting systems not only include devices, but involve multiple links from chips, devices, light source modules, drivers, various assembly interface processes to lighting systems. In this section, the LED system reliability framework was introduced. The framework includes failure trees, Bayesian networks, and Markov chains. These models are used to predict the life of semiconductor lighting systems, while showing the failure modes of key components. The failure modes of key components were showed at the same time.

10.3.1 LED System Reliability

The lifetime of lamp-level products not only depends on the LED, but also the reliability of other parts inside the lamp. Thus, when the reliability of a lamp was analyzed, the lifetime and reliability of not only the LED but also the remaining components should be considered [23]. Here, a lamp-level LED product is divided into the following major components: LED light source (including LED chips, packages, bonding wires), LED power drivers, optical materials (including phosphors, paste caps, etc.), machinery, and heat dissipation modules.

The failure modes of the lighting products mainly include two types: the sudden failure and the optical degradation. It is generally considered that the time, at which the product suffers the sudden failure or the luminous flux of the product reduces to 70% or 50%, is the value of the lifetime of the product.

The relationship between the main components and the system reliability is shown in Fig. 10.6, where more "★" means the stronger interaction.

It can be seen that there are many factors involved in the reliability of LED system products, which made the failure mechanism complicated due to the interplay among these factors. Therefore, methods such as failure tree, Bayesian network and Markov chain need to be introduced to analyze prediction models of LED system products. The concepts of failure trees, Bayesian networks, and Markov chains will be introduced next.

Failure tree (FT) is a system analysis method. The maximum number of trusted methods can be found according to the environment and operation by using FT. The system will show an unexpected state due to the influence of these methods.

As shown in Fig. 10.7, the system reliability consisting of three partial failures (A1, A2, and A3) is considered. From the characteristics of the system, the system

	LED package and bonding wires	Optical materials	Electronic materials	Cooling system	Mechanical part
Optical degradation	★★★	★★★	★	★	★
Sudden failure	★		★★★	★★★	★★★

Fig. 10.6 The relationship between components of LED system and the system reliability

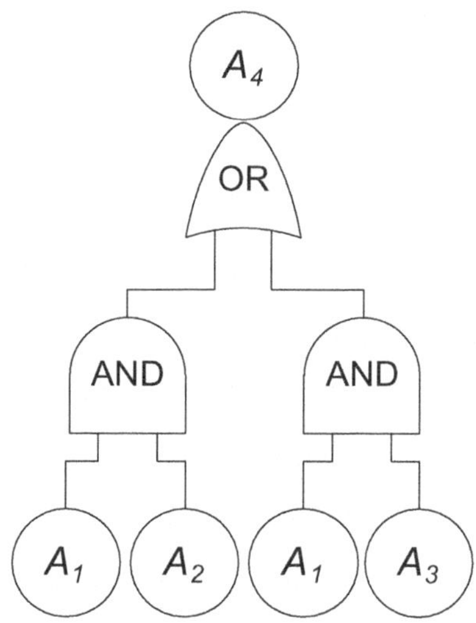

Fig. 10.7 An example of a system failure tree

will fail as long as A1 fails. Therefore, the failure tree can be simplified. The simplified failure tree is shown in Fig. 10.8. The bottom event of the failure tree can often be represented by a Boolean variable, whose purpose is to build the system failure tree more effectively. Thus, FT can be seen as a graphical Boolean variable. Boolean variables are usually 0 or 1. If A1 and A2 are Boolean variables, the Boolean operations A1 $\otimes$ A2 and A1 $\oplus$ A2 are also Boolean variables.

In the logical formulas, $\otimes$ and $\oplus$ represent AND gates and OR gates, respectively. Therefore, A1 $\otimes$ A2 means that only if A1 and A2 fail at the same time, the system will fail. A1 $\oplus$ A2 means that if A1 or A2 fails, the system will fail. System failure can be described by logical formulas, and the system failure tree can be simplified by using Boolean operations.

In Bayesian networks, cause and effect variables are represented by nodes. Each node has its own probability distribution. Nodes can be an abstraction of any problem,

Fig. 10.8 A simplified system failure tree

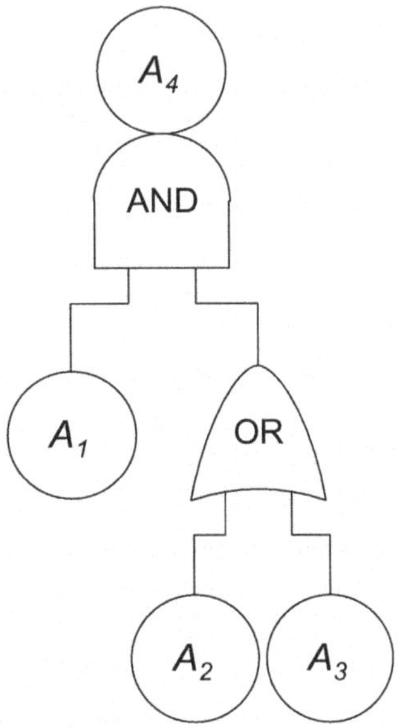

and arrows indicate the causal relationship between nodes. The network contains the important conditional independence assumption. A simple Bayesian network is showed in Fig. 10.9. In the figure, X1 X3 indicates the direct relationship between the two variables: X1 is the parent node of X3 and X3 is the child node.

Its joint probability distribution function is.

$$P(x_1, x_2, x_3) = P(x_1)P(x_2|x_1)P(x_3|x_1). \tag{10.17}$$

Once an observation is obtained for one variable, the probability distributions of other variables can be obtained through network updates. For example, a result of

Fig. 10.9 A simple Bayesian Network

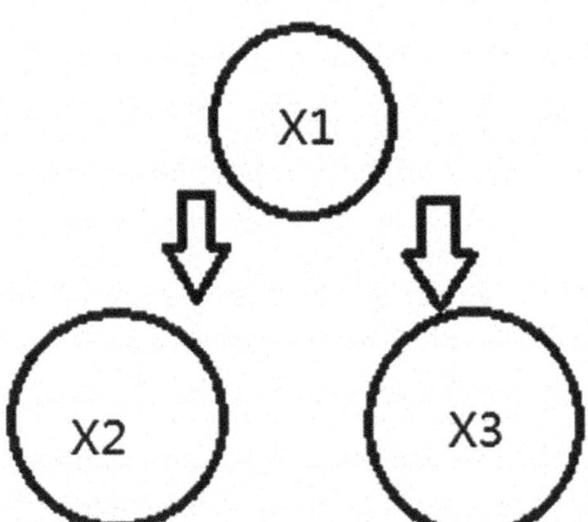

X2 is observed as e, which will propagate through the network to update the prior probability. The probability of X1 and X3 posterior is.

$$P(x_1, x_3|e) = \frac{P(x_1)P(e|x_1)P(x_3|x_1)}{\sum\limits_{x_1} P(x_1)P(e|x_1)}. \tag{10.18}$$

For a specific problem, the general steps of Bayesian network modeling are as follows: (1) determine the key variables involved in the problem and the possible values of the variables, and use one node to represent one variable in the Bayesian network. (2) Judge the correlation or independence between different nodes, and use graphs and directed arcs to represent them in Bayesian networks. (3) Determine the probability parameters required for Bayesian network calculations and calculate the probability. In general, detailed analysis about the key technologies and difficulties of the problem to be solved is required to construct a reasonable Bayesian network model. Only in this way can Bayesian networks effectively solve practical problems, rather than just probabilistic calculations.

The reliability of semiconductor lighting systems depends on the individual components. For example, the light output power of the LED chip is reduced to 80% of initial, and the light transmission performance of the optical material is reduced to 80% of initial. That means the overall luminous performance is reduced to 64%, which is less than 70%. Thus, the LED product is considered to have failed. In addition, a sudden failure of a component, power module for example, can cause the overall product to fail. Its Bayesian network can be represented by the following figure (Fig. 10.10):

A dynamic Bayesian network (DBN) is an extended concept of a static Bayesian network combined with (discrete) time. Therefore, a dynamic Bayesian network is a model of discrete-time stochastic processes. A DBN represents a set of static Bayesian networks which connect arc points between different time nodes in each time tangent.

Markov chain

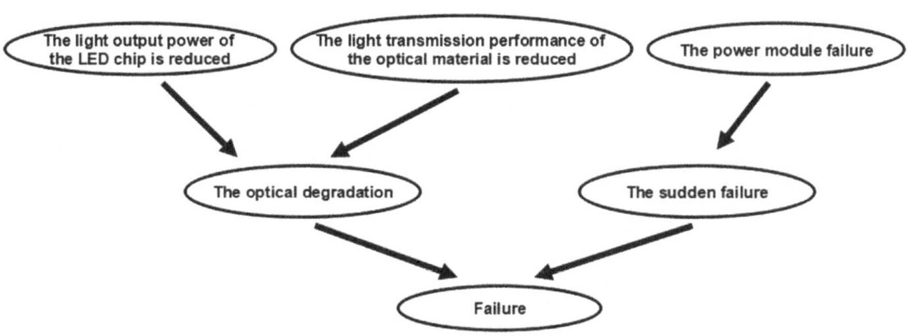

Fig. 10.10 Bayesian network in the reliability analysis of LED system products

Fig. 10.11 A simple MC

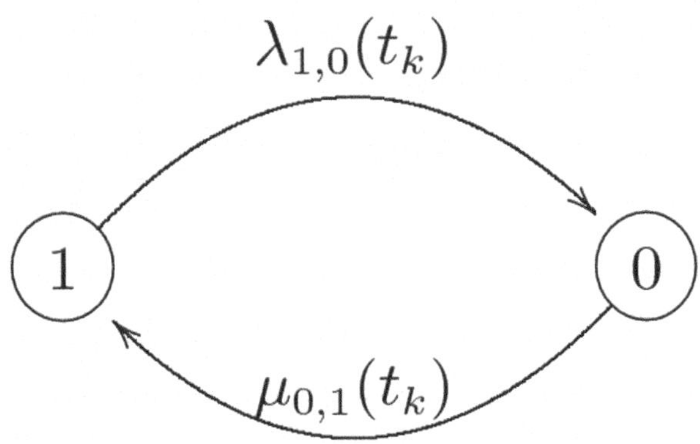

The Markov Chain (MC) is a discrete state which describes a stochastic process during the transition times. For example, $X(t_k)$ (X_t, for short) is expressed as the state of the process k at time t, and the first-order property of the Markov chain is.

$$P(X_{k+1} = x_{k+1}|X_k = x_k, \ldots, X_0 = x_0) = P(X_{k+1} = x_{k+1}|X_k = x_k). \quad (10.19)$$

The sequence of the process k in the Eq. (10.19) can be set to the non-negative integers and finite fields. If the transition of states in a MC process is independent of the system time, the process will be described as static state or time-uniformity. The Fig. 10.7 is a simple MC.

Different from Bayesian networks, the graph of the MC is an acyclic graph, and these nodes do not represent random variables, but rather a change state with time. The Fig. 10.11 shows a system including two process state.

- 1 represents normal working state.
- 0 represents do not working properly.

Supposing that a subsystem is in the normal working state at the beginning of time, if the failure probability and repair probability in this system is λ and μ, respectively, the probability of the subsystem failure within the discrete time will be given by.

$$P(X_k = 0) = \frac{\lambda}{\mu + \lambda} + \frac{\mu}{\mu + \lambda}(1 - \mu - \lambda)^k \quad (10.20)$$

when the number of states k → ∞, the Laplace transform is used to represent the probability of continuous time, the formula is simplified to

$$P(X_t = 0) = \frac{\lambda}{\mu + \lambda} + \frac{\mu}{\mu + \lambda}e^{-(\lambda+\mu)t} \quad (10.21)$$

when the time t → ∞, and the period of time is simplified to a standard function, the subsystem failure probability is reduced to

$$P(X_k = 0) = \exp\{\overline{Q}\Lambda(t)\} = \sum_{k=0}^{\infty}(Q(t)/f(t) + I)^k \frac{\Lambda(t)^k}{k!} e^{-\Lambda(t)} \tag{10.22}$$

where $Q(t) = \overline{Q}f(t), \Lambda(t) = \int_0^t f(s)ds$, $\overline{Q}$ is the transition-strength matrix, A is a matrix exponent in the exp{A}.

10.3.2 The Cases of Reliability Analysis in the LED Lighting System

Case one

A simple LED lighting system includes a circuit board, three LED devices which are soldered on it and a common drive power. By analyzing the failure modes and interaction effects, the potential system failures mainly includes the following three aspects.

The failure of light source: considering the aging and mutagenic failure, it causes the luminous efficiency output less than 70% of the initial luminous efficiency; the welding points: the welding points are out of order; the drive failure: the drive failure is represented by aluminum electrolytic capacitor.

Figure 10.12 shows the three LED failure trees diagram of the semiconductor lighting system. It has four main failure modes, including the LED aging, the LED abrupt failure, the solder failure, and the drive failure.

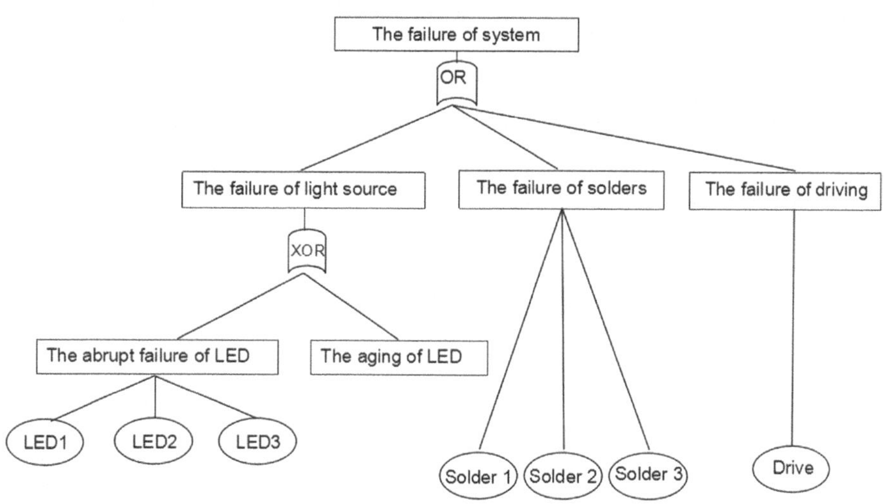

Fig. 10.12 The three LED failure trees of the semiconductor lighting system

It should be noted that, the two failures of the light source can affect each other. For example, after inducing the failure caused by the aging of the light source, the light source can no longer induce the failure caused by the abrupt failure, and vice versa. Therefore, the failure of the light source uses the NAND gate to describe the relationship between the abrupt failure and the aging failure.

According to the regression method of least squares for each critical failure mode, it can be seen that the failure data that changes over time is fitted into (and conforms to) the Weibull distribution.

$$f(t) = \frac{bt^{b-1}}{a^b} e^{-\left(\frac{t}{a}\right)^b}$$
$$h(t) = \frac{bt^{b-1}}{a^b}. \tag{10.23}$$

when b > 1, the failure rate increases. Since the failure rate is not constant, the probability function can be expressed as:

$$P(t) = \exp\{\overline{Q} \cdot t\} = \sum_{k=0}^{\infty} \overline{Q}^k \frac{t^k}{k!} \tag{10.24}$$

where Q is the transition-strength matrix.

Using the regression method of failure mechanism in each component fitting to the failure data that changes over time in each component with uncertainty distribution of 5, 50 and 95%, respectively, assuming each failure mode meet the MC (Markov Chain) shown in Fig. 10.11, the reliability of the simplified system can be calculated without abrupt failure as shown in Fig. 10.13, which will fail after approximately 120,000 h.

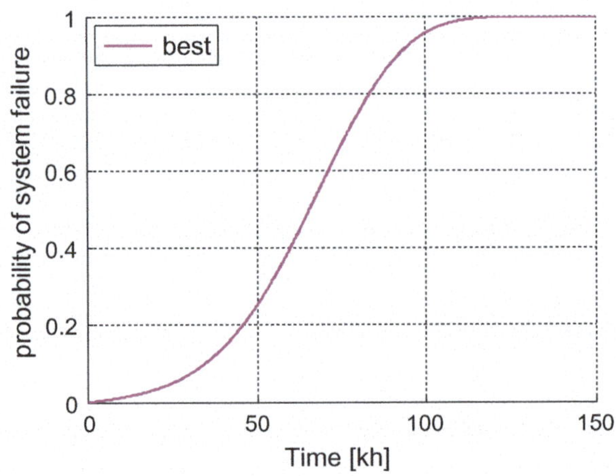

Fig. 10.13 The failure probability of the LED system that changes over time

Case two [24]

Consider the luminous flux probability of the single LED at a specific time rather than the failure probability. The total luminous flux probability of the single LED can be obtained by the Bayesian network algorithm. To obtain the total luminous flux of all LEDs light sources, it can simply multiply the number of LEDs in it and the luminous of individual devices. By setting the light failure criterion, you can get the probability of the light source and calculate the reliability when it embedded into system. The probability distribution of the estimated luminous flux can be expressed as

$$P(LED\_Failure) = P(3\_led\_output \leq 0.70) = \int_0^{70} f_{3\_led\_output}(z)dz \tag{10.25}$$

For the semiconductor lighting system with three LEDs, Fig. 10.14 shows the probability distribution of the estimated total luminous flux at 44,000 and 114,000 h, respectively.

It is worth noting that the probability distribution of total luminous flux can be observed to shift to the left with the time increasing as shown in the Fig. 10.14. This is because the total luminous flux can decay caused by the aging of the light source with time increasing as well as the increase in the mutagenic probability.

Figure 10.15 shows the probability of the system failure. In both cases, the probability of the system failure at 50,000 h is approximately 0.22. This is basically consistent with the estimated result by the earlier method.

With the popularization of LED lighting applications, the establishment of reliability estimated methods for LED lighting systems is becoming more and more urgent. This chapter presents a reliability estimated method that can be applied to LED lighting systems. The presented cases describe the difference in estimation

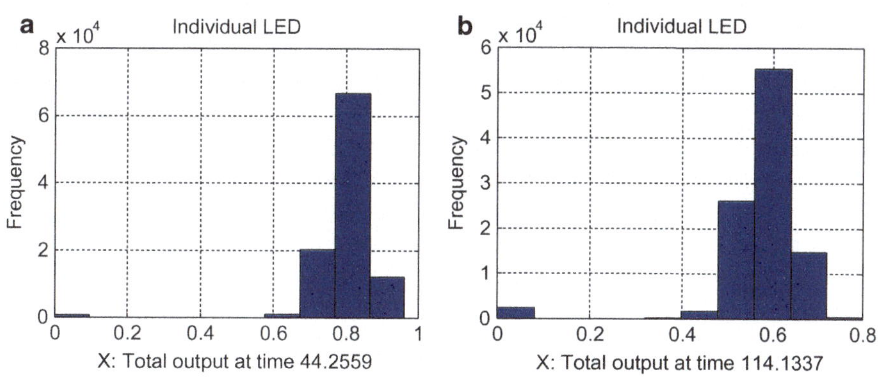

Fig. 10.14 The probability distribution of the estimated total luminous flux at **a** 44,000 h and **b** 114,000 h

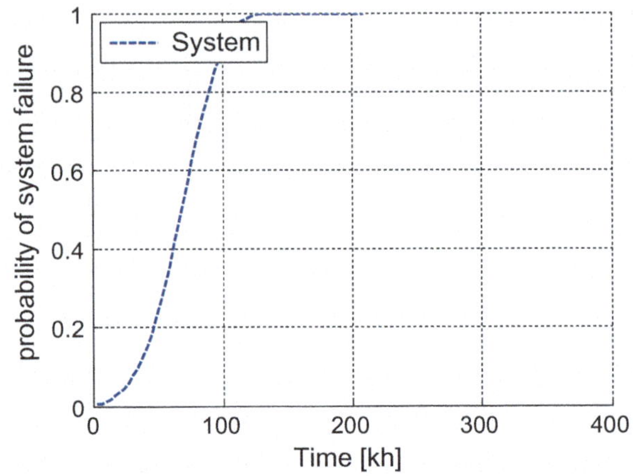

Fig. 10.15 The estimated lifetime of the semiconductor lighting system

methods. But in practice, the estimated reliability for the LED lighting application systems is a daunting task. Not only is there a wide variety of components in the LED system, in addition to a large number of unknown failure modes and failure mechanisms, but also the failure mechanisms can affect each other. In order to know the reliability of the entire LED system, it is not only necessary to test the reliability performance of the entire system, but also to test the reliability performance of each component. For the long-life LED systems, it is estimated to extrapolate its reliability under the normal conditions by accelerating aging tests. For the accelerated-life experiments, if you simply increase stresses such as temperature, electrical power and so on, it will lead to new failure models in some components. Therefore, the accelerated-life experiments should be carried out under certain estimation and evaluation conditions. Firstly, the failure of each component is not equal to the system failure. Each component in the system exhibits its own different failure mode, which require three different aging conditions for accelerated testing. Furthermore, the analysis model has to obeys the physical mechanism of the failure. When determining the stress of the accelerated aging experiment in the system and components, the implementation plan should be set according to the actual research results. For the LED lighting systems, not only temperature, humidity, electrical power and mechanical stress have to be considered, but also the effect of light on them. In addition, it is necessary to know the failure modes of the connection interface for the LED systems. Accelerated reliability testing methods involving LED modules and driving systems need to be further explored and improved.

References

1. M. Meneghini, A. Tazzoli, G. Mura et al., A review on the physical mechanisms that limit the reliability of GaN-based LEDs. IEEE Trans. Electron Devices **57**, 108–118 (2010)
2. L. Zhao, E. Thrush, C. Humphreys et al., Degradation of GaN-based quantum well light-emitting diodes. J. Appl. Phys. **103**, 024501 (2008)
3. R. Xie, N. Nirosaki, K. Sakuma et al., White light-emitting diodes (LEDs) using (oxy)nitride phosphors. J. Phys. D Appl. Phys. **41**, 144013 (2008)
4. H.W. Jang, J.K. Kim, S.Y. Kim et al., Ohmic contacts for high power LEDs. Phys. Status Solidi A Appl. Mater. Sci. **201**(12), 2831–2836 (2004)
5. R. Dugnani, M. Wu, Fracture mechanisms for silicon dice, in *Proceedings of the 35th ISTFA* (2009), pp. 309–313.
6. G. Meneghesso, S. Levada, R. Plerobonet al., Degradation mechanisms of GaN-based LEDs after accelerated DC current aging, in *IEEE International Electron Devices Meeting* (2002), pp. 103–106.
7. K. Mizuishi, H. Kurano, H. Sato et al., Degradation mechanisms of GaAs MESFETs. IEEE Trans. Electron Devices **ED-26**, 1008–1014 (1979).
8. J. Arnold, DFR solutions (2004)
9. H. Kim, H. Yang, C. Huh, S. Kim et al., Electromigration-induced failure of GaN multi-quantum well light emitting diode. Electron Lett. **36**, 908–910 (2000)
10. K. Kohler, T. Stephan, A. Perona et al., Control of the Mg doping profile in III-N light-emitting diodes and its effect on the electroluminescence efficiency. J. Appl. Phys. **97**, 1049141–1049144 (2005)
11. DL Barton, M Osinski, P Perlin et al., Life tests and failure mechanisms of GaN/AlGaN/InGaN light emitting diodes, in *IEEE 35th Annual Proceedings of Reliability Physics Symposium* (1997), pp. 276–281
12. E. Meissner, M. Haeckel, J. Friedrich, A Practical Example of GaN-LED Failure Cause Analysis by Application of Combined Electron Microscopy Techniques. Materials (Basel) **10**(10), 1202 (2017)
13. F. Oldervoll, F. Strisland, Wire-bond failure mechanisms in plastic encapsulated microcircuits and ceramic hybrids at high temperatures. Microelectron. Reliab. **44**, 1009–1015 (2004)
14. P. Jacob, A. Kunz, G. Nicoletti, Reliability and wearout characterisation of LEDs. Microelectron. Reliab. **46**, 1711–1714 (2006)
15. S.J. Chang, C.H. Chen, Y.K. Su et al., Improved ESD protection by combining InGaN-GaN MQW LEDs with GaN schottky diodes. IEEE Electron Device Lett. **24**, 129–131 (2003)
16. D. O'Mahony, W. Zimmerman, S. Steffen et al., Free-Standing gallium nitride Schottky diode characteristics and stability in a high-temperature environment. Semicond. Sci. Technol. 24(12), 1–8 (2009)
17. Y. Su, S. Chang, S. Wei et al., ESD engineering of nitride-based LEDs. IEEE Trans. Device Mater. Reliab. **5**, 277–281 (2005)
18. G. Meneghesso, M. Meneghini, E. Zanoni, Recent results on the degradation of white LEDs for lighting. J. Phys. D Appl. Phys. **43**, 354007 (2010)
19. W. Driel, G. Wisse, Influence of material combinations on delamination failures in a cavity-down TBGA package. IEEE Trans. Compon. Packag. Technol. **27**, 651–658 (2004)
20. D. Jia, W. Jia, Y. Jia, Long persistent alkali-earth silicate phosphors doped with Eu^2+, Nd^3+. J. Appl. Phys. **101**, 023520 (2007)
21. B. Dierre, R. Xie, N. Hirosaki et al., Blue emission of Ce^3+ in lanthanide silicon oxy-nitride phosphors. J. Mater. Res. **22**, 1933–1941 (2007)
22. L. Zhou, B. An, Y. Wu et al., Analysis of delamination and darkening in high power LED packaging, *IEEE International Symposium on the Physics and Failure Analysis of Integrated Circuits* (2009), pp. 656–660
23. S. Tarashioon, *An Introduction to Driver Reliability* (Springer, 2013), pp. 207–230
24. S. Koh, SSL System Reliability TNO report (2014)

Chapter 11
Applications of LEDs

Solid-state lighting (SSL) market is huge, being worth hundreds of billions of dollars. In 2014, output value of SSL product in China was worth more than 350 billion RMB [1]. SSL products consist of LED backlight, outdoor lighting, indoor lighting, LED display, etc. This chapter focuses on key areas related to LED applications, including an overview of technologies such as lighting, backlighting, display, communications, and biological applications.

11.1 New Light Environment Technology

The primary role of SSL technology is energy saving. In contrast with conventional artificial light sources, SSL has unique characteristics such as monochromaticity and adjustability, which enables the expansion of its application horizon. This section mainly introduces the main application fields of SSL, especially in general lighting area where traditional lamps are being replaced. This could be ascribed to the fact that LED has higher luminous efficiency, lower power consumption, longer lifetime. Meanwhile, the device is safe, reliable and environment friendly. In recent years, with the rapid development of industrial modernization and urbanization in China, the relationship between supply and demand of energy and resources has become increasingly tense, which puts new demands on China's power supply and capacity. Due to its outstanding features such as environmental protection and power saving, LED is gradually replacing traditional light source, and more widely used.

J. Li et al., *III-Nitrides Light Emitting Diodes: Technology and Applications*,
Springer Series in Materials Science 306,
https://doi.org/10.1007/978-981-15-7949-3_11

11.1.1 LED Lighting Technology Background

Saving energy, protecting the environment, and developing green lighting products have been the focus of China's lighting industry since the early 1990s. Therefore, when high-efficiency LEDs appeared, they immediately attracted the attention of relevant domestic funding agencies. In 2003, Ministry of Science and Technology of China established the "National Semiconductor Lighting Project Coordination Leading Group", which started the domestic LED lighting industry. At the beginning of 2009, Ministry of Science and Technology of China launched the pilot project of SSL application, which set off the development climax of domestic LED road lamps. In 2009, in view of the development process of domestic LED lighting, the National Development and Reform Commission, together with the Ministry of Housing and Construction and other six ministries and commissions, issued the "opinions on the development of LED lighting energy-saving industry", which clarified the development goal of China's LED lighting industry by 2015. In 2010, the National Development and Reform Commission (NDRC), together with the Ministry of Housing and Urban-rural Development and the Ministry of Transportation, launched bidding for 50 LED demonstration projects. The relevant ministries and commissions of China have carried out a series of promotion activities and relevant policies for LED lighting products. Such activities greatly accelerated the development pace of Chinese SSL industry and encouraged the application of LEDs.

At present, major application fields of SSL include (1) outdoor lighting, such as street lamps, tunnel lights, etc. These have been widely used, and energy saving is remarkable; (2) indoor lighting such as bulbs, and flat panel lamps. Tube lamps have also been widely used. Other applications, such as buildings and landscape lighting, which are easy to install, adjustable and design, have considerable influence on the lighting technique. It can be combined with streets, leisure space in urban lighting. LED-based signal or indicator with comfortable and flexible surface brightness can be used for signs where space is limited. In display lighting, such as museum lighting and shopping mall lighting, the use of LED lamps has great value since it has no ultraviolet infrared radiation. Its spectrum can be adjusted as well. Stage lighting and display screen, thanks to the dynamic and digital control ability of LED, open up new lighting forms for these places. SSL also has great application potential in vehicle, thanks to the higher efficacy and low consumption of electricity.

11.1.2 Basic Principles of LED Lighting

LED lighting system is the LED application carrier. With the expansion of LED lighting products and acceleration of industrialization, LED lighting systems are also evolving. It is one of the development trends to further simplify the function of the lighting system. In LED lighting systems, luminaire construction, heat dissipation, secondary optics, driver, intelligent control, reliability and cost control are key technical elements.

LED light distribution includes primary optical design and secondary optical design. Among them, the primary optical design is completed in the packaging process, which mainly determines the light output, the luminous flux size, the light intensity, and the light intensity distribution of the light emitting device. The primary light distribution design is to ensure the light output quality of each LED. It is considered to extract as much light as possible from the LED chip. Due to the small area of the LED chip, the large luminous flux, and the light-emitting characteristics of the half-space light, the use of the LED chip for direct illumination application has the problem of unreasonable use of light energy, thereby reducing the energy utilization efficiency of the LED lamp. Therefore, the secondary light distribution of the lighting fixture becomes a conventional scheme for improving the optical performance of the LED light source and improving the energy utilization rate. Secondary light distribution design mainly considers luminous flux, light intensity, illuminance, illuminating angle and brightness, which are the research scope of non-imaging optics.

The purpose of LED thermal management technology is to derive and dissipate the heat generated by the LED device. Heat-dissipating materials should have high thermal conductivity, such as new plastics, ceramics, graphite, metal, thermal conductive adhesives, heat-dissipating coatings, etc. The development and design of new luminaire structures with efficient heat dissipation is a key factor in reducing the thermal resistance of LED lamps. In addition to developing high-performance heat-dissipating materials and improving the heat dissipation performance of materials, LED lamp heat dissipation technology can also be divided into passive heat dissipation and active heat dissipation according to the heat exchange mode with ambient environment. Passive heat dissipation refers to the self-dissipation of heat in an air environment without the need for additional ancillary facilities. Traditional passive heat dissipation includes natural convection heat dissipation, uniform temperature plate heat dissipation, heat pipe and loop heat pipe heat dissipation, etc. The new type has micro channel heat dissipation substrate and the use of new thermal conductive materials. Active cooling of LED lighting refers to the consumption of a certain amount of electric energy, using a fan, a pump, etc. to drive the heat-dissipating medium to be forced to flow through the LED lighting device or to cool the LED light source by means of semiconductor cooling, and bring the heat to the heat-dissipating module. Compared with passive heat dissipation technique, active cooling technology has the advantages of high cooling strength and good cooling effect. It is especially suitable for high-power density or ultra-large LED lighting devices. It mainly includes fan-type, water-cooled, thermoelectric cooling, ion wind cooling and synthetic jet.

The main function of the LED driver is to convert ac voltage into dc voltage and simultaneously match the LED voltage and current. LED drive technology has the following features: DC control, high efficiency, pulse-width-modulation (PWM) dimming, overvoltage protection, load disconnection, compact size and ease of use. LED drive power is not only simple control and drive, but also has strict requirements on energy efficiency, life, power factor, constant current accuracy, electromagnetic compatibility, etc. Its development has been generally driven by industrial power

Fig. 11.1 LED lamp with dimming control to fit natural spectrum (from the left to the right are: ordinary LED, fitting daylight spectrum, the spectrum fitting incandescent lamp, candle fit spectrum)

supply, constant voltage drive, constant current drive and intelligent control. Efficient, low-cost, high-reliability LED driver power supply development is the future development direction of LED lighting driver power technology. On this basis, the modularization and standardization of the drive power products, with high integration, high intelligent power supply and control technology, will help to further improve the quality of LED drive power. Non-electrolytic LED drive power, low power and low-cost drive chip, ac direct drive LED, all solid-state large capacity capacitor and intelligent power technology based on user demand are expected to be the breakthrough of LED drive power technology in the future.

With a large number of LED lighting products entering the road, business, home and other fields, the LED application market and intelligent lighting system integration will lead to system innovation era. As shown in Fig. 11.1, LED intelligent control technology for dimming and change the design of the color provides greater flexibility, very suitable for construction, lighting, indoor lighting and dimmer, energy-efficient lamps and outdoor lighting applications. These applications can remotely control the lighting, but also improve the lighting product and add value and technology content, which further improve the lighting quality.

These features can bring great value to the users. How to integrate intelligent control and cost organically and to realize the optimal cost performance of LED lighting products are the main points of recent work. LED intelligent lighting control can be divided into wired control and wireless control.

11.1.3 Lighting and Display and Construction of Fusion

Since the LED light source in SSL can be free of ultraviolet and infrared radiation, it generally does not cause damage to the object to be illuminated. Compared with the conventional light source, the color generation does not require an additional filter device and is easy to change. The illumination system is simple, inexpensive, and easy to install. Its precise light distribution can be used as a substitute for fiber optic lighting.

LED's dynamic, digital control of color, brightness and dimming, saturated tone can create static and dynamic lighting effects. In other words, almost any color can be realized. LED opens up new ideas in spatial illumination. Characteristics of

LEDs such as long life, high lumen maintenance lead to reduction of maintenance costs and the frequency of replacement of the light source. The device performance may still suffer from color drift. The color shift is relatively small compared to the conventional decorative lamp. LED light source is small and thin. New projection lamp will undoubtedly become the major highlight of projector lamps because there is no place to put conventional cast light for many buildings. The combination of SSL and the building's surface brings creative space for lighting design. It will also have an impact on the lighting practices of modern and historic buildings.

At present, LED color decorative wall surface has been widely used in architectural design. In the future, SSL is expected to replace almost all traditional lighting products. In addition, with further LED performance improvement, lighting and display technologies will gradually merge and form new industrial platforms.

11.1.4 Lighting and Outlook

LED lights are widely used in major events such as the Beijing Summer Olympic Games and the World University Games. Such examples have fully demonstrated the SSL technology. Meanwhile, the technical level of LED products in China has increased and the cost has decreased in recent years. At present, LED products have occupied a large number of mainstream supply chain and the general lighting market. The application scope of LED lights in China has entered the mainstream lighting market such as commercial and household lighting market from professional fields such as transportation and landscape. Applications in special fields such as vehicle, ship-based and military will need the support of new technology. LED is a green light source in the 21st century. With its huge energy saving potential and good illumination, it opens up a completely new technical field for us. The continuous innovation of LED technology is creating an energy-saving and high-quality lighting environment for us, which requires the joint efforts of manufacturers, designers, government and other participants.

11.2 Visible Light Communication Application System

LED has a fast switching speed and a high frequency response compared with conventional light source. The nitride-based LEDs can be modulated at high speed (on the order of megahertz). Therefore, white LED is able to work as communication source, which has both lighting and communication functions. Compared with wireless communication technologies such as infrared communication and radio frequency communication, visible light communication (VLC) technology has the advantages of safety, energy saving, green environmental protection, and high transmission power. It can be widely used in various fields such as smart home, LED broadcasting system, navigation, intelligent transportation, robot positioning, etc. In particular, some specific applications are aircraft, hospitals and other radio frequency sensitive areas, military bases and other information security areas. The white light LED-based visible light communication technology does not directly replace the traditional communication technology, but is a good complement to these wireless communication technologies.

The United States is at the forefront of research in semiconductor lighting, but research on semiconductor lighting information networks started late. Boston University (responsible for LED communications, computer network system technology research), Rensselaer Institute of Technology (responsible for new material device technology and system applications), and the University of New Mexico (construction platform for nanomaterials, devices, bioimaging and display) center have achieved data transmission between two laptop computers by two LED lights flashlight demonstration system. Boeing is also engaged in research for multimedia entertainment system on the aircraft's study and development [2].

Europe has also carried out a lot of research and development work in this field. The representative research institutions include the University of Oxford, University of Cambridge, Imperial College London, Siemens, France Telecom and so on. Lubin Zeng from University of Oxford and his colleagues proposed a communication system based on white-light LED arrays and detectors with data rates of up to 1Gbps [3]. Researchers at Fraunhofer Institute of Communications in Germany together with Siemens has realized the data transmission rate of 513 Mbps. The bit error rate is lower than 2×10^{-3} [4]. The University of Oxford put forward the method of improving the data transmission rate by using pre-equalization and post-equalization, and developed the point-to-point music playback demonstration system (data transmission rate less than 1 Mb/s) and the one-way communication of 2.5 mbit/s with the transmission distance of 2 m. Siemens realized point-to-point data transmission with a speed of over 10 Mb/s using white LED. At the beginning of this century, the Nakagawa research group of Keio University in Japan proposed the concept of constructing an indoor optical wireless network using illuminated white LEDs, and made a preliminary theoretical analysis of the parameters and influences of the visible light communication system [4]. In 2003, Japan established visible light communications Consortium (VLCC). The members include telecommunications, lighting, LED aspects of manufacturing, power electronics, electronics manufacturing firms. VLCC not only emphasizes the research and development of advanced technologies, but also pays great attention to the research work of industry standards. In 2007, two standards were proposed adopted by the Japan Electronics and Industrial Technology Association (visible light communication system standard CP-1221 and visible light ID system standard CP-1222). South Korea is also very active in the research of semiconductor lighting information networks. Researchers from Samsung and Korea Electronics and Telecommunications Research Institute have implemented point-to-point communication between mobile devices and wavelength division multiplexing using RGB tri-color light to achieve one-way communication between fixed facilities and mobile devices as well as two-way communication from fixed facilities to mobile devices.

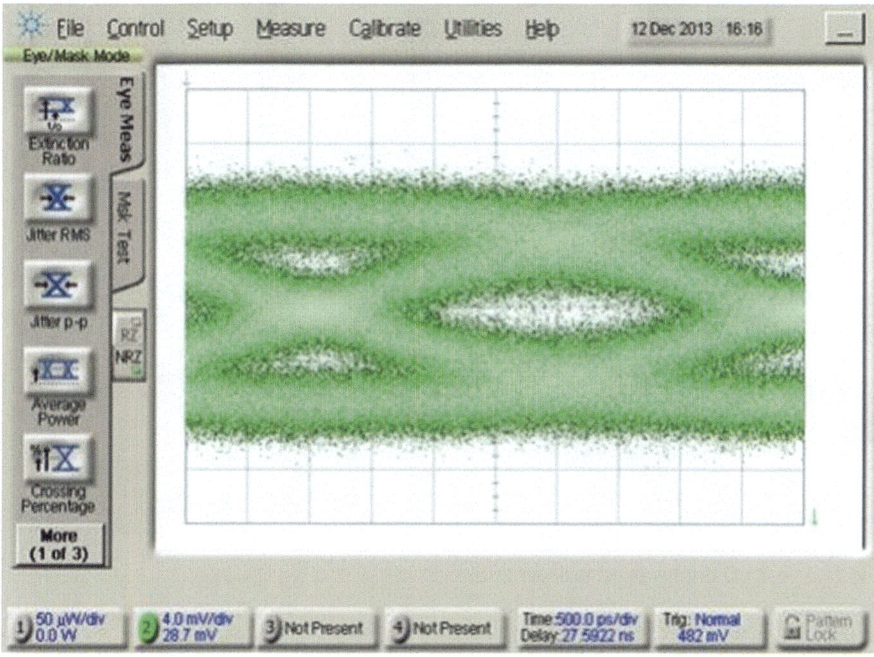

Fig. 11.2 Eye diagram of a delta-doped LED devices with transmission speed of 260 Mbits/s

The research on white LED wireless communication technology in China started relatively late. Representative units include the Institute of Semiconductors of the Chinese Academy of Sciences, Fudan University, PLA Engineering University, Jinan University, Xi'an Institute of Technology, Changchun University of Science and Technology, Beijing University of Posts and Telecommunications, Zhejiang University, Southwest Jiaotong University, etc. [5–14]. In 2009, the Institute of Semiconductors, Chinese Academy of Sciences studied the application of VLC in home control and broadband access, which was exhibited at the 2010 Shanghai world expo. Figure 11.2 shows the device eye diagram by using a delta-doping LED light source.

11.3 LED Display

11.3.1 LED Display Overview

Display becomes one of the most important ways of information exchange. The information conveyed by the display includes text or images. Electronic display

Fig. 11.3 A LED display panel is under test

devices can be divided into two categories: active light-emitting and passive light-emitting [15, 16]. The former mainly performs direct display by modulating the brightness and color of each pixel in the display. The LED display technology belongs to this category. The LED scrolls and full-color displays during the opening ceremony of the 2008 Beijing Olympics impressed us all. At present, the demand for LED displays also comes from sports, urban construction, command, advertising, and culture. Application environment of LED display mainly includes indoor and outdoor. According to the output color classification, the LED display can be divided into monochrome, multi-color and full-color display. In addition, it can also be classified by the center distance or dot pitch of two adjacent pixels in the LED display, such as P2.5 (point spacing is 2.5 mm), P3, P10, and so on (Fig. 11.3).

In order to obtain high quality display, the following parameters are mainly considered in LED display technology.

The unit of brightness is candela per square meter (cd/m^{-2}). The brightness requirements of the display device need to be considered in combination with the ambient light intensity. For example, the brightness requirements for the screen of the cinema and the LED display operated outdoors are obviously different.

Contrast refers to the ratio between the maximum brightness and the minimum brightness of the picture presented by the display device. Considering the ambient light level in the real environment, the display device must have a sufficiently high brightness to have sufficient contrast in the environment. It needs to be pointed out that LED is an active type light emitting device. Therefore, even if the maximum brightness of LED display device is low, its contrast can still be high. Grayscale refers to the black and white brightness level of an image. The larger value leads to more distinct of the image hierarchy.

The resolution describes the ability of the display device to show the smallest detail of the image and is a sign of the clarity of the image display. For LED display devices, increasing the resolution means increasing the number of pixels per unit area. Currently, LED displays on the market mainly use monochromatic or multi-color LED as unit pixels in the display screen. The center-to-center or dot pitch of two adjacent pixels in the LED display reflects the pixel density of the screen. The smaller the dot pitch, the higher the pixel density and the resolution are. In order to achieve high-definition resolution, it is necessary to continuously optimize the chip and packaging technology of LED to minimize the distance between each pixel.

The response time, also known as the rise time, is the time from the application of the voltage to the appearance of the image display. The time from the cutting off the power to the disappearance of the image, also known as the afterglow time, becomes the fall time. If the response time is too long, there will be smearing on the screen, especially for high-speed motion pictures which will become blurred. In general, active light-emitting display devices have better response time than non-active light-emitting display devices. For instance, LEDs have response times on the order of microseconds, while the response time of liquid crystal display (LCD) is on the order of milliseconds.

The above parameters are essential for achieving high quality image display. Therefore, an ideal display device should have high brightness, high contrast, high resolution, high color gamut coverage, low power consumption, high reliability, long lifetime and light weight. Compared with traditional display technologies such as cathode ray tube (CRT), liquid crystal display (LCD), and plasma TV, LED display technology has advantages in brightness, contrast, color gamut coverage, power consumption and response time.

11.3.2 Outdoor LED Display

The general LED display system mainly includes three parts: LED dot matrix structure (display screen), drive control and heat dissipation.

The LED screen of the display system is composed of a large number of LED devices, where each LED package is an individual pixel. The packaging form of the LED used for display panel is generally surface mount package, i.e. SMD-LED. According to the application requirements, there are one or more LED chips of the same kind or multiple colors in the package. For a full-color LED display, each of the pixels contains a number of LED chips of different light colors as sub-pixels. The pixel density of an LED display is described by dot pitch or pixel period, which is the center-to-center distance of two adjacent pixels. The smaller the dot pitch, the higher the pixel density (or the finer and the clearer the display) is. As the level of LED packaging continues to increase, the pixel pitch of high-density indoor LED displays improves continuously. The pixel center distance of indoor LED display screens in research and development has reached 1 mm or less. The key technology to improve pixel density and realize high-definition display is to realize ultra-small illuminating pixels, that is, smaller-sized LED full-color illuminating units. By properly arranging the internal structure of the LED chip and packaging structure, the size of the pixel can be effectively reduced.

A display module is formed by encapsulating a number of LED lamp beads on a substrate as illustrated in Fig. 11.4. Several modules are assembled to form an LED matrix unit. The back side of each LED display matrix unit has a corresponding drive circuit and components for attachment to the connection. According to the required LED screen area, a certain number of LED matrix units are assembled into a whole LED display screen as shown in Fig. 11.5.

LED display drive control: The current injected into each pixel unit of the LED display can be individually controlled. The pixels synthesize various characters, numbers, graphics and images on the display screen under the action of the driving signal [17]. LED display driver control system mainly includes data transmission, processing module, scanning control and display driver module. In order to improve the display quality and user comfort, the brightness correction of LED display screen and the light color consistency of the display screen are the key issues to be considered in the driver control part.

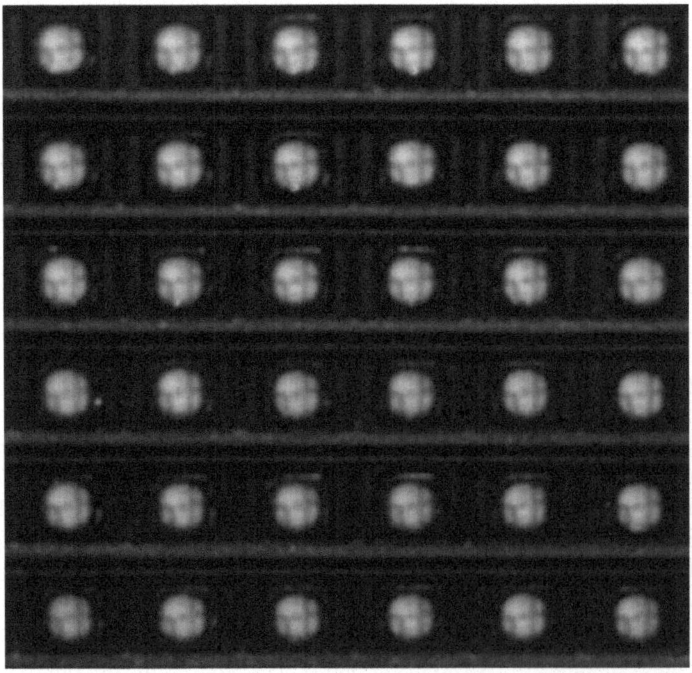

Fig. 11.4 A display module consisting of multiple LEDs

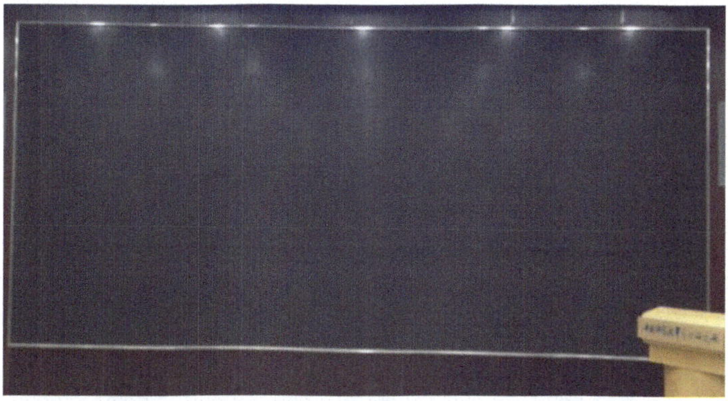

Fig. 11.5 Display screen consisting of a certain number of LED matrix units

The LED display system dissipates heat. The heat source of the LED display system mainly includes the LED display screen, the driving circuit and the switching power supply. If the heat dissipation system has a problem, the electronic components in the LED display system work in a high temperature environment for a long time, which may cause the reliability of the display to decrease. Therefore, it is necessary to analyze and design the heat dissipation of the whole display system. The main purpose of this part of the work is to provide a low thermal resistance path between the heat source and the external environment to ensure that the heat inside the system is quickly and smoothly transmitted.

When designing the heat dissipation method of the LED display system, factors such as power consumption, surface area, volume, heat flux density, volume power density, and working environment conditions of the LED display should be considered. According to the mechanism and actual demand of heat conduction, heat dissipation methods such as natural heat dissipation cooling, forced convection cooling heat dissipation, and heat-pipe heat transfer can be adopted. For outdoor display systems, a variety of protection measures needs to be considered. For instance, waterproof, dustproof, anti-corrosion, lightning protection, anti-reflection, anti-electromagnetic interference can be implemented to protect the display system.

11.3.3 Small Pitch Display and Indoor Applications

The small-pitch LED display technology is the focus and hot topic of the current development of LED display technology. It is regarded as a stepping stone for the LED display technology to open up the market for the commercial display. Compared with the traditional indoor large-screen splicing, the small-pitch LED display has great advantages. For example, small-pitch LED display has the advantage of seamless splicing compared with the LCD splicing screen; Compared with DLP (digital

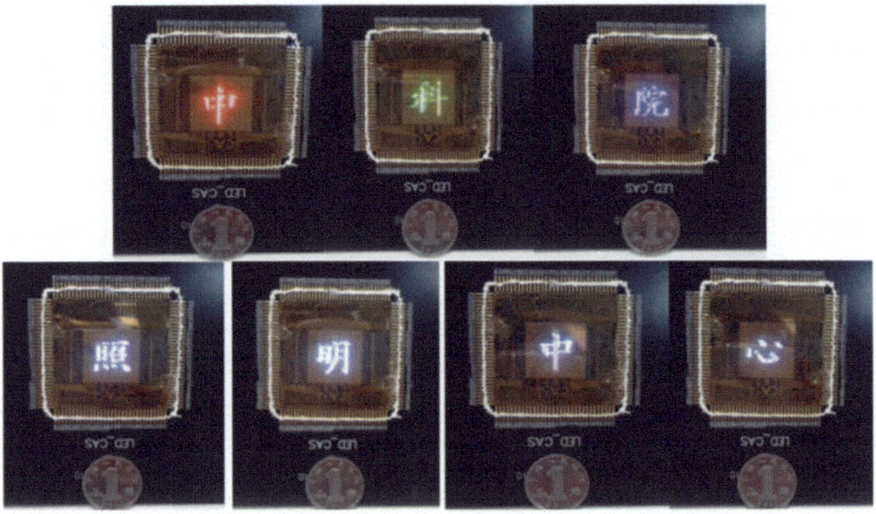

Fig. 11.6 A LED display sample with pixel size of P0.5 based on COB packaged devices

light processing-based projection technology) mosaic screen, its application platform is more flexible and reliable, easy to maintain, and flexible application environment. Manufacturers from both China and foreign countries usually use surface mount packaging technology to complete high-density LED dot matrix screen which involves PCB board, SMD and driver IC content. With the gradual reduction of the pitch size of LED full-color display screens, SMD LEDs are gradually replaced by ultra-small RGB integrated devices. In 2014, Everlight launched their ultra-small RGB LED products. Based on such product, pixel size of the LED screen was close to the 1.0 mm. In the field of small-size RGB full-color packaging, we have proposed LED matrix based on COB packaging for decreasing pixel size as illustrated in Fig. 11.6.

As an emerging technology, LED display technology also has certain shortcomings such as the need to improve display uniformity, especially at the edge of each module to ensure the flatness of large-area screen.

11.3.4 Wide Color Gamut LED Back Light Technology

With the progress of compound semiconductors, especially the rapid development of wide-bandgap compound semiconductors in recent years, the development of materials, device processes and their application technologies have ensured the semiconductor light-emitting diodes to cover the entire visible spectra range. As we know, LED has the advantages of rich color, high efficiency, small size and long lifetime.

As an ideal color point light source, LED can be integrated into an array that functions as a surface light source with uniform chromaticity. Such a technology has been widely used in various applications. It has been noticed that the light source has become the mainstream backlight of liquid crystal display. The main function of the backlight module of the liquid crystal panel is to provide a uniform, high-brightness light source. The basic principle is to convert a point or line type light source into a high-brightness and uniform luminance surface light source through effective light modulation.

At present, LED backlight technology can be divided into straight type and side-in type. The former is simple and without the light guiding plate, but the mixed light distance is not easy to reduce. It generally supports the backlight area adjustment technology, which is easy to achieve brightness adjustment, and can achieve extremely high dynamic contrast. This is very important for improving the image quality of the television. The latter backlight method sets the LED on the side of the specially designed light guiding plate as a backlight. The cost of such design is low. The thickness is also easily reduced, but the light efficiency is also reduced. Different LED backlights have their own characteristics. The LEDs also have a lot of room for improvement in terms of the design of point light sources into planar light sources. The technology is, therefore, continuously evolving.

11.4 LED for Plant Breeding

11.4.1 Overview

British scientist Joseph Priestley discovered photosynthesis in the 18th century. Dutch scientist Jan Ingenhousis proved that photosynthesis could only occur under the condition of light [18]. At the end of the 19th century, Russian researchers carried out a variety of artificial light source irradiation experiments on different plants such as cucumber, tomato, spinach, two gourds, broccoli, onions, winter rape, etc. It has been noticed that the plants grown speed is fastest under orange red light exposure, followed by blue light and green light. In 1996, Mitsubishi Chemical developed a Western lettuce cultivation device using red LEDs as a light source, which promotes the increase of sugar content in the plant. In 2002, Japan's CCS company Morimoto Miho made a plant growth system including LED light source, growth control software, signal conditioner and data collection software.

As far as the research in China is concerned, fluorescent lamps were used for light cultivation in the 1980s. The results showed that under red light lumination the leaf area expanded and the photosynthetic rate was fast. The chlorophyll content, soluble sugar and total sugar content were increased. Chlorophyll a/b value and total nitrogen content were low. It also has been noticed that blue light can thicken seedling stem, regulate stomatal opening, promote cell membrane permeability and cytoplasmic circulation. Seedling growth was better under red light mixed with blue light. In the 21st century, the obvious advantages of LED have attracted the attention of researchers from the Netherlands, Japan, America and China. The investment in research and development has been enthusiastic. In particular, the successful development of high-power LEDs in recent years has laid the foundation for the application of LEDs in greenhouse. The sugar content of komatsuna cultivated by Japanese researchers increased 1.3 times thanks to the application of LEDs operating at 660 nm, and the vitamin content in the crops is increased by 54–72% when the plants was irradiated by 730 nm near-infrared LED. This study shows the feasibility of using LEDs in plant factories. In recent years, researchers from Chinese Academy of Sciences, Chinese Academy of Agricultural Sciences and Nanjing Agricultural University compared plant growth rate under LEDs and natural light. Results indicate that LED exposure is superior in terms of leaf area, leaf number, leaf growth speed, etc. Therefore, it shows that plant growth quality is better with suitable LED exposure.

Light is the prime factor for plant growth. Photosynthesis, driven by sunlight, is the basis of most life on earth. The fundamental function of agriculture is plant production, which is the largest conversion of light energy into stored chemical energy on our planet, as well as the synthesis of inorganic matter into organic matter and the release of oxygen. Generally speaking, light with different spectra has different function on the plants growth [19]. For instance, light with wavelength greater than 1000 nm is not involved in photosynthesis, but converted into heat; light between 700 and 1000 nm can promote stem elongation; 610–720 nm is the strongest absorption band of chlorophyll, with strong photosynthetic effect and strong photolytic effect in many cases; 510–610 nm is the low efficiency zone of photosynthesis; 400–510 nm is the strong absorption zone of chlorophyll and lutein, which is also the sub-peak area of photosynthesis. Light with spectra range of 320–400 nm is capable of making the plants shorter and the leaves thicker. For even shorter spectra range, such as 260–320 nm, the light has strong sterilization effect. Light with wavelength less than 260 nm has harmful and lethal effect on plants. At present, human beings are faced with such major problems as food, energy, resources, environment and population, which are all closely related to plant production. Light is the primary factor affecting plant growth and development. Therefore, the characteristics, rules and light control standards of plant growth and development should be further studied to develop new types of plants suitable for high yield. It is a meaningful and innovative work to provide efficient light environment for plant growth. The shortage of land resources and the increase of population determine the scarcity of land for plant cultivation. At the same time, with the continuous improvement of the quality of life, it is increasingly urgent to supply green, safe and healthy food plants for human beings. The demand for devices and technologies used in the industrial production of refined and engineered cultivation equipment for agricultural plants, including LED artificial lighting technology, is increasing.

At present, LED light source can be adjusted with uniform light intensity and quality, and can be precisely configured with luminescence spectrum to meet the needs of different environments of plants and organisms which cannot be realized by traditional light sources. LED is capable of emitting monochromatic light with

Fig. 11.7 Field experiment of rice seedlings at the Institute of Semiconductors, Chinese Academy of Sciences

accurate wavelength for plant growth. Moreover, the combination of LEDs with different spectra would ensure the formation of the light spectrum basically consistent with plant photosynthesis and morphological formation. The light energy utilization efficiency is high as shown in Fig. 11.7.

Besides plants breeding, LEDs can be used in aquaculture. Limited by environmental factors such as offshore pollution and shrinking far-sea fisheries, aquaculture has grown rapidly worldwide. As an ecological factor, light plays an important role in the feeding, metabolism, reproduction and endocrine of fish. Artificial lighting not only increase the growth efficiency of aquaculture, but also obtain the weight of commercial fish as much as possible. By precisely control the fish growth at critical time and stage, we could obtain healthy and quality food.

At present, compared with water system treatment, nutritional bait research and development, seedling technology research and disease prevention, the effects of light on fish growth and reproduction need to be further studied. Most of the Chinese fish breeding factories still use traditional fluorescent lamps and the function is limited in light cycle and intensity. If the light source is replaced by LEDs, it is possible to select appropriate light color, intensity and cycle to control the growth and reproduction of fish, which will greatly increase the production of commercial fish, as shown in Fig. 11.8.

Fig. 11.8 Experimental site of the effect of light source on fish in collaboration with Institute of Semiconductors, Chinese Academy of Sciences

The research of LED light source in the agricultural field is promising. Due to the wide spectra coverage and compact size of LED, the device will be widely used in agriculture in the future.

11.4.2 Alternative Plant Lighting

At present, the most commonly used artificial light source in greenhouse mainly includes fluorescent lamps, high pressure sodium lamps, low pressure sodium lamps, metal halide lamps and the like. The disadvantages of these light sources are high energy consumption, and high operating costs. The energy costs accounting for 50–60% of the total operating costs. Therefore, improving luminous efficiency and reducing energy consumption have always been an important issue in artificial light applications in the agricultural field. Ordinary fluorescent lamps provide more green light, about 50%. The rest part is mostly red and blue light, which account for about 25% of the total spectrum. The infrared spectrum is very low. Fluorescent lamps have high luminous efficiency and long lifetime with up to about 12,000 h, but its power is relatively small (28 and 36 W). Because the spectra of such light contain a lot of green light, it is easy to cause crops to grow in vain. Therefore, such light is generally used in plant tissue culture. The sodium lamp is a gas-emitting light source that emits mercury and sodium vapor. Its luminescence spectrum is concentrated at 589 nm. The luminous efficiency is as high as 120 lm/W. Standard operating sodium vapor pressure of the lamp is about 10 kPa. There are more red orange light and less blue green light in the luminescence spectrum. However, high-pressure sodium lamp is a heat source. The surface temperature is high and the amount of heat generated during operation is large. Therefore, the crop cannot be irradiated by such light source at a close distance. Metal halide lamp is a gas illuminating light source in which a mixed vapor of mercury and a rare metal halide generates an electric arc to discharge light. Different spectra can be exhibited by changing the composition of the metal halide. It has high luminous efficiency, high power and long lifetime between 5000 and 20,000 h. However, the filler in the lamp also contains mercury and will cause pollution. Its spectrum contains a lot of infrared light, which generates a large amount of heat and cannot illuminate the crop at close range. Most of the test light sources use filters or colored polyvinyl chloride films to obtain light quality. It is impossible to quantitatively and precisely modulate the spectral energy distribution and regulate the light environment, thus affecting the reliability and comparability. For a long time, the traditional artificial light source applied in the facility has uncontrollable wavelength, high energy consumption and high operating cost, which has become an obstacle to the development of application in the agricultural field.

Light has a regulating effect on photosynthesis, growth, morphogenesis and material metabolism of plants. At present, the light sources in greenhouses used to add up more light are generally fluorescent lamps, high-pressure sodium lamps and incandescent lamps. Spectral energy diffusion is designed according to the needs of the human eye for light. The spectrum required for plant growth is different from the

Fig. 11.9 LED greenhouse fill light system experiment

needs of the human eye. According to the research, LEDs, regardless of the GaN, gallium arsenide and indium phosphide output, are all monochromatic. They can be used alone or in combination. The bio-efficiency of such light source is high and can be used to concentrate light of a specific wavelength. Evenly illuminating plants can regulate crop flowering and fruit, and can control plant height and plant nutrient composition as shown in Fig. 11.9. Since LED has higher efficiency, less heat is generated during operation. Plant will not get burnt even at close range. In addition, LED is a compact light source and can designed in various shapes, which will occupy less space. It can be multi-layered and cultivated in a three-dimensional combination, and is convenient to install. Therefore, the plant factory size can be miniaturized. Extremely durable and energy efficient characteristics of such light source would ensure the reduction of operating costs.

Plant tissue culture, that is, plant aseptic culture technology, is based on the theory that plant cells have pluripotency, organs (such as roots, stems, leaves, shoot tips, flowers, fruits, etc.), tissues (such as forming layers, Epidermis, cortex, medullary cells, endosperm, etc.) or cells (such as megaspores, microspores, somatic cells, etc.) and protoplasts. Induced by sterile and suitable artificial medium and artificial conditions such as light and temperature, callus, adventitious buds, adventitious roots finally form the intact plants. The artificial light source used in the tissue culture room is generally a multi-layer rack mounted fluorescent lamp. Since the fluorescent lamp has high heat, it cannot be too close to the plant. This would lead to higher power consumption and heat load in the room, which prompted researchers to use LED instead of fluorescent lamp as shown in Fig. 11.10. Seedling cultivation is an important part of fruit and vegetable production. Since the seedling formation is an irreversible process, the health of seedling growth will directly affect the growth and development of the plant and affect the yield and quality of the crop. The use of light technology to cultivate strong seedlings is a new method of energy saving,

Fig. 11.10 LED plant tissue culture optical system experiment

environmental protection, economic efficiency and simplicity. It has outstanding advantages and is of great significance for cultivating strong seedlings. Studies have reported that light environment regulation has a significant impact on the growth and development of cucumber, tomato, sweet pepper, rapeseed and other seedlings. In addition, the gender performance of melons is susceptible to environmental factors and chemical regulation. Since the sex differentiation of melons occurs in the seedling stage, the gender performance of melons can be artificially controlled during the seedling period by controlling the light irradiation. LED has unparalleled superiority in the application of light environment regulation in vegetable seedling stage, and will play an important role in vegetable plantation.

11.4.3 Lighting Design Features

According to the light quality required by specific plant, the design of the luminaire and the artificial light supplement in the greenhouse have three main elements such as light quality, light intensity and photoperiod. The light intensity, i.e. the illuminance requirement, is based on the daylight compensation point of the main crop. The light compensation points of tomato, cucumber and pepper are 3000 lx, 2000 lx and 1500 lx, respectively, and the light saturation points are 70,000 lx, 55,000 lx and 30,000 lx, respectively. Therefore, the light intensity of greenhouse light supplement is generally required to reach 1000–3000 lx. Secondly, as far as the optical quality requirements is concerned, the plant uses sunlight for photosynthesis. Its absorption spectrum lies between 380 and 760 nm [17]. Hence according to the type of the crop, there are specific requirements for spectral morphology. Finally, the requirements of the photoperiod, the natural day and night, alternating and recurring phenomena form the cycle of light and dark and spectral morphology. Crops adapt to this change during the long evolutionary process. However, before or after the winter solstice or even cloudy days, the illumination time often fails to meet the growth and development needs of the crop. Artificial light is therefore needed to increase the illumination time. In recent years, artificial light supplementation in greenhouses has become an important means of facility agricultural production. Various artificial light sources have also been developed rapidly.

The LED light source system includes a plant growth system that is an LED light source, growth control software, signal conditioner, and data collection software. The system can control the LED illumination through the growth control software, and collect the growth environment information measured by the sensor into the database through the data collection software, and at the same time realize automatic control of different light qualities.

11.4.4 Systematic Design Trend

Light is one of the most important environmental factors for crop growth. Some scholars have pointed out that "1% of light is 1% of production". This statement clearly indicates the importance of visible light in crop production. The illumination of the greenhouse itself is much lower than that of the open field. Especially in the winter and spring seasons and under rain and snow, the situation of insufficient light will become more obvious. It has become an important factor limiting the yield of vegetables. Artificial light supplementation in the greenhouse will be an inevitable choice.

A large number of studies has successfully proved that LED can be used as an artificial light source for tissue culture, greenhouse, and closed plant factories. One can utilize various light qualities of various LED radiations to study the light quality response characteristics of different crops. One can also combine different proportions and different light spectrum of the LED to generate special lighting condition. Due to the easy selection of LED wavelengths and compact structure [20–23], it will be more and more widely used in the field of facility agriculture in the future. However, further improvement of light efficiency and cost reduction of the device need to be improved urgently. With the continuous maturity of LED technology, the reduction of manufacturing costs, LED will be the most promising artificial light source in the agricultural field in the 21st century, and show good development prospects.

11.5 Medical Applications

Thanks to the unique characteristics of LEDs, the device can be used in medical lighting, surgical lighting and endoscope light source by virtue of its characteristics of high intensity and high color rendering performance. Meanwhile, with its biological effects recognized by people, LED is also being gradually applied in medical treatment. In contrast with existing laser treatments, such as weak laser treatment (physiotherapy, light needle), spectroscopy (photodynamic therapy), light knife treatment (vaporization, coagulation, cutting), LED are anticipated to be used in these fields thanks to the performance development [24–27]. With the improvement in LED's photoelectric conversion efficiency, monochromatic, emission angle, LED beam characteristics will move closer to the laser beam. Although one is coherent and the other is incoherent, studies have shown that the coherence and polarization of the beam have little effect on the results when interacting with organisms. LED will become the main light source for light therapy.

11.5.1 Treatment of Neonatal Jaundice

Neonatal hyperbilirubinemia (jaundice) is a common disease that can easily lead to hearing damage. In severe cases, it can cause bilirubin encephalopathy and affect the quality of life. Conventional light source in the treatment includes mercury vapor blue light or incandescent lamp. However, the application of such light sources introduces thermal effects and the addition of ultraviolet rays, which would bring side effects to the patients. Blue LED produces narrow wavelength exposure, which matches the absorption spectrum of bilirubin and promotes the decomposition of bilirubin into biliverdin and non-toxic water-soluble substances out of the body [23].

11.5.2 Treatment of Hemorrhoids

Acne is a common disease in adolescents. The main cause of the disease is the vigorous sebum division and the rapid proliferation of propionibacterium acnes. Drug treatment has side effects and the efficacy is limited, while the LED light with a wavelength of 415 nm produces photochemical reaction and oxygen, kills bacteria, and achieves therapeutic effect. Treatment of acne with LED is therefore safe for the patient [24].

11.5.3 Treatment of Wound Healing

Low-energy laser treatment of wound healing has accumulated a lot of clinical experience [25]. Many foreign companies have introduced LED wound healing products. The effect is very good because LED red light can make human muscle and skin cells grow at 5 times of the normal speed. The wound healing is promising, and the US FDA has approved clinical applications several years ago.

11.5.4 Treatment of Oral Ulcer Inflammation

After high-dose chemotherapy and chemotherapy for leukemia patients, oral mucosal cells will be severely killed. Patients often suffer from acute oral ulcers, which causes difficulty in swallowing and produces severe gastrointestinal reactions. 688 nm LED light is used after chemotherapy. It can significantly relieve oral pain and better control oral ulceration.

11.5.5 Treatment of Joint Pain

Blue light can be used to illuminate the painful joints to alleviate the pain. The longer of the light exposure, the more obvious the pain is alleviated. Therefore, LED is combined with sheath for the treatment [26].

11.5.6 Application in Medical Beauty

The treatment of baldness with red light 660 nm irradiation allows cell activation, promotes hair growth and toughness, and avoid the hair root premature atrophy [27]. Using 625 nm red light irradiation, one can increase the skin fibroblasts activity, promote protein production, recover the elastic, makes the skin smooth. Combined with the photosensitizer, it can also remove color pigment deposition, reduce the damage caused by light. LED red light, combined with local repair paste catalase, can effectively stimulate the secretion of melanocyte black pigment, reduce the concentration of hydrogen peroxide in the skin for therapeutic purposes vitiligo [28].

References

1. China Solid State Lighting Alliance Annual Report 2014 (in Chinese)
2. H. Chen, J. Duan, X. Chen, Progress of solid-state lighting information network, in *Proceedings of the 7th National Photonics Conference*, July 11, 2010, Changchun, China (2010)
3. Z. Lubin, D. O'brien, M. Hoa, et al., High data rate multiple input multiple output (MIMO) optical wireless communications using white LED lighting. IEEE J. Sel. Areas Commun. **27**(9) (2009)
4. M. Nakagawa, Y. Tanaka, T. Komine et al., Indoor visible light data transmission system utilizing white LED lights. IEICE Trans. Commun. **86**, 8 (2003)
5. L. Zhu, B. Liu, Y. Yang et al., Wireless communication systems based on semiconductor lighting. High Technol. Lett. **20**(8), 863–867 (2010)
6. S. Zhu, L. Zhao, H. Yang, Device technology of LED used in visible light communication. ZTE Technol. J. **20**, 6 (2014)
7. C. Wang, LiFi: light bulb turned into a router; *China Economic Weekly* (2013), p. 42 (in Chinese)
8. F. Wang, J. Rao, X. Xiang, Research of performance of circular array light source in underwater wireless LED optical communication. Laser Technol. **38**(4), 527–532 (2014)
9. H. Zhao, C. Chen, X. Chen, et al., Analysis and solution on noise and interference of white light LED communication system. Opt. Commun. Technol. **1**, 60–62 (2011)
10. D. Ding, Z. Kexi, Design and simulation of wireless LAN based on visible light communication. J. Xi'an Univ. Technol. **23**(1), 29–32 (2007)
11. J. Zang, Y. Piao, Z. Song et al., Indoor visible light communication system based on white LED. Chin. J. Lumin. **30**(6), 877–881 (2009)
12. J. Ding, Z. Huang, Y. Ji, Utilization of white LED for wireless communication system performance analysis, in *Proceedings of CIE-YC2009*, Chinese Institute of Electronics (2009)
13. Y. Wang, Research of indoor wireless visible light communication system. Master Dissertation, Zhejiang University (2008) (in Chinese)
14. X. Wang, On the performance of space shift keying modulation for visible light communications. Master Dissertation, Southwest Jiaotong University (2014) (in Chinese)
15. G. Ying, W. Hu, Y. Qiu, *Flat Panel Display Technology* (Posts & Telecom Press Co., Ltd., Beijing, 2002)
16. M. Tian, *Electronic Display* (Tsinghua University Press, Beijing, 2001) (in Chinese)
17. T. Wu, Overview of LED display and drive methods. J. Guangdong Ocean Univ. **27**(4), 104 (2007). (in Chinese)
18. R. Michael, Matthews, science and worldviews in the classroom: Joseph Priestley and photosynthesis. Sci. Educ. **18**(6), 929–960 (2009)
19. H. Gao, D. Zhu, Relationship between plant growth and illumination. Light & Lighting **5**(4) (2005) (in Chinese)
20. C. Nie, Y. Yang, Development of LED plant growth illumination system. Practical Electron. **8** (2014) (in Chinese)
21. F. Shi, H. Wang, R. Zhang, et al., Research progress on accelerating plant-growth with LED. China Light & Lighting **3** (2011) (in Chinese)
22. C. Miao, B. Hu, The application of LED spectroscopic lighting in plant growth. Adv. Display **9** (2012), p. 196
23. X. Yan, L. Ding, Y. Ding et al., The application of LED light source in plant tissue culture. China Agr. Sci. Bull. **25**(12), 42–45 (2009). (in Chinese)
24. J. Chai, K. Chen, J. Zhou et al., Time-effect relationship of neonatal high unconjugated bilirubinemia blue light therapy. J. Practical Med. **26**(19), 3535–3537 (2010). (in Chinese)
25. Y. Li, J. Shi, Y. Wang, et al., narrow-band blue light treatment of acne vulgaris clinical observation. Youjiang Med. J. **38**(3), 289 (2010) (in Chinese)
26. F. Zhao, J. Qu, R. Zheng et al., Acceleration of wound healing of rabbits with topical oxygen therapy combined with near infrared LED irradiation. Negative **28**(21), 1976–1978 (2007). (in Chinese)

27. L. Wang, F. Wu, X. Shen, M. Wang, Comparing the effects of acupoint irradiation with 650 nm-10.6 μm compound laser or RLED phototherapy on patients with knee osteoarthritis. China J Tradit. Chin. Med. Pharm. **25**(2), 217–220 (2010)
28. I.V. Meglinski, S.J. Matcher, Quantitative assessment of skin layers absorption and skin reflectance spectra simulation in the visible and near-infrared spectral regions. Physiol. Meas. **23**(4), 741–753 (2002)

Chapter 12
Novel Nitride LED Technology

At present, the light efficiency of the high-power white LED has reached 250 lm/W in the industry. The light efficiency reported in the laboratory has exceeded 300 lm/W. The innovation of LED technology and application of LEDs have far exceeded expectations. However, the luminous efficiency of the traditional planar LED with InGaN/GaN multiple quantum well structure is difficult to rise significantly due to some inherent problems. At the same time, the light efficiency is still far away from the theoretical limit of 400 lm/W. In addition, with the demand of some special applications, some new LED technologies such as nanorod LEDs, quantum dot LEDs, polarized LEDs, etc., have been explored. The exploration of the growth, preparation and application of these new LED structures not only contribute to the research of basic physics but also to solve the bottleneck problem in the development of semiconductor lighting by improving the material and device performance and by approaching the theoretical limit of LED lighting. Thus, it has important research value.

12.1 GaN-Based Nanorod LED

The research on GaN nanorods began in the late 1990s. The initial work utilizes MBE to grow different morphologies of GaN nanorods. In 2004, Kim of Dongguk University in Korea first developed the nanorod LED by MOCVD method, which officially opened the research of nanorod LED, and has become a research topic in recent years.

J. Li et al., *III-Nitrides Light Emitting Diodes: Technology and Applications*,
Springer Series in Materials Science 306,
https://doi.org/10.1007/978-981-15-7949-3_12

12.1.1 Advantages of Nanorod LEDs

There are three main reasons for the bottleneck of lighting efficiency of LED with traditional structures.

First, high-quality GaN bulk single crystal homogeneous substrates are difficult to prepare. Current GaN-based LEDs are mainly heteroepitaxially grown on sapphire, 6H-SiC or Si substrates. However, large lattice mismatch and thermal mismatch in heteroepitaxial growth will lead to high-density threading dislocation in the GaN epitaxial layer, which will result in a large non-radiative recombination probability of electrons and holes in the active region. This will reduce the internal quantum efficiency [1].

Second, the excessive lattice mismatch (up to 11%) of InGaN/GaN multiple quantum wells in GaN-based LEDs can also cause severe polarization effects in InGaN. The stress-induced polarization field in the InGaN quantum well layer causes the tilt of energy band. Then the wave function overlapping electrons and holes in space is reduced, which is the so-called quantum confinement Stark effect (QCSE). The probability of radiative recombination electron-hole pairs will be reduced, which eventually lead to a decrease in the quantum efficiency of the LED [2].

Third, since the refractive index difference between the GaN material (refractive index of 2.5) and air (refractive index of 1) is large, the light emitted from the active region cannot be extracted from the interface due to too small critical angle of total reflection (only 23.5°). This will result in low light extraction efficiency (only 4%). Light that cannot be extracted will repeatedly propagate through the dielectric material until all of the light energy is dissipated into thermal energy, which also adversely affects the performance of the device [3].

Researchers have done a lot of work in reducing dislocations, eliminating polarization and improving light extraction by using lateral epitaxy, homogenous growth, etc. to suppress dislocations and to improve the internal quantum efficiency of LEDs. People also use non-polar and semi-polar growth to reduce the polarization effect, or use surface roughening, photonic crystals, graphic substrates, etc. to increase light extraction efficiency. However, it is difficult to take care of all aspects of relevant issues, which further limit the development of high efficiency LEDs. In recent years, nanorod LED has become the hot topic in the semiconductor lighting since it shows many advantages. As shown in Figs. 12.1 and 12.2, it can effectively solve the problem that the internal and external quantum efficiency of conventional planar structure LEDs are not high [4, 5]. The specific advantages are listed as follows:

① The growth free energy of nanorod is relatively low. It is possible to achieve near-defect-free growth, thereby greatly improving the crystal quality of GaN materials;

② The special geometry and large specific surface area make the nanorod LED a much larger illuminating area than the planar structure. Simultaneously, the nano-scale scattering effect also has an anti-reflection effect, which will greatly improve the light extraction efficiency;

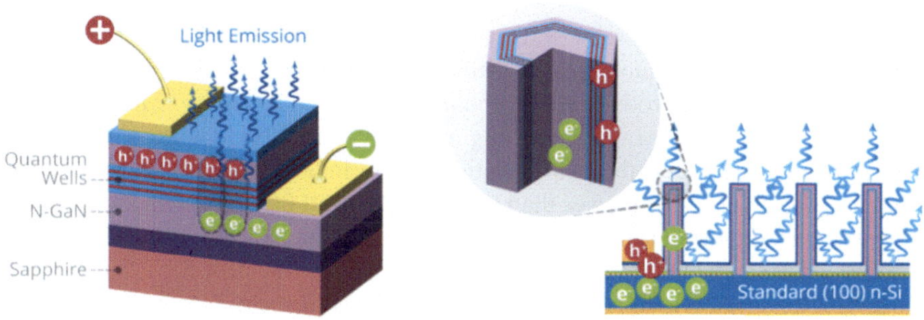

Fig. 12.1 Structural illustration of the traditional planar 2D structure and nanorod 3D structure LED

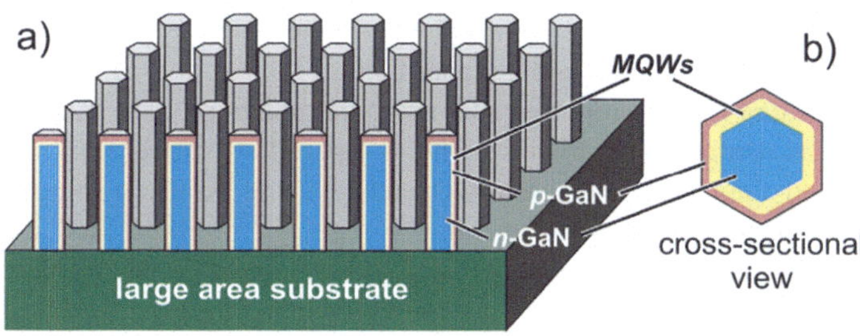

Fig. 12.2 Structural illustration of nanorod LED structure [5]

③ The nanorod LED can realize the growth of the active region along the non-polar or semi-polar plane, which effectively reduces the quantum-confined stark effect and thereby increases the internal quantum efficiency;

④ The saturation current of the nanorod LED can be greatly improved with a weakened droop effect and is suitable for large injection current. This is also the development general trend of LED lighting in the future.

12.1.2 Preparation Method of Nanorod LED

There are two main methods for preparing nanorod LEDs. One is the "top-down" method, and the other is the "bottom-up" method [6]. The "top-down" is a method of performing ICP etching of a planar GaN LED with a mask. Common pattern transfer methods include electron beam exposure, nanoimprint, holographic lithography, nanosphere lithography, etc. The materials used as the mask mainly include Ni, SiO_2, and polystyrene PS sphere. "Bottom-up" is a method of epitaxial growth on nanostructures, which is mainly divided into self-organized growth and selective Area growth.

"Top-down" method:

In 2005, Tao-Hung Hsueh et al. of Taiwan Jiaotong University reported that the InGaN/GaN MQW nanorods could be obtained by RIE-ICP etching, where contractive Ni sphere formed by rapid annealing at 850 °C under N_2 was used as a mask. The PL results show that the peak position of the InGaN/GaN MQW nanorod is 446.8 nm, which is 5.1 nm lower than the quantum well of the bulk material. They believe that the blue shift is the result of strain relief and quantum confinement effects in MQW. At the same time, the light emission intensity of the nanorod is 5 times that of the bulk material, which is ascribed to the decrease of the piezoelectric field, the increase of the superposition of electron and hole wave functions, and the increase of the radiation recombination rate. Seung Hwan Kim et al. also found that the nanorod structure acted as photonic waveguide and scattering center, which can effectively alleviate the effect of total reflection in the device. J. Bai et al. used the same method to verify that the PL intensity of nanorod LEDs is increased by 8 times and the IQE by 50% compared with conventional planar LEDs [7].

In 2011, Qiming Li et al. reported a two-step method to prepare nanorods using a silica sphere as a mask on a planar LED. A tapered structure was firstly etched by plasma, and then was selectively wet etched using KOH to mitigate surface mechanical damage caused by dry etching. A regular "flashlight-like" dislocation-free nanorod LED was obtained as shown in Fig. 12.3. Liang-Yi Chen et al. of Taiwan University also used a spin-coating method with silica nanospheres as a mask, and then etched the nanorods by ICP-RIE. In order to reduce defects caused by mechanical damage and avoid short circuits, they deposited 100 nm thick SiO_2 layer at surface of the nanorod by PECVD. This can effectively reduce the leakage [8].

However, using top-down methods also has some drawbacks. A significant portion of the GaN material is etched away, which no longer contributes to the generation of light and adds additional cost for the bottom-up approach during device fabrication. In addition, the increased defects and leakage current are also problems that need to be addressed.

"Bottom up" approach:

There are two main bottom-up methods. One is self-organized growth such as catalyst-assisted self-organized growth and catalyst-free self-organized growth. The other is selective area growth [5].

① The catalyst-assisted self-organized growth

Catalyst assisted GaN nanorods self-organized growth method is also referred to as the VLS (Vapor-Liquid-Solid) synthesis since such a process uses gaseous reactant where the catalyst is in a liquid state during the reaction and the final reaction product is solid. The growth mechanism of catalytic reaction growth method based on the gas-liquid-solid (VLS) [6], as shown in Fig. 12.4, was proposed by Wagner in the study of a large number of single crystal whiskers grown in 1960. Since the diameter of the nanowire is determined by the diameter of the particles of the catalyst, the length of the nanowire can be controlled by the reaction time. This method gives

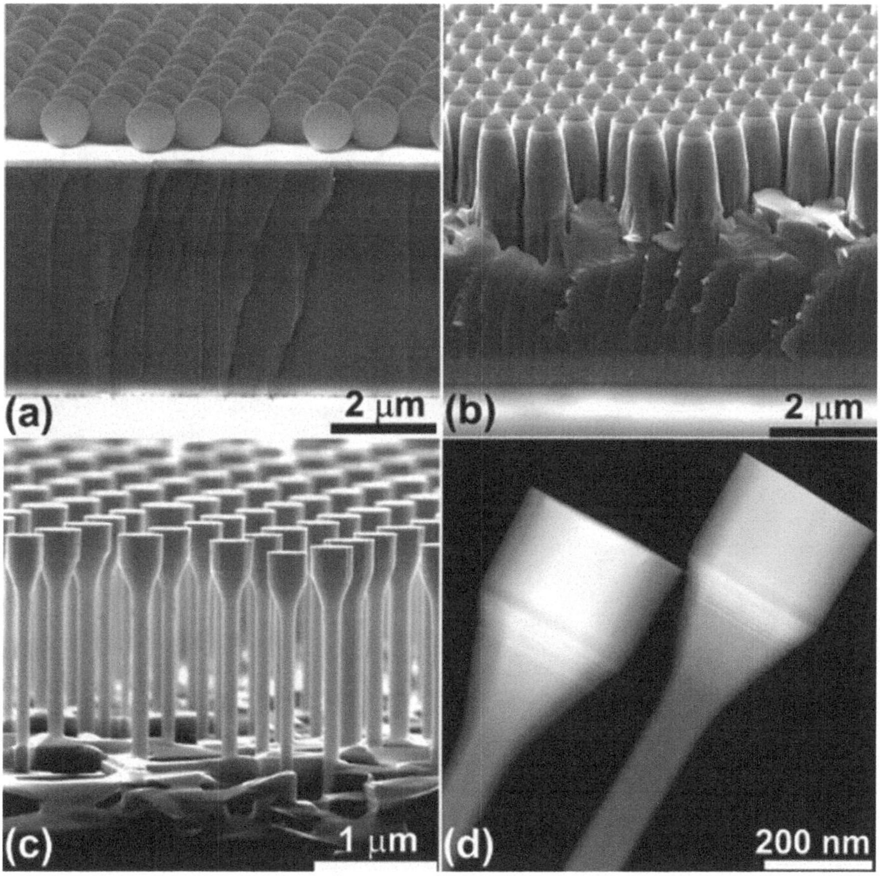

Fig. 12.3 SEM image of **a** the coated and self-assembled monolayer silica spheres on wafer of planar LED; **b** tapered nanorod the LED formed by plasma etch; **c** "Flashlight-like" nanorod LEDs array formed by wet etch; **d** STEM bright field fringe pattern of "flashlight-like" nanorod LEDs shows InGaN MQWs position

an effective means for obtaining uniform nanowires. Commonly used catalysts are Fe, Ni, Au, and Ta. By controlling the size and growth time and temperature of the catalyst, nanostructures of different diameters and lengths can be obtained. Although metal plays an important catalytic role in the growth process, it can diffuse into the nanostructure during growth, introduces defect levels, enhances non-radiative recombination, reduces internal quantum efficiency, and thus impacts the properties for materials and devices. In addition, the GaN nanorod crystal orientation is highly diverse and inconsistent, which also causes troubles in the device process, and is therefore not suitable for mass production.

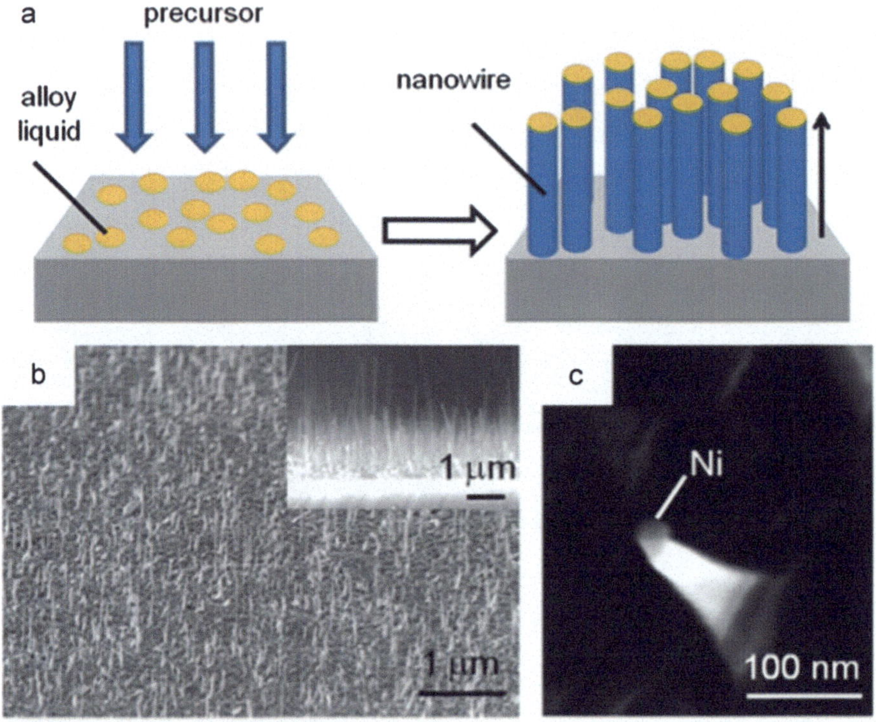

Fig. 12.4 **a** Schematic illustration of the VLS process; **b** SEM oblique (inset) and image of GaN nanowires by the VLS method using Ni catalyze with tilted view; **c** SEM image showing Ni catalyst on top of a single nanowire [6]

② Catalyst-free self-organized growth

Catalyst-free self-organized growth usually does not suffer the problem of metallic impurities. Commonly used methods are MOVPE, HVPE, PAMBE, and so on. By this method, nanorod structure is mainly formed by nucleation island growth as shown in Fig. 12.5. The growth method, however, requires strict control of V/III ratio, temperature and other conditions in the growth process. Studies have shown that high V/III ratio can control the density of nucleation and inhibit the coalescence of the nucleation islands by reducing the diffusion length of Ga atoms in N-rich atmosphere. Thus it is beneficial for columnar growth of GaN, while low V/III ratio is conducive to layered growth. Doping also affects the growth and morphology of the nanorods. Koester et al. reported a method for GaN nanorods epitaxial growth on a thin SiN_x layer grown in situ using MOCVD. They found that higher SiH_4 flux promotes vertical growth of the material. In contrast, increasing the flow rate of Mg will lead to larger diameter and smaller height of GaN nanorods. That is to say, using Mg for p-type doping can increase the tendency of lateral growth of nanorod.

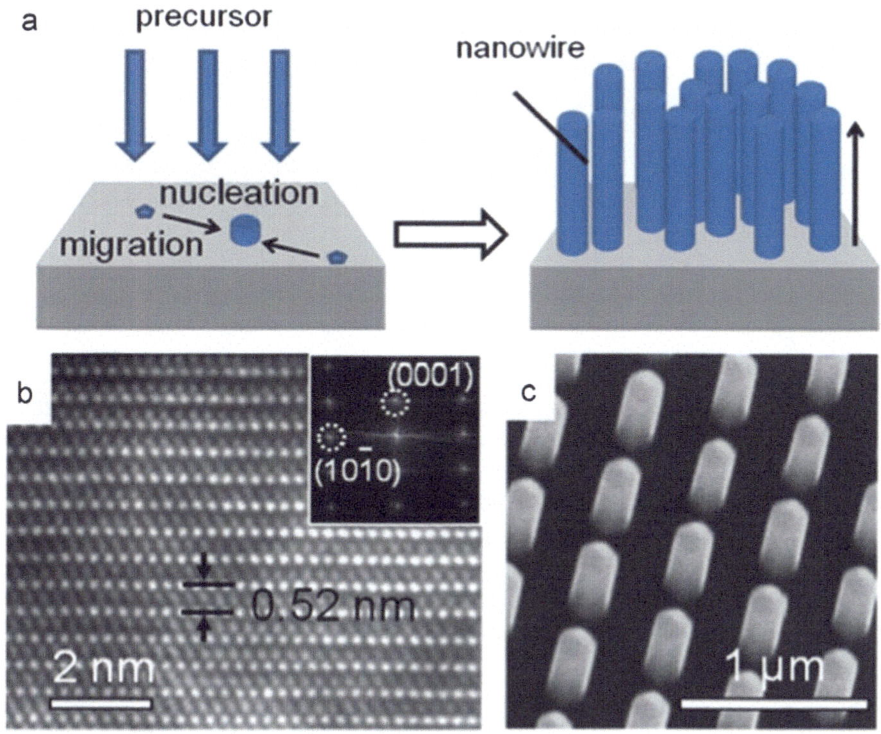

Fig. 12.5 **a** Schematic illustration of catalyst-free MOVPE growth; **b** HRTEM image of GaN nanorods grown by catalyst-free MOVPE, inset shows the diffraction pattern of GaN nanorods; **c** SEM image of GaN nanorod array with tilted view [6]

③ Selective Area Growth

The selective area growth is the development trend of the future micro-nano structure LED. Although GaN nanorods by self-organized growth have demonstrated good results, the wavelength of the nanorods grown by this method is difficult to control because the quantum well size, position, and material parameters that determine the wavelength of the nanorod are uneven. However, the choice of growing GaN nanorods on a patterned substrate is a good solution to this problem. In 2006, S. D. Hersee et al. first reported the use of pulse growth mode to achieve MOCVD growth of GaN nanorods on SiO_2 patterns. During vertical growth, GaN nanorods with uniform dimensions and morphology were obtained by alternately supplying Ga flow and NH_3 flow. In 2010, W. Bergbauer et al. achieved the selective area growth of GaN nanorods in continuous gas flow mode. They found that when H_2/N_2 ratio is 1:2, the columnar growth mode is very significant as shown in Fig. 12.6. It is because that larger H_2 ratio can suppress the coalescence between single GaN nanorod. However, when the H_2/N_2 ratio is too large, the GaN nanorod will decrease as the H_2/N_2 ratio increases [9].

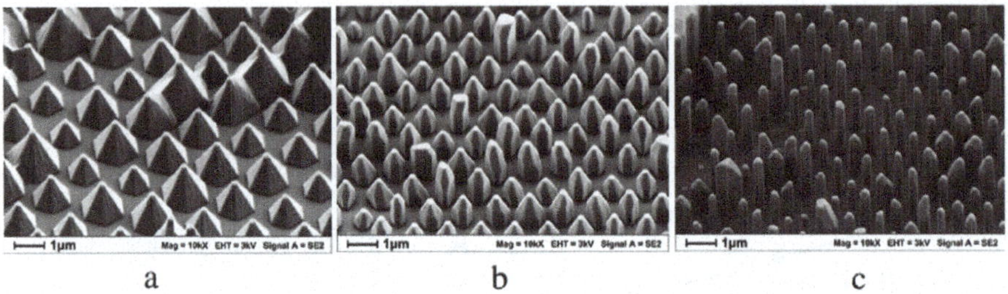

Fig. 12.6 **a** Pyramidal nanostructures of GaN formed by using pure N_2 as carrier gas during growth process; As-grown GaN nanorods with **b** the H_2/N_2 ratio being 1/2 **c** the H_2/N_2 ratio being 2/1

In addition to these traditional methods, Chul-Ho Lee et al. reported a new method for epitaxially growing GaN/ZnO coaxial LEDs directly on graphene films and obtained a blue LED with good performance as shown in Fig. 12.7 in 2011 [10]. Since graphene films can be transferred to any flexible substrate, this approach opens up new directions for the preparation of inorganic flexible optoelectronic devices.

In short, the "top-down" process is relatively simple and can be implemented directly based on planar structure LEDs, which can effectively release the stress in the planar structure LED and increase the light extraction. However, the etching process

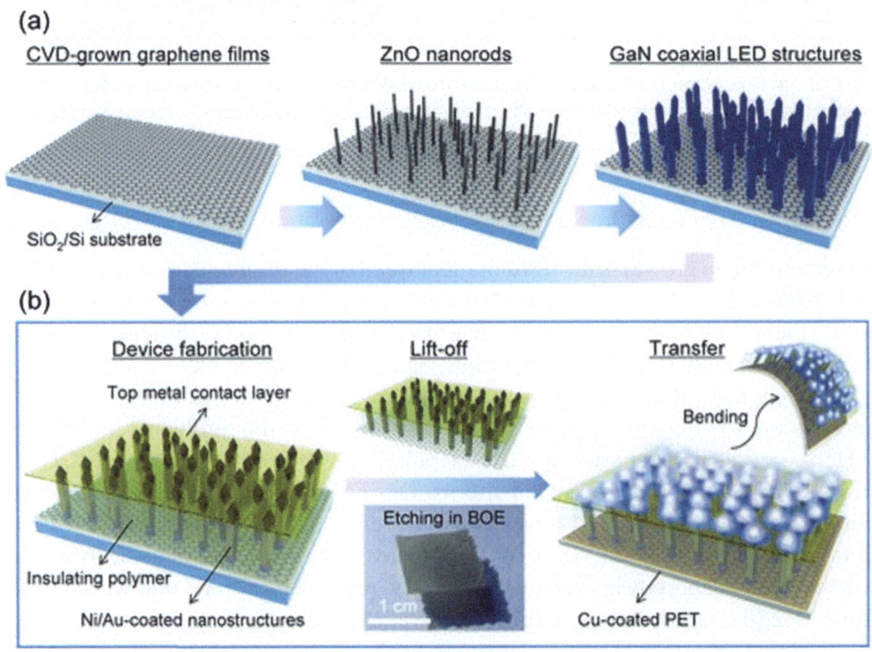

Fig. 12.7 Schematic illustration of preparation process of graphene–based flexible GaN/ZnO coaxial nanorod LED

will inevitably cause structural and surface damage, which leads to an increase of surface defects. In addition, the reduction of the active area, the waste of GaN materials, and the increase of leakage also limit its application. In contrast, the driving force of the "bottom-up" approach is the reduction of Gibbs free energy, making nanostructures and nanomaterials closer to thermodynamic equilibrium. That is to say, "bottom-up" method is easier to obtain a GaN nanorod structure with uniform chemical composition and reduced defects. Among them, the nanorod structure obtained by self-organized growth in the "bottom-up" method is not uniform in size and diverse in crystal orientation. The wavelength of the nanorod LED is strongly dependent on the size of the nanorod, the position and the alloy component of the quantum well. Thus, this method is difficult to control the emission wavelength of LED, and is not suitable for mass production. In contrast, "bottom-up" method of selective area growth possesses almost all the advantages of nanorods LED due to its orderly and controllable growth characteristics, which is the trends of nanorods LED and has become the future mainstream R&D technology of nanorod LED.

12.1.3 Application of Nanorod LED

Monochromatic light LED:

Since the quantum well of the nanorod LED grows radially, the quantum-limited Stark effect caused by the polarized electric field can be effectively alleviated. Thus, the In composition can be more efficiently incorporated into the quantum well of the nanorod LED, which makes the nanorod LEDs possible by realizing visible light full-band monochromatic LEDs with InGaN material systems. At present, the Swedish GLO company has begun to produce products in this regard as shown in Fig. 12.8.

Phosphors-free white LED:

In addition to monochromatic light LEDs, GaN nanorods can also be used to achieve single-chip white LEDs without phosphors. At present, there are three main methods for realizing GaN-based micro-nano structure phosphors-free single-chip white LED [11–14].

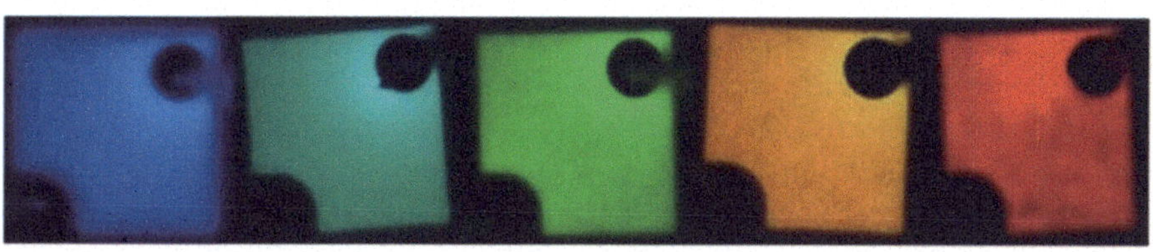

Fig. 12.8 Glo's monochromatic light nanorod LED chip

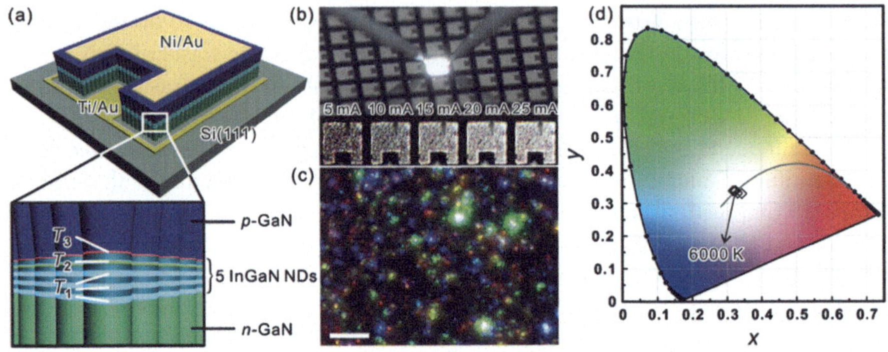

Fig. 12.9 **a** Schematic illustration of the nano-array LED structure with multiple InGaN nanodisks in the active region; **b** image of luminescence of the nanorod LED array under injection current of 20 mA; **c** micro-EL CIE emission spectrum under injection current of 20 mA

① *Adjust the quantum well width.* In 2010, Hon-Way Lin et al. in Taiwan Tsinghua University reported a method for white light LED achieved by adjusting the width of the quantum well [11]. They grow nanorods along the c-axis orientation using PA-MBE method on the 3 inch n-type Si (111) substrate. By changing the number, location and thickness of quantum wells in the pn junctions of GaN nanorods, single-chip white light LED can be implemented as shown in Fig. 12.9. S. Albert et al. also optimized the In/Ga and total V/III ratios and growth temperature in order to tune the well width. The adequate structure tailoring and monolithic integration in a single nanocolumnar heterostructure of three InGaN portions emitting in the red-green-blue colors leaded to white light emission [13].

② *Adjusting the diameter of the Nanorod.* In 2010, Hiroto Sekiguchi et al. reported a new method for controlling the In composition in InGaN QWs on the same wafer [12]. They grow InGaN/GaN MQW nanorod array with diameter of 137–270 nm on the same substrate using Ti as a mask by rf-PAMBE selective area growth. The results show that the emission wavelength gradually increases from 479 (blue) to 632 nm (red) with the increased size of nanorods as shown in Fig. 12.10.

③ *Adjusting the potential field distribution.* In 2011, Young Joon Hong et al. grew GaN nanorod array on n-GaN/Al_2O_3 (0001) substrate using catalyst-free selective area growth with porous SiO_2 as a mask by MOVPE [14]. The results show that the illuminating color changes from red to blue as the bias voltage increases. In contrast, thin film LEDs grown on the same substrate do not have the phenomenon of EL peak shift under different bias voltages. To explain this phenomenon, they used STEM to obtain a cross-sectional view of the sample. It was found that the quantum wells of the nanorods are distributed in the three regions including the platform, the slope and the side which is corresponding to the c-plane, r-plane and m-plane of the wurtzite GaN, respectively. EDX spectra show that the In component of $In_xGa_{1-x}N$ QWs on the top of the nanorod was four times that of the side. Subsequently, they measured the wavelength of the electroluminescence peak of the nanorod LED with different size of p-electrode,

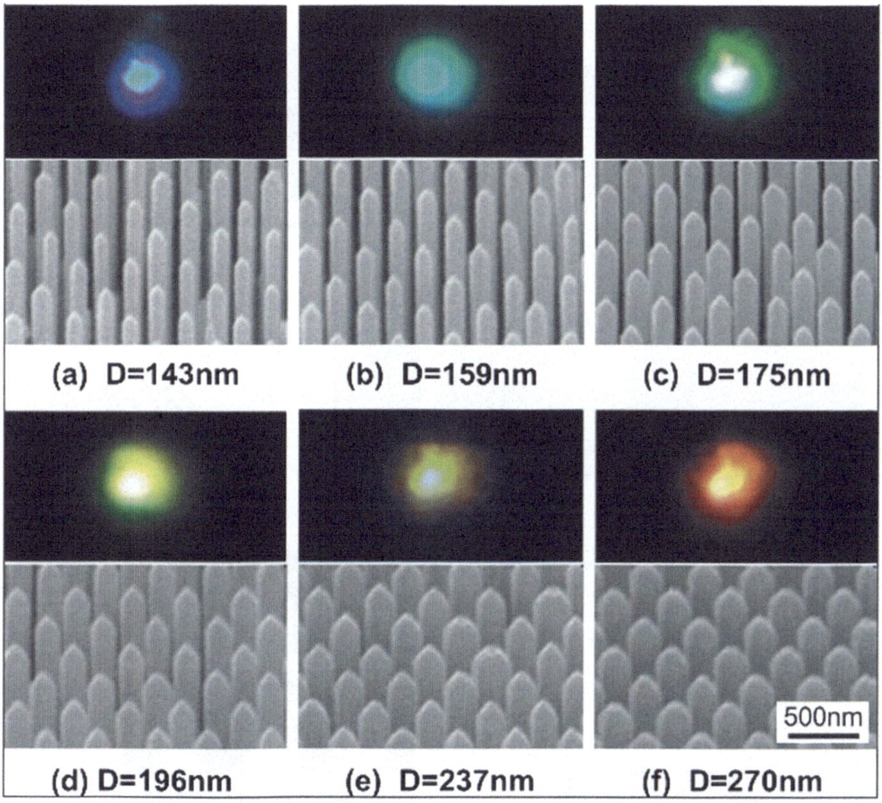

Fig. 12.10 Nanorod morphology and illuminating diagrams at different diameters

and found that the color of electroluminescence is closely related to the current density [14]. This is used to explain such phenomenon by the field distribution model as shown in Fig. 12.11.

In short, GaN-based nanorod LEDs have unparalleled advantages compared to traditional surface LEDs and have attracted much attention in the international LED industry. However, the main work remains to study the effects of nanorod structure, shape and composition on device performance. Due to the difficulties in the growth of 3D structures, efficiency of nanorod LED is far below that of traditional LEDs. The large surface area of the nanorod LED causes strong surface adsorption. The 3D structure also increases the complexity of the device process. Also, the current will concentrate on the bottom of the nanorod LED, resulting in a very high current density at the bottom of the rod. It is also a challenge to achieve uniform current spreading. In addition, the novel physical properties, synergy and particle motion of nanorod LED materials and devices will lead to a series of fundamental problems such as the influence of nanostructures on interface states, carrier transport and composite mechanisms in nanorods, etc. These issues have not been clarified yet and remain to be explored.

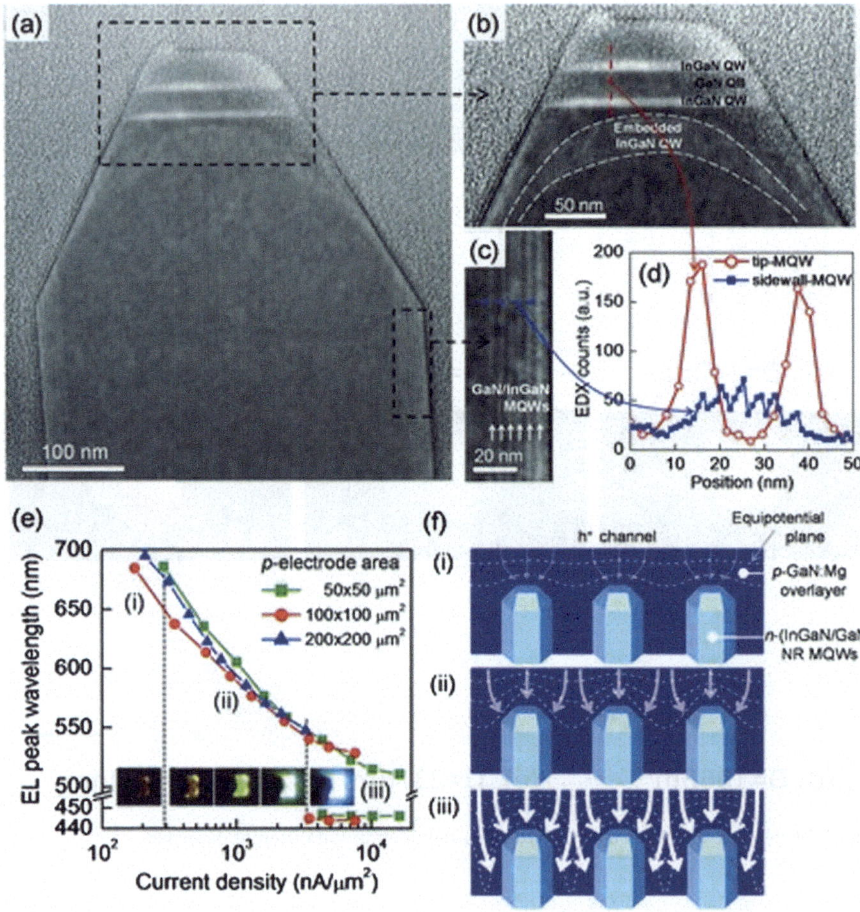

Fig. 12.11 STEM image and field distribution model for explaining dimmable LEDs

12.2 Quantum Dot LED

Quantum dots are quasi-zero-dimensional nanomaterials with a typical size of 1 to 100 nm. Essentially, the size of quantum dots needs to be smaller than de Broglie wavelength of electron. The movement of electrons in quantum dots in all directions is limited, and the quantum confinement effect is very significant. The three-dimensional confinement of quantum dots can lead to some unique physical properties such as discrete energy level distribution of atom-like atoms, quantum size effect, high radiation efficiency, etc. These characteristics make quantum dots high potentials in important application of nano optoelectronic devices.

Quantum dot LEDs using quantum dots as luminescent media can improve LED performance or achieve some unique optical properties, which has attracted great concern. This section will focus on semiconductor quantum dot LEDs to introduce their fabrication methods, optical properties, advantages of quantum dot LEDs over traditional quantum well LEDs, and the current state of research on quantum dot LEDs.

12.2.1 Preparation Method of Quantum Dots

There are many methods for preparing quantum dots, among which lithography and self-organization methods are common. There are also some special preparation methods. The quantum dots can be prepared by photolithography to obtain a uniform array. As quantum dots generally require small dimensions (on the order of nanometers), they are typically prepared using fine exposure techniques such as electron beam exposure and focused ion beam techniques. However, the size of quantum dots obtained by photolithography is still too large due to the resolution of current photolithography processes. Furthermore, subsequent etching and other processes can also introduce some contaminations and damages.

In order to avoid the above problems of lithography method, some methods for forming quantum dots without photolithography and etching steps have been developed. In other words, quantum dots can be prepared by epitaxy on a patterned substrate. For example, SiN_x square openings are formed on a GaAs substrate by lithography and wet etching. Then, AlGaAs is epitaxially grown by MOVPE, leading to a spontaneously formed AlGaAs pyramid in the SiN_x opening. Subsequently, monolayer GaAs is epitaxially grown. GaAs quantum dots can be formed on top of the AlGaAs pyramid, while GaAs quantum wells are formed on the four side walls of the AlGaAs pyramid [15]. By adopting this method, a uniformly arranged quantum dot array with uniform size can be obtained. At the same time, contamination and etching damage in the process of preparing quantum dots by the photolithography method are avoided. However, the as-grown quantum dots still have some problems. First, the quantum dot pitch is defined by lithography, which is generally large. Thus, it is difficult to obtain dense quantum dot alignment, which is disadvantageous for high-power LED or LD devices. Second, there are also losses of illumination at side wall.

Self-organized growth of defect-free high density quantum dot can be achieved by Stranski-Krastanow (SK), which leads to a great concern. The SK growth mode refers to a film growth mode where a two-dimensional continuous film (so-called "wetting layer") is formed in the epitaxial growth process of the material, followed by formation of a three-dimensional island (i.e., quantum dots). The transition from 2D growth to 3D growth stems from the lowest energy principle. Because the strain energy and surface energy should be considered during the growth process, the strain energy will increase to a certain extent when the thickness of 2D layer is above a certain value. Despite the increased surface energy when changed into 3D growth, the strain energy can be reduced to maintain the lowest energy of the system. However, the position and size of the 3D island formed in the SK mode are generally random. This implies that it is difficult to achieve uniform growth.

Stress-induced formation of quantum dots is an interesting method for preparation of quantum dot. The formation of a horizontally structure with modulated stress on the quantum well, such as a nano-island, can regulate the stress state of the below quantum well. Thus, characteristics of quantum dot can be obtained. For example, an InP self-organized quantum dot structure is grown on an InGaAs/GaAs quantum well to form GaAs stress-induced quantum dots [16].

12.2.2 Optical Properties of Quantum Dots

Quantum dots have unique luminescent properties due to their unique three-dimensional limitations. Next, we will introduce the unique optical properties of quantum dots compared to quantum wells.

First, the density distribution of states of quantum dots is significantly different from that of quantum wells as shown in Fig. 12.12. The density of states refers to the number of electronic states in a unit energy interval near the photon energy E per unit volume. The state density of a quantum well is stepped, while the density of states of a quantum dot is a series of delta functions. The quantum well exhibits a discrete energy level in the z direction and a continuous energy level in the xy plane. That is to say, the energy is continuous above the ground state. For quantum dots, as they are constrained in three dimensions, they present discrete sub-levels similar to atoms (quantum dots are therefore called "artificial atoms").

The relationship between the intensity of spontaneous emission and the photon energy is

$$I(hv)\alpha\left|M^2\right|g(hv) * \text{level occupancy factors} \quad (12.1)$$

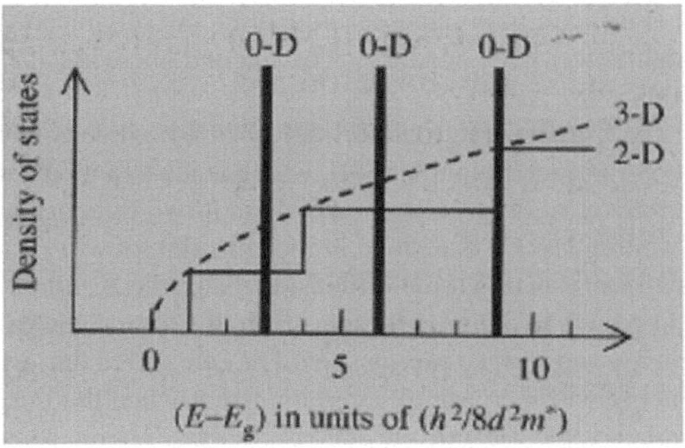

Fig. 12.12 Comparison of density of states for quantum wells with 0D degrees of freedom, quantum wells with 2D degrees of freedom, and bulk materials with 3D degrees of freedom

Among them $|M^2|$ is the transition matrix element, and $g(hv)$ is the density of states. For quantum well light-emitting devices, there is usually only one luminescence peak. For quantum dot light-emitting devices, as the carriers may occupy several discrete levels of quantum dot, the quantum dots correspondingly have a plurality of sub-level peaks.

The spontaneous emission spectrum of quantum dots is determined by the density distribution of states and the level of carrier occupancy [17]. Simultaneously, there is also a unique influencing factor: phonon relaxation [18]. When electrons are excited by light to a high energy level or injected into the energy level of the barrier region, they need to rely on the scattering of phonons (generally releasing LO phonons) to relax to the discrete sub-level of the quantum dot. Phonon scattering requires conservation of energy and momentum. For a quantum well, the conservation condition is easy to satisfy due to the existence of a continuous energy level distribution. However, for a quantum dot, the energy level is discrete and the energy difference between the occupied high energy level and the discrete energy level of quantum dot is very high. Thus, it is hard for the energy difference to be equal to an integer multiple of the phonon energy. Therefore, it is believed that quantum dots lack an effective carrier relaxation path, which is called "phonon bottleneck".

The phonon bottleneck problem of quantum dots has been verified in experiments such as PLE. It is found that as the photon energy of the light source changes, the PLE signal intensity of the quantum dot exhibits an oscillating change. The oscillation period equals to the energy of an LO phonon as shown in Fig. 12.13. It shows that the phonon bottleneck problem is indeed an important issue which affects the relaxation process of quantum dot. However, the phonon bottleneck problem is not as serious as imagined. On the one hand, quantum dots are uneven. On the other hand, the Auger process can release excess energy. However, the mechanism of the Auger relaxation is still controversial and needs further verification.

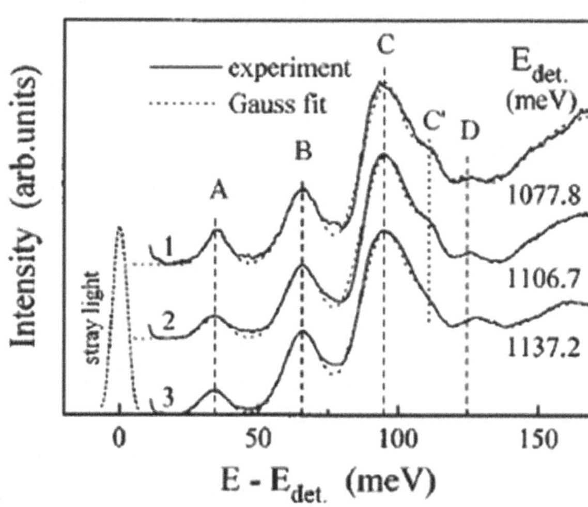

Fig. 12.13 PLE spectrum InAs/GaAs quantum dots [18], wherein 1, 2, 3 three curves obtained at different probing photon energy, 1 at low-energy side of the PL peak, 2 at the PL peak, 3 at high energy side of PL peaks

12.2.3 Advantages and Research Status of Quantum Dot Light-Emitting Diodes

The unique optical properties of quantum dots can be applied to light-emitting diodes to improve the performance of light-emitting diodes.

Compared to conventional quantum well LED, LED with nitride quantum dot based active region has many advantages. First, it can reduce the formation of dislocations. When the quantum dots are formed, the surface energy is enhanced and the strain energy is reduced. The stress can be also released to reduce the formation of dislocations. The reduction of dislocations not only reduces dislocation-related non-radiative recombination, but also improves LED efficiency. It can also reduce the leakage of device, improve the antistatic capability of LED, reduce the light decay of LED, and improve the reliability and stability of LED. Second, the polarization effect can be attenuated. The polarization electric field caused by the polarization charge can cause the spatial separation of electron and holes in the quantum well, which reduces the radiation recombination rate. At the same time, the energy band bending caused by the polarization electric field is not conducive to carrier transport, which increases the LED operating voltage, the electron leakage current, and drop of efficiency at the large injected current (Droop). The reduction in strain in the quantum dots helps to reduce the polarization effect and improve the LED performance. Third, the strong local effect of quantum dots on carriers is beneficial to increase the overlap of wave functions of electron and hole, thereby enhancing the radiation recombination rate. At the same time, the lateral limitation of quantum dots also reduces the lateral diffusion of carriers and the recombination resulted from the dislocations. In addition, wide spectral emission and spectral modulation are easily achieved using quantum dots. Quantum dots exhibit discrete sub-levels similar to those of a single atom. When a higher carrier density is injected, the carrier will continue to fill the excited state due to the energy level filling effect after filling the ground state. Thus, the multi-level luminescence can cause the broadening of the luminescence spectrum, depending on the energy level (or density of states) distribution of the quantum dots and the filling level of the carriers. In addition, due to the random position and size of the island at the beginning of the quantum dot formation from the two-dimensional to three-dimensional growth mode, the size and composition of the self-organized quantum dots are generally not uniform, which also leads to a certain uneven spread of luminescence spectrum. Due to the strong quantum confinement effect, the sub-level distribution of the quantum dots can be adjusted in a large range by adjusting the size of them, which further adjusts the luminescence spectrum of the quantum dots. In summary, quantum dot LEDs have many advantages over quantum well LEDs, and have potentials in application of improving LED luminescence performance and regulating LED spectroscopy.

The preparation, structure and optical properties of quantum dots and related device research have received great attention in GaAs and InP material systems, and have achieved fruitful research results. However, research on nitride quantum dots based materials and devices are still in its infancy. At present, research in this field in China and abroad mainly focuses on the study of tunable growth and optical properties of quantum dots. Wu et al. systematically calculated the optical properties of InGaN quantum dots. The results show that the wavelength can be adjusted in the entire visible range. The electron-hole overlap integral in the quantum dots is larger than the quantum well, which is conducive to radiative recombination [19]. Schulz et al. calculated the polarization electric field in InGaN quantum dots, showing that the quantum dot structure can reduce the polarization effect [20]. Park et al. found that the quantum dots with small size have strong luminescence than that of quantum dots with large size, which is attributed to the weak polarization effect in small-sized quantum dots [21]. Zhang and Xu et al. prepared blue, green and red quantum dot LEDs based on InGaN quantum dots [22]. The yellow-green InGaN quantum dot multilayer stack structure was successfully prepared in Tsinghua University in China [23]. The InGaN quantum dot with green emission was successfully prepared by the Institute of Semiconductors of the Chinese Academy of Sciences, with anomalous dependence between PL intensity and temperature [24]. There are few studies on nitride based quantum dot LEDs. The Chua research group prepared InGaN quantum dot single-chip white LEDs [25]. However, the luminous efficiency was quite low. In general, although InGaN quantum dots initially show unique luminescence properties and potentials in applications, the controllable growth of InGaN quantum dots is still difficult. In addition, the luminescence efficiency is still low, which requires to further improve the quality of material to achieve high-efficiency and high-performance nitride based quantum dots LED.

In addition to nitride-based quantum dot LEDs, the coating of monolayer of colloidal CdSe/ZnSe quantum dots on the nitride LED as a fluorescent conversion layer as shown in Fig. 12.14 has attracted attention in recent years [26].

In addition to applications in light-emitting diodes, quantum dots are also widely used in lasers, photodetectors, solar cells, and single-photon sources. Compared with traditional quantum well lasers, quantum dot lasers can reduce threshold current density, improve temperature stability, and easily achieve single mode lasing. As the free dimension of the active region is reduced, from bulk materials, quantum wells, quantum wires, to quantum dots, the threshold current density Jth of the laser can be effectively reduced, and the temperature stability of Jth can be improved. For photodetectors, compared with quantum well, carriers in quantum dots have longer lifetime due to stronger quantum constraints. This is beneficial to improve the detection rate, responsiveness, photoconductivity gain of photodetectors, and dark current. For solar cells, in addition to the long carrier lifetime of quantum dots, this can facilitate the extraction of photogenerated carriers. The quantum dot absorption spectrum can be adjusted in a large range with the quantum dot size, which is also beneficial to enhance the light absorption and improve the solar conversion efficiency. Besides, single photon source based on single quantum dot luminescence plays an important role in quantum information processing and is widely studied at present.

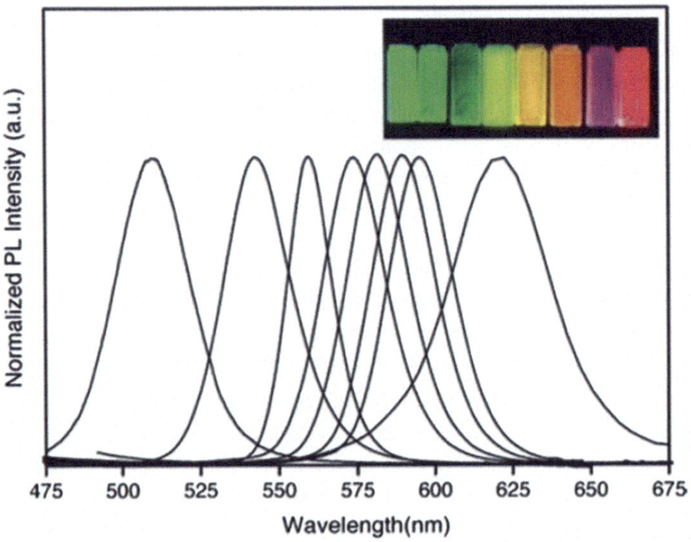

Fig. 12.14 Normalized PL spectra of CdSe-ZnSe quantum dots with different size (2.2, 2.7, 3.2, 3.4, 3.7, 3.8, 4.0, and 4.8 nm from left to right) [26]

12.3 Surface Plasmon Enhanced GaN-Based LED

Surface plasmon enhanced GaN-based LEDs are light-emitting devices that increase internal quantum efficiency through the coupling of surface plasmon and electron-hole pairs. The coupling is induced by the metal film or particles placed near the light-emitting region. The metal film and particles can generate surface plasmons. The coupling can increase the photon state density of the electron hole pair, thereby increasing the radiation recombination rate and quantum efficiency of the electron hole pair without changing the material properties and the band structure.

12.3.1 Basic Properties of Surface Plasmons

12.3.1.1 Surface Plasmon Polariton (SPP) and Localized Surface Plasmon (LSP)

Surface plasmons are divided into two types: surface plasmon polariton (SPP) and localized surface plasmon (LSP). SPP is the elementary excitations of surface electromagnetic wave propagating at the interface between the metal and the dielectric, caused by the continuous longitudinal collective oscillation of electrons at the metal boundary. The SPP field is mainly restrained within the region several nanometers away from the metal surface, which renders the surface plasmons some unique properties such as high electromagnetic field strength and DOS (density of state). Figure 12.15a shows the distribution of electromagnetic fields and charges. Considering the properties of the SPP at the metal-dielectric interface, to simplify the problem we assume that the dielectric substrate with ideal plane shows no light absorption, and the metal with ideal plane is thick enough that the interaction of the plasmons at the upper and lower interfaces is negligible. Then, the electric field intensity of the SPP propagating along the metal interface (z = 0) in x direction can be expressed as

$$E_{SP}(x, z) = E_0 \exp(ik_{SP}x - k_z|z|) \tag{12.2}$$

The following expression corresponds to the surface mode with wave vector of propagating along the surface, its field strength decays exponentially in the direction perpendicular to the interface.

$$k_z^2 = k_{SP}^2 - \left(\frac{\omega}{c}\right)^2 \begin{cases} \varepsilon_i \\ \varepsilon_m \end{cases} \text{Dielectric metal} \tag{12.3}$$

Generally, metal's dielectric constant is greater than the dielectric's dielectric constant. The penetration depth of SPP's electric field in the metal is smaller than that in the dielectric. Figure 12.15b shows the penetration depth of the SPP. The dispersion relationship of the SPP is expressed as follow:

$$k_{SP}^2 = \left(\frac{\omega}{c}\right)^2 \frac{\varepsilon_i \varepsilon_m}{\varepsilon_i + \varepsilon_m} \tag{12.4}$$

where ε_i, ε_m is the dielectric constant of dielectric the metal, ω/c is the light's wave vector. Figure 12.17c shows the dispersion curve of the SPP and photons in the dielectric. Since the SPP's wave vector is always larger than the photon's wave vector, the SPP at the ideal interface cannot radiate photons, which means SPP can't

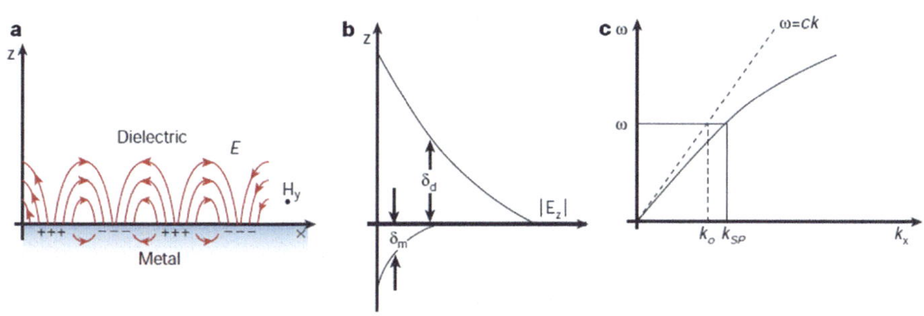

Fig. 12.15 The SPP at the metal-dielectric interface combine the properties of surface charge and electromagnetic field, **a** the distribution of surface electromagnetic field and charge, **b** penetration depth of the interface's evanescent field in the metal and dielectric, **c** dispersion relationship of the SPP

be stimulated by the radiation to the interface. The coupling of the SPP and photons can be enhanced by using rough metal surface or grating structure which can increase the scattering of the light by SPP.

In the case of metal particles, the surface plasmons cannot propagate along the interface in the form of waves, but are localized near the surface of the particles. This is called "LSP", and the electrons will collectively oscillate when they resonate with the electromagnetic waves. LSP is another different excited state compared to the SPP. The SPP can propagate at the interface and obey the dispersion relationship. While the LSP is restrained near the metal particles, it has one or more discrete resonance modes and resonance energy. The energy of the LSP is determined by the shape, size, and medium environment of the metal. LSP can resonate with energy-matching photons without too many concerns about the wave vector of the excitation light. Therefore, LSP can easily dissipate in the form of radiated photons. While only the SPP satisfying the energy-momentum conservation can radiate photons.

LSP can also be seen as the feature of metal surface. If there are nano metal particles on the metal surface or the metal surface is rough, SPP and LSP can coexist. When the resonance frequency of the two is close, the LSP on the rough surface will strongly influence the behavior of SPP. The LSP can both dissipate and excite the SPP. Therefore, LSP can enhance the scattering and excitation of the SPP.

12.3.1.2 LSP's Energy Dissipation

LSP's energy dissipation can be divided into three categories: scattering, absorption, and pure dephasing dissipation. Scattering is to radiate surface plasmons' energy in the form of photons. Absorption is the process of energy relaxation which converts energy into heat energy through internal dissipation. Pure dephasing dissipation refers to the energy dissipation caused by the elastic scattering of SP's energy quantum. The scattering causes the collapse of the phase relationship between SP and the excited electromagnetic field. Experiments show that this kind of energy dissipation in the metal particles is rare. In the quasi-static approximation, the polarizability of the spherical metal particles, the scattering cross sections and the absorption cross sections are expressed as follow:

$$\alpha = \frac{4\pi}{3} V \frac{(\varepsilon_m - \varepsilon_e)}{2\varepsilon_m + \varepsilon_e} \tag{12.5}$$

$$C_{abs} = kIm(\alpha) \tag{12.6}$$

$$C_{sca} = \frac{k^4}{6\pi} |\alpha|^2 \tag{12.7}$$

$$C_{ext} = C_{abs} + C_{sca} \tag{12.8}$$

Wherein, $\varepsilon_{m,}$ ε_e represent the dielectric constant of the surrounding dielectric and metal environment. Im(α) and $|\alpha|$ are the imaginary part and the real part of the susceptibility. It can be seen from the formula that the scattering cross section is proportional to the square of the polarizing rate, and the absorption is linear with the polarizing rate, and the polarizing rate is proportional to the volume of the particles. Therefore, as the particle size increases, the scattering rate of the LSP increases.

The absorption and scattering of different materials' LSP can be analyzed through metal's band structures. For metal nanoparticles, the absorption of LSP mainly derives from the excitation of electrons and holes in the metal's band. For Drude-like metal, such as Au and Ag, electrons show low inter-band transition activity when lower than the interband transition threshold, scattering is the main part of dissipation; for Pd and Pt which have large interband transition activity, absorption is the main part of dissipation.

The thresholds of strong s-d interband activity for Au and Ag are 2.4 and 3.8 eV. When the energy of LSP is less than the threshold, the interband transition activity of electrons is low so that the dissipation mainly derives from scattering. When the energy of LSP is greater than the threshold, dissipation mainly comes from the absorption caused by the interband transition. For Pt and Pd, the s-band near the Fermi level overlaps with the d-band so that the activity of the interband transition is very high through the entire wave band from ultraviolet to near-infrared, which is why strong non-radiative electron-hole transitions are observed in these metals. Al is a very interesting metal that combines the characteristics of the above two types of metals. Al has a strong narrow band transition near the 1.5 eV. When it is lower or higher than this value, Al shows low interband transition activity, hence relatively high scattering efficiency.

12.3.1.3 Factors Affecting LSP Resonance Energy

The resonance energy of LSP is affected by factors such as metal material, particle size, particle shape, dielectric environment and particle spacing. Variation rules may be generally summarized as follows: (1) generally, as the particle size increases, the LSP dipole resonance peaks show red shift, and high-order dipole's resonance peak appears; (2) the LSP's resonance peak is very sensitive to the particles' shape change. The general rule is that the stronger the circular symmetry of the particle, the larger the LSP energy is; (3) the larger the refractive index of the dielectric environment, the smaller the resonance energy of the LSP is; (4) as the particle spacing decreases, the LSP resonance peak has a red-shift tendency; (5) under the same dielectric environment, LSP's resonance energy of three commonly used metals Ag, Au and Al which have the same particle morphology has following relationship: Al > Ag > Au.

12.3.2 Principles of SP Coupling Enhanced LED

Since the DOS (density of state) of SP resonance is very high compared to vacuum or dielectric, the energy of the radiation sources matches the SP resonance energy when the radiation sources (the luminescent molecule or the electron-hole pair in the LED) is in the SP's evanescent field. The radiation source will transfer energy to the SP through spontaneous emission at a very high speed, and then the energy will be radiated by the SP. This process increases the radiation sources' radiative recombination rate, which adds the competitiveness of the radiative process to the non-radiative process, thereby improving the spontaneous emission efficiency. Figure 12.16 shows the electron-hole pairs' radiation process for conventional LED and surface plasmon-enhanced LEDs [27].

The transfer rate of the SP coupling's energy from radiation source to the SP mode is

$$\Gamma_p = \frac{2\pi}{h} \langle d \cdot E(a) \rangle^2 \rho(\hbar\omega) \tag{12.9}$$

where d represents the momentum of the electron-hole pair, a is the position of the quantum well relative to the metal/semiconductor interface, and $E(a)$ is SP's electric field strength at a. $\rho(\hbar\omega)$ is SP's density of state.

Whether or not the radiation efficiency of the radiation source can be improved ultimately depends on the scattering efficiency of the SP and the radiation efficiency of the radiation source. Taking the LED's quantum well as an example, the following equation gives the internal quantum efficiency of the LED after SP coupling.

$$\eta_{\text{int}}^{*}(\omega) = \frac{k_{\text{rad}}(\omega) + c'_{\text{ext}}(\omega) k_{\text{SP}}(\omega)}{k_{\text{rad}}(\omega) + k_{\text{non}}(\omega) + k_{\text{SP}}(\omega)} \tag{12.10}$$

where $k_{\text{rad}}(\omega)$, $k_{\text{non}}(\omega)$, $k_{\text{SP}}(\omega)$ are the quantum well's radiation recombination rate, the non-radiative recombination rate and the rate at which the quantum well transfer

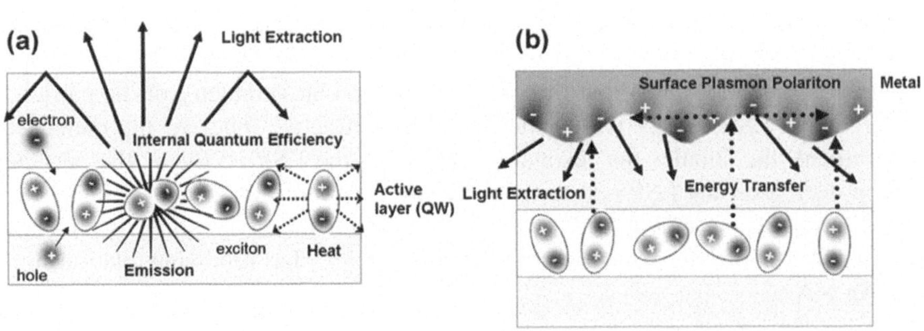

Fig. 12.16 Electron-hole pairs' radiation process for conventional LED (**a**) and surface plasmon enhanced LED (**b**)

to SP energy, $c'_{ext}(\omega)$ is the SP's scattering efficiency. It can be seen from the equation that the $k_{rad}(\omega)/k_{non}(\omega)$ and $c'_{ext}(\omega)$ are the two key factors determining whether the SP can improve the quantum efficiency in the quantum well. If the extreme case is that c'_{ext} is equal to 0 or k_{non} is far less than k_{rad}, then the quantum efficiency of the quantum well cannot be improved. Only when $C'_{ext}(\omega) > k_{rad}(\omega)/(k_{non}(\omega) + k_{rad}(\omega))$ can the internal quantum efficiency of quantum well be likely to increase.

Methods to improve the SP scattering efficiency include increasing surface roughness of the metal film, fabricating metal grating structure, using LSP coupling and selecting metal with low absorption such as Ag, Au or Al.

12.3.3 Coupling Methods for SP Coupling Enhanced GaN-Based LED

In the early stage, people explored how to improve emission efficiency of InGaN quantum wells mainly using light injection. In 2004, Okamoto et al. deposited Ag, Al and Au on three different InGaN quantum well respectively to systematically study the SP coupling enhanced luminescence [28]. A 50 nm thick metal is deposited on the cladding layer of the GaN-based blue quantum well. The distance between the metal film and the quantum well is 10 nm. Then, the PL methods with back incidence and back light emission are used to compare the luminous efficiency of different sample with or without metal. The PL intensity of the sample with Ag at the peak position increase by 14 times compared to the sample with no metal, while the sample with Al shows a 8-times increase compared to the sample with no metal. At the same time, Okamoto also studied how the thickness of the cladding layer between different metal films and quantum wells affects the sample's PL intensity. As the distance between the metal film and quantum well increases, the PL intensity's multiplication factor decreases and eventually drops to 1 at the thickness of 150 nm. This verifies that the SP coupling is a near-field effect and exponentially decays with distance. Temperature-dependent PL can be used to directly calculate the internal quantum efficiency of quantum wells, to distinguish the effect of SP on internal and external quantum efficiency. The internal quantum efficiency of quantum wells with Ag surface plasmon increased by 6.8 times from 6 to 41%. At the same time, results from time-resolved PL show that the spontaneous emission rate of quantum wells is 32 times higher than that of conventional quantum well structures due to the SP coupling [29].

Compared to metal films, metal particles are more advantageous in terms of SP enhancement because LSP has high scattering rate and adjustable SP resonance frequency. In 2006, Pompa increased the PL efficiency of CdSe/ZnS by 30 times using Au particle array with triangular prism shape, and nanostructures were prepared by electron beam exposure [30].

Undoubtedly, the SP coupling can greatly increase the luminous efficiency of quantum wells or other radiators. However, the application of SP to LEDs, especially

GaN-based LEDs, faces serious challenges. The reason is that the p-type region of the GaN-based LED must ensure enough thickness to keep sufficient hole injection and reduce leakage current caused by defects such as dislocations. The thickness is generally above 200 nm. This makes it hard for the metal film or particles on the surface of the LED chip to couple with the quantum well. Despite this, a variety of methods have been tried to fabricate the SP enhanced GaN-based LEDs.

Yang et al. [31] reduced the distance between the surface metal and the active region by growing a 70 nm p-type GaN in the green LED, and preparing a layer of silver nanoparticles on the surface of p-type GaN using annealing. Although the coupling distance has exceeded the penetration depth of Ag in gallium nitride (47 nm), LSP still has a certain degree of coupling enhancement to the quantum well's luminescence. The external quantum efficiency is improved by about 120%.

In order to further shorten the distance between the metal particles and the active region, embedding metal particles in the vicinity of the quantum well region in the process of material growth is another option. The C.Y Cho et al. fabricated a 0.6 nm layer of Ag in the growth of p-type GaN near the blue-light quantum well, and then continued the growth of p-type GaN after annealing. This keeps the distance between Ag particles and the active region within 30 nm, greatly reducing the distance between the metal and active region. Eventually, the LED's optical power at 20 mA is increased by 38% [32]. However, the addition of a metal layer in the middle of the growth process reduces the GaN's material quality and degrades the optical and electrical properties of the device. This also increases the fabrication cycle of the LED epitaxy. In order to embed Ag particles into the GaN without degrading the material quality, Chu-Young Cho et al. improved the above method by locally preparing the Ag nanoparticles and then growing a layer of silica nanodisks on the particles. This method can improve the thermal stability of Ag and the quality of the material. Eventually, Chu-Young Cho further increased the optical output power by 72% using this method [33]. Although this method can prepare metal particles close to the active region, it has many defects such as interrupting the growth process when depositing metal, reducing the quality of GaN, and increasing the unintentional doping of Au and Ag in GaN.

To solve this problem, C. H. Lu proposed to prepare a two-dimensional nanopore array structure in the p-GaN layer of conventional LED epitaxial wafer using a top-down method and deposit metal particles at the hole bottom in 2012 [34]. Then in 2013, C. C. Yang et al. prepared an electro-injected LED device with a two-dimensional nanopore array structure [35]. Although this coupling method does not have to consider many complicated issues in the material growth process, it also causes current leakage at the hole bottom and etching damage to the material. In 2014, Z. G. Yu et al. successfully solved the hole bottom's leakage problem by introducing a sidewall passivation layer. He further extended the nanopore into the quantum well to increase the coupling intensity, achieving 5 times of luminescence gain and 4.5 times of radiation lifetime reduction [36].

So far, the preparation of SP coupling enhanced LED is still very challenging. However, the superior properties of surface plasmons still attract people's attention to design new device structures to achieve this goal.

12.3.4 Surface Plasmon Application in Improving LED's Modulation Bandwidth

As already mentioned, surface plasmons can increase electron-hole pairs' radiation recombination rate. With this feature, high-speed, high-power LEDs show great potential in the field of visible light communication.

Bandwidth refers to the frequency bandwidth occupied by the signal. When used to describe the channel, bandwidth refers to the maximum frequency bandwidth of a signal that can effectively pass through the channel. LED's modulation bandwidth refers to the maximum frequency bandwidth of the loaded signal that the LED can carry. When the LED's AC optical power drops to half (−3 dB) at a certain reference frequency, this corresponding frequency is defined as LED's modulation bandwidth. The LED's modulation bandwidth can be affected by many factors such as the actual modulation depth. The I-V characteristics of the device determine the channel capacity and transmission rate of visible light communication system.

Surface plasmons increase the LED's modulation bandwidth through the reduction of the carrier's radiation lifetime. Okamoto et al. first studied how surface plasmon affects the radiation recombination rate [29]. Figure 12.17 shows the PL decay curves for the two samples at different wavelengths. It can be seen from Fig. 12.17 (a) that the PL decay time at different wavelengths is equivalent; while in Fig. 12.17b, the PL decay rate is the fastest at 440 nm, and gradually becomes slower as the wavelength increases. This means that the carrier's radiation lifetime is greatly reduced by the SP coupling. Therefore, surface plasmons will play an important role in high-speed LEDs for visible light communication.

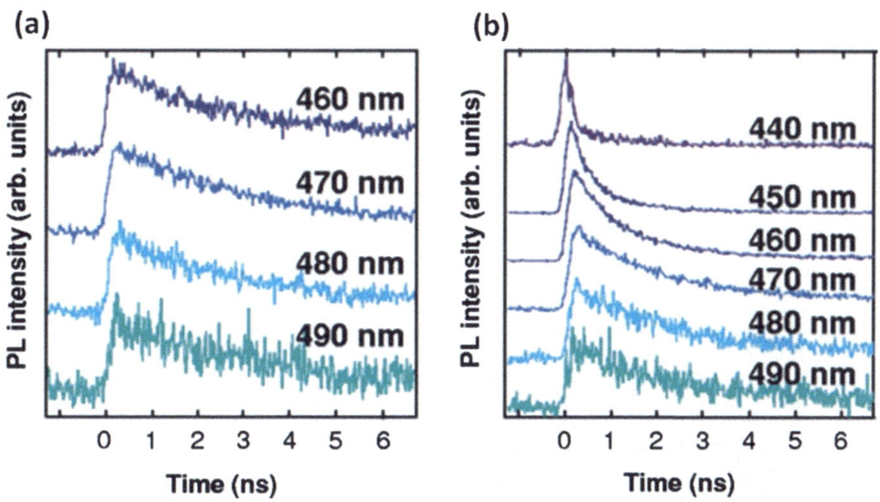

Fig. 12.17 PL decay curve of **a** samples without Ag covered, and **b** samples with Ag covered at different wavelength

12.4 GaN-Based Polarizing LEDs

Light is an electromagnetic wave with electric and magnetic fields perpendicular to each other. The polarization light occurs when the electric field is polarized for an electromagnetic wave. Natural light is non-polarized, and will exhibit a polarization phenomenon only when its symmetry is broken. Some methods is used to fabricated polarizing LEDs, including secondary optical design, LEDs plus linear polarizer, non-polar GaN LEDs (or semi-polar GaN LEDs), side-emitting LEDs and surface plasma coupled LEDs etc. Polarizing LEDs have been applied in automotive headlights, 3D movies, biological physiology, LCD display backlighting and medical treatment.

12.4.1 Secondary Optical Design

An optical design is generally carried out when the LED chip is packaged, including the escaping angle, distribution and color temperature. Especially, secondary optical design is imperative in order to improve in light extraction efficiency. Schubert et al. obtained the polarizing LEDs using a secondary optical design in 2007 [37]. Based on the Bourus law, they designed a non- symmetrical lens to enhance the proportion of the polarized light. The polarization ratio was increased to 1.083 by optimizing the shape of the lens. Then, they fabricated InGaN-based LEDs on (0001) sapphire substrates [38]. The polarized light from the sides was mostly parallel to the m orientation. Furthermore, the polarized light is modulated to the same direction by the special mirror as shown in Fig. 12.18. The polarization light emitted from the

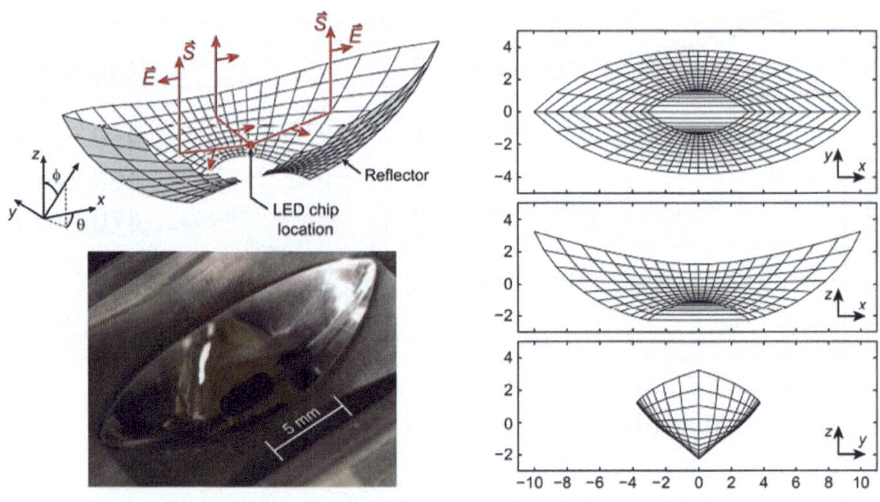

Fig. 12.18 Schematic view of the mirror [39]

Table 12.1 Effects of the metal gratings placed on different positions on the polarization ratio

Configuration	Extraction efficiency (%)	Loss compared to the unpolarized LED (%)	Extinction ratio	
			Uncollimated	Collimated
Unpolarized LED	23.0	–	0.97	0.96
LED + external polarizer	9.6	58.3		2063.27
WGP on all sides of the chip	5.6	75.7	2.04	5.38
WGP on the encapsulation	12.9	43.9	2.72	5.01
Optimized WGP on encapsulation	12.6	45.2	2.37	76.86

sides of the LED is achieved using a polarizer. The polarization ratio is up to 2.5, or it is 1.9 when the light emitted from the top surface of the LED is computed. It is a low cost approach to obtain polarized light by secondary optical design because only the lens is fabricated. Furthermore, they may also be applied to other light sources. Unfortunately, this method would lead to the energy loss and reducing total light efficiency.

12.4.2 The LEDs Plus Metal Grating

Polarized light can also be produced by applying a metal grating to the LEDs, where the metal grating plays the role of a polarizer due to the different TM and TE mode transmittance. The polarization ratio of LEDs is enhanced using a metal grating by Sepsi et al., where aluminum metal grating with a period of 150 nm, a line width of 60 nm, the height of 150 nm is of high polarization selectivity [40]. Furthermore, the effects of the metal gratings placed on different positions on the polarization ratio were compared by RCWA (Rigorous coupled wave analysis) and Monte-Carlo tracing method as shown in Table 12.1. In order to obtain high light extraction efficiency, the LEDs should be encapsulated along with a metal grating compared with an external polarize. However, an almost half of the light is lost even using the method, which will lead to reduced efficiency.

12.4.3 Nonpolar LEDs

Dingle firstly discovered that c-plane GaN is of optical polarization characteristics in 1971. Gardner found that m-plane GaN also exhibits polarization characteristics by electroluminescent in 2005 [41], and the polarization ratio is up to 0.8, which depends on the band structure. As shown in Fig. 12.19, biaxial stress in x and y directions is isotropic for c-plane polar InGaN/GaN heteroepitaxy, which did not affect the crystal symmetry [41]. The band structure is shown in Fig. 12.19a, where |HH > and |LH > are |X>, |Y> uniformly hybrid heavy hole bands and light hole bands, |Z> splits the hole band for the crystal field. However, biaxial stress is anisotropic for nonpolar and semi-polar InGaN/GaN heteroepitaxy, which leads to the bands split into |X >, |Y>, |Z> three zones. The polarization direction of CB-|X > composite luminescence is parallel to "a" axis due to electric dipole of |X> parallel to it, as well as the polarization direction of CB-|Z> luminescence parallel to "c" axis. According to Fermi's rule, the probability of radiation recombination is proportional to the transition matrix, corresponding to wave functions of the hole band and conduction band electron. The stronger luminescence derives from the lower energy CB-|X> complex and its polarization direction along "a" axis.

Kubota and other researchers prepared m-plane GaN LEDs with a polarization ratio of up to 90% in 2008 [43]. However, the light extraction efficiency is reduced because the total reflection happens at the interface due to the large difference in refractive index difference between GaN and air. Conventional light extraction methods including surface roughening and patterned substrates are ineffective because the polarized light will become irregular through several Fresnel reflections. Fortunately, Matioli proposed to use the embedded photonic crystal to improve the extraction efficiency in 2011 [44]. The photonic crystal can enhance light extraction efficiency, and the polarization of the light did not change. G diffraction of the photonic crystal can be modulated by the refraction, and the diffracted light can be entered into the air when the vector $k0$ is smaller than air cone as shown in Fig. 12.20.

Masui and Yamaguchi predicted the longer wavelength of the LEDs with the higher polarization ratio according to the band theory [46, 47]. The valance band

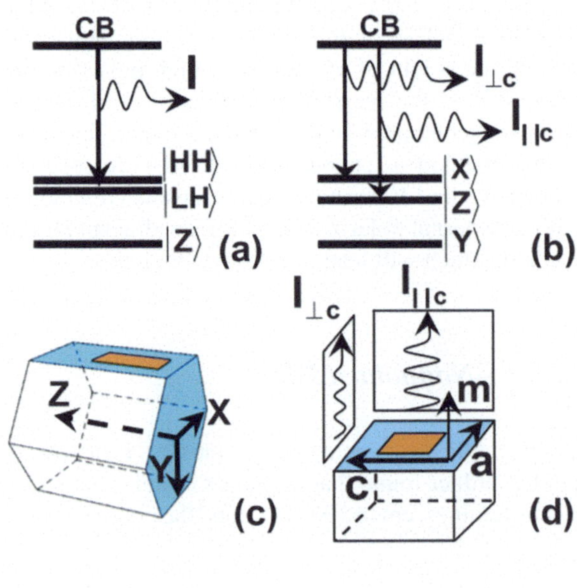

Fig. 12.19 Energy band and structure of GaN on m plane [42]

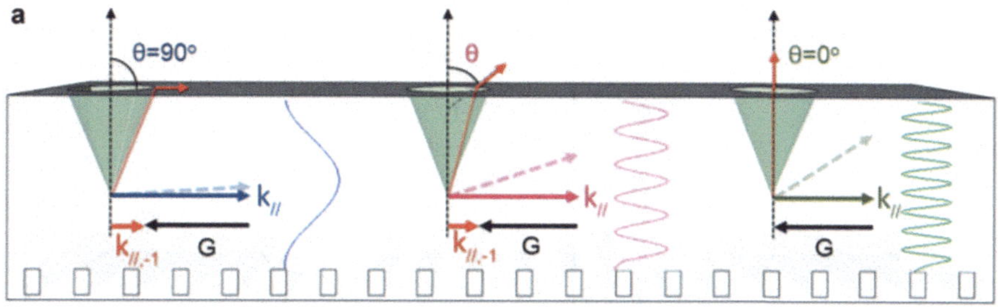

Fig. 12.20 Diffraction of the photon crystal to the wave vector [45]

further separates due to the larger tension stress with the increase of the wavelength in the InGaN. Then, the probability of CB-|X> recombination was enhanced due to further reduction of CB-|Z> transition matrix. The theory was confirmed by Brinkley in 2011 [42]. They produced Blue LEDs on m-plane GaN, and the higher polarization ratio was obtained with the increase of In composition.

The m-plane GaN-based LEDs are of the higher polarization ratio. This is one of the very promising methods. Unfortunately, it is difficult to grow the high-qualify material. It is not suitable for commercialize on a large scale. At first, m-GaN substrates are costly. The photonic crystal with the size of 100 nm is a challenge. Therefore, the process can't be realized at present. Secondly, the quality of the material should be improved at the wavelength in green band. Unfortunately, it is very difficult due to high dislocation density. Of course, A-plane GaN and semi-polar GaN show the similar properties as discussed above.

12.4.4 The Edge-Emitting Polarized LEDs

The edge-emitting polarization was found by J. Shakya in blue and ultraviolet LEDs in 2002 [48]. The polarization ratio was up to 1.8–2.3, which is related with the light confinement of the multiple quantum wells because the polarization light with low-order mode was more easily escaped from the sides. Chuanyu Jia in Peking University found that the polarization ratio decreases from 3.2 to 1.9 due to the competition between the quantum dot and the quantum well with rich-In zone when the wavelength of InGaN-based LEDs changed from 395 to 455 nm [49]. A 460 nm wavelength of LED was prepared by Schubert In 2007, and the polarization ratio was up to 7 as shown in Fig. 12.21. The side-emitting polarization LEDs was manufactured using conventional process with simple and low-cost advantages. However, the light extraction efficiency is low because only side-emitting light is polarized. This is a small portion of total light.

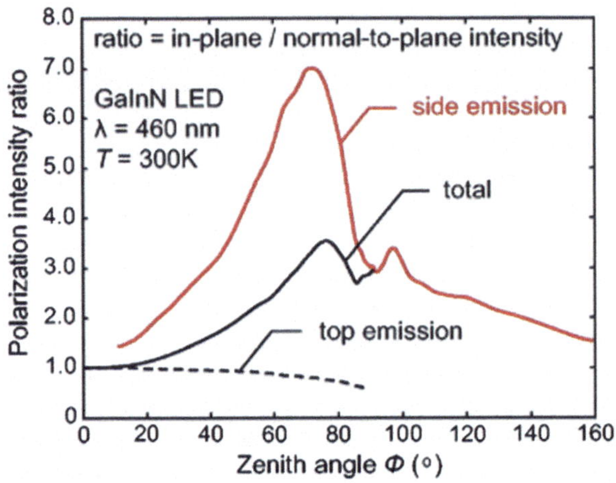

Fig. 12.21 Polarization of the side-emitting LEDs, showing the polarization ratio of edge and top emission [49]

12.4.5 Surface Plasmon Coupled Polarized LEDs

Surface plasma is collective oscillations of excited state electrons at the interface between the metal and the dielectric. Such energy entities include the electromagnetic field in the surface and the dissipative field in the dielectric. At an extended metal dielectric interface, such an energy entity propagates along the interface, called surface plasmon polariton. In addition, such an energy entity exists a localized state of electron oscillation in the vicinity of metal nanoparticles or nanostructures, called a localized surface plasmon. The energy is enhanced because both SPP and LSP can be coupled with the dissipative field that could provide a radiation recombination channel.

In 2008, Shen KC prepared the surface plasmon coupled LEDs with the polarization characteristic [50]. They grew a one-dimensional grating of metallic Ag at 15 nm on the LED quantum well. The polarization light was perpendicular to the groove of the grating as show in Fig. 12.22. However, lot of light lost due to the absorbing from the metal used in the structure. In order to overcome the problem, a layer of SiO_2 was inserted between the metal and quantum well by Shen. KC in 2010 year, and the polarization ratio of the LEDs is up to 1.82 [51]. The effective volume between SP and quantum wells increased effectively due to SP dissipative field extended, which led to the increasing emitting light from coupling channel.

In summary, there are five methods to obtain the polarized light using the LEDs. Second optical design is equivalent to using a polarizer. Only special lens are required in the way, and it is low cost and good compatibility with others light sources. However, the light extraction efficiency is reduced due to a significant portion of the light loss. When the mirror is designed to change the polarization of the sides, it is inefficient due to further light loss. This is the same as the LEDs plus the polarizer, and the overall efficiency of the LED was greatly reduced to about 50%. The m-plane GaN-based LEDs are of the higher polarization ratio. It is one of the very

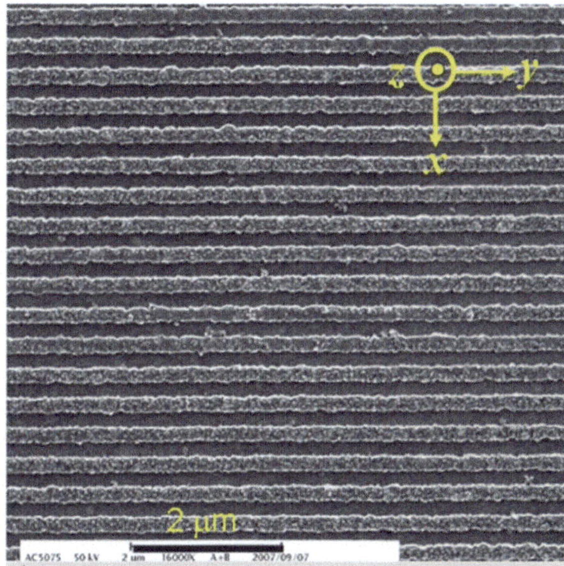

Fig. 12.22 Ag metal grating SEM image [49]

promising methods. Unfortunately, it is difficult to grow high-qualify material. The photonic crystal is also a challenge. So this is not suitable for commercialize on a large scale. For the polarized light by surface plasmon coupling the LED, there are two competitive radiation channels including quantum well and SP. Only the polarization light is derived from the SP channel. Unfortunately, this is a small portion of total light. However, it is a promising method if further development of the technology can be done in the future.

References

1. S. Nakamura, The roles of structural imperfections in InGaN-based blue light-emitting diodes and laser diodes. Science **281**, 956–961 (1998)
2. F. Bernardini, V. Fiorentini, D. Vanderbilt, Spontaneous polarization and piezoelectric constants of III-V nitrides. Phys. Rev. B. **56**, 10024–10027 (1997)
3. M.H. Chang, D. Das, P.V. Varde et al., Light emitting diodes reliability review. Microelectron. Reliab. **52**, 762–782 (2012)
4. F. Qian, Y. Li, S. Gradecak et al., Gallium nitride-based nanowire radial heterostructures for nanophotonics. Nano Lett. **4**, 1975–1979 (2004)
5. S.F. Li, A. Waag, GaN based nanorods for solid state lighting. J. Appl. Phys. **111**, 071101 (2012)
6. M.S. Kang, C.H. Lee, J.B. Park et al., Gallium nitride nanostructures for light-emitting diode applications. Nano Energy **1**, 391–400 (2012)
7. J. Bai, Q. Wang, T. Wang, Greatly enhanced performance of InGaN/GaN nanorod light emitting diodes. Phys. Status Solidi A-Appl. Mater. Sci. **209**, 477–480 (2012)

8. G.T. Wang, Q.M. Li, J.J. Wierer, D.D. Koleske, J.J. Figiel, Top-down fabrication and characterization of axial and radial III-nitride nanowire LEDs. Phys. Status Solidi A-Appl. Mater. Sci. **211**, 748–751 (2014)
9. W. Bergbauer, M. Strassburg et al., Continuous-flux MOVPE growth of position-controlled N-face GaN nanorods and embedded InGaN quantum wells. Nanotechnology **21**, 305201 (2010)
10. C.H. Lee, Y.J. Kim, Y.J. Hong, S.R. Jeon, S. Bae, B.H. Hong, G.C. Yi, Flexible inorganic nanostructure light-emitting diodes fabricated on graphene films. Adv. Mater. **23**, 4614–4619 (2011)
11. H.W. Lin, Y.J. Lu, H.Y. Chen, H.M. Lee, S. Gwo, InGaN/GaN nanorod array white light-emitting diode. Appl. Phys. Lett. **97**, 073101 (2010)
12. H. Sekiguchi, K. Kishino, A. Kikuchi, Emission color control from blue to red with nanocolumn diameter of InGaN/GaN nanocolumn arrays grown on same substrate. Appl. Phys. Lett. **96**, 231104 (2010)
13. S. Albert, A. Bengoechea-Encabo, X. Kong, M.A. Sanchez-Garcia, E. Calleja, A. Trampert, Monolithic integration of InGaN segments emitting in the blue, green, and red spectral range in single ordered nanocolumns. Appl. Phys. Lett. **102**, 181103 (2013)
14. Y.J. Hong, C.H. Lee, A. Yoon, M. Kim, H.K. Seong, H.J. Chung, C. Sone, Y.J. Park, G.C. Yi, Visible-color-tunable light-emitting diodes. Adv. Mater. **23**, 3284–3288 (2011)
15. K. Kumakura, K. Nakakoshi, J. Motohisa, T. Fukui, H. Hasegawa, Novel formation method of quantum-dot structures by self-limited selective-area metalorganic vapor-phase epitaxy. Jpn. J. Appl. Phys. Part 1-Reg. Pap. Short Notes Rev. Pap. **34**, 4387–4389 (1995)
16. J. Tulkki, A. Heinamaki, Confinement effect in a quantum-well dot induced by an InP stressor. Phys. Rev. B. **52**, 8239–8243 (1995)
17. S. Raymond, X. Guo, J.L. Merz, S. Fafard, Excited-state radiative lifetimes in self-assembled quantum dots obtained from state-filling spectroscopy. Phys. Rev. B. **59**, 7624–7631 (1999)
18. R. Heitz, M. Grundmann et al., Multiphonon-relaxation processes in self-organized InAs/GaAs quantum dots. Appl. Phys. Lett. **68**, 361–363 (1996)
19. Y.R. Wu, Y.Y. Lin, H.H. Huang, J. Singh, Electronic and optical properties of InGaN quantum dot based light emitters for solid state lighting. J. Appl. Phys. **105**, 013117 (2009)
20. S. Schulz, E.P. O'reilly, Theory of reduced built-in polarization field in nitride-based quantum dots. Phys. Rev. B. **82**, 033411 (2010)
21. I.K. Park, J.Y. Kim, M.K. Kwon, C.Y. Cho, J.H. Lim, S.J. Park, Phosphor-free white light-emitting diode with laterally distributed multiple quantum wells. Appl. Phys. Lett. **92**, 091110 (2008)
22. M. Zhang, P. Bhattacharya, W. Guo, InGaN/GaN self-organized quantum dot green light emitting diodes with reduced efficiency droop. Appl. Phys. Lett. **97**, 011103 (2010)
23. W.B. Lv, L. Wang, J.X. Wang, Z.B. Hao, Y. Luo, InGaN/GaN multilayer quantum dots yellow-green light-emitting diode with optimized GaN barriers. Nanoscale Res. Lett. **7**, 617 (2012)
24. J. Ma, X.L. Ji et al., Anomalous temperature dependence of photoluminescence in self-assembled InGaN quantum dots. Appl. Phys. Lett. **101**, 131101 (2012)
25. S.J. Chua, C.B. Soh, W. Liu, J.H. Teng, S.S. Ang, S.L. Teo, Quantum dots excited InGaN/GaN phosphor-free white LEDs. Phys. Status Solidi C Curr. Top. Solid State Phys. **5**, 2189–2191 (2008)
26. H.S. Chen, C.K. Hsu, H.Y. Hong, InGaN-CdSe-ZnSe quantum dots white LEDs. IEEE Photonics Technol. Lett. **18**, 193–195 (2006)
27. K. Okamoto, Y. Kawakami, High-efficiency InGaN/GaN light emitters based on nanophotonics and plasmonics. IEEE J. Sel. Top. Quantum Electron. **15**, 1199–1209 (2009)
28. K. Okamoto, I. Niki, A. Shvartser et al., Surface-plasmon-enhanced light emitters based on InGaN quantum wells. Nat. Mater. **3**, 601–605 (2004)
29. K. Okamoto, I. Niki, A. Scherer, Surface plasmon enhanced spontaneous emission rate of InGaN/GaN quantum wells probed by time-resolved photoluminescence spectroscopy. Appl. Phys. Lett. **87**, 071102 (2005)

30. P.P. Pompa, L. Martiradonna, A.D. Torre et al., Metal-enhanced fluorescence of colloidal nanocrystals with nanoscale control. Nat. Nanotechnol. **1**, 126–130 (2006)
31. D.M. Yeh, C.F. Huang, C.Y. Chen, Y.C. Lu, C.C. Yang, Localized surface plasmon-induced emission enhancement of a green light-emitting diode. Nanotechnology **19**, 345201 (2008)
32. C.Y. Cho, M.K. Kwon, S.J. Lee et al., Surface plasmon-enhanced light-emitting diodes using silver nanoparticles embedded in p-GaN. Nanotechnology **21**, 205201 (2010)
33. C.Y. Cho, K.S. Kim, S.J. Le et al., Surface plasmon-enhanced light-emitting diodes with silver nanoparticles and SiO_2 nano-disks embedded in p- GaN. Appl. Phys. Lett. **99**, 041107 (2011)
34. C.H. Lu, C.C. Lan, Y.L. Lai et al., Enhancement of green emission from InGaN/GaN multiple quantum wells via coupling to surface plasmons in a two-dimensional silver array. Adv. Funct. Mater. **21**, 4719–4723 (2011)
35. H.S. Chen, C.F. Chen, Y. Kuo et al., Surface plasmon coupled light-emitting diode with metal protrusions into p-GaN. Appl. Phys. Lett. **102**, 41108 (2013)
36. Z.G. Yu, L.X. Zhao, X.C. Wei et al., Surface plasmon-enhanced nanoporous GaN based green light-emitting diodes with Al_2O_3 passivation layer. Opt. Express **22**, A1596–A1603 (2014)
37. M.F. Schubert, A. Noemaun, S. Chhajed et al., Encapsulation shape with non-rotational symmetry designed for extraction of polarized light from unpolarized sources. Opt. Express **15**, 10452–10457 (2007)
38. M.F. Schubert, S. Chhajed, J.K. Kim et al., Polarization of light emission by 460 nm GaInN/GaN light-emitting diodes grown on (0001) oriented sapphire substrates. Appl. Phys. Lett. **91**, 051117 (2007)
39. M.F. Schubert, S. Chhajed, J.K. Kim et al., Linearly polarized emission from GaInN light-emitting diodes with polarization-enhancing reflector. Opt. Express **15**, 11213–11218 (2007)
40. O. Sepsi, I. Szanda, P. Koppa, Investigation of polarized light emitting diodes with integrated wire grid polarizer. Opt. Express **18**(14), 14547–14552 (2010)
41. N. Gardner, J. Kim, J. Wierer et al., Polarization anisotropy in the electroluminescence of m-plane InGaN-GaN multiple-quantum-well light-emitting diodes. Appl. Phys. Lett. **86**, 111101 (2005)
42. S. Brinkley, Y. Lin, A. Chakraborty et al., Polarized spontaneous emission from blue-green m-plane GaN-based light emitting diodes. Appl. Phys. Lett. **98**, 011110 (2011)
43. M. Kubota, K. Kuniyoshi, T. Taketoshi et al., Temperature dependence of polarized photoluminescence from nonpolar m-plane InGaN multiple quantum wells for blue laser diodes. Appl. Phys. Lett. **92**, 011920 (2008)
44. E. Matioli, S. Brinkley, K.M. Kelchner et al., Polarized light extraction in m-plane GaN light-emitting diodes by embedded photonic-crystals. Appl. Phys. Lett. **98**, 251112 (2011)
45. E. Matioli, S. Brinkley, K.M. Kelchner, Y.L. Hu, S. Nakamura, S. DenBaars, J. Speck, C. Weisbuch, Light Sci. Appl. **1**, e22 (2012)
46. H. Masui, H. Yamada, K. Iso et al., Optical polarization characteristics of m-oriented InGaN/GaN light-emitting diodes with various indium compositions in single-quantum-well structure. J. Phys. D-Appl. Phys. **41**, 225104 (2008)
47. A. Yamaguchi, Anisotropic optical matrix elements in strained GaN quantum wells on semipolar and nonpolar substrates. Jpn. J. Appl. Phys. Part 2-Lett. Express Lett. **46**, L789–L791 (2007)
48. J. Shakya, K. Knabe, K. Kim et al., Polarization of III-nitride blue and ultraviolet light-emitting diodes. Appl. Phys. Lett. **86**, 091107 (2005)
49. C. Jia, T. Yu, S. Mu et al., Polarization of edge emission from III-nitride light emitting diodes of emission wavelength from 395 to 455 nm. Appl. Phys. Lett. **90**, 211112 (2007)
50. K. Shen, C. Chen, C. Huang et al., Polarization dependent coupling of surface plasmon on a one-dimensional Ag grating with an InGaN/GaN dual-quantum-well structure. Appl. Phys. Lett. **92**, 013108 (2008)
51. K. Shen, C. Liao, Z. Yu et al., Effects of the intermediate SiO_2 layer on polarized output of a light-emitting diode with surface plasmon coupling. J. Appl. Phys. **108**, 113101 (2010)

The manufacturer's authorised representative in the EU is Springer Nature Customer Service Centre GmbH, Europaplatz 3, 69115 Heidelberg, Germany. If you have any concerns regarding our products, please contact ProductSafety@springernature.com

Printed and bound by CPI Group (UK) Ltd, Croydon, CR0 4YY
24/07/2026
02175924-0007